The JCT Major Project Form

The JCT Major Project Form

Neil F. Jones
LLB(Hons), *FCIArb, Solicitor*
Pinsents

Editorial offices:
Blackwell Publishing Ltd, 9600 Garsington Road, Oxford OX4 2DQ, UK
Tel: +44 (0)1865 776868
Blackwell Publishing Inc., 350 Main Street, Malden, MA 02148-5020, USA
Tel: +1 781 388 8250
Blackwell Publishing Asia Pty Ltd, 550 Swanston Street, Carlton, Victoria 3053, Australia
Tel: +61 (0)3 8359 1011

First published 2004 by Blackwell Publishing Ltd

Library of Congress Cataloging-in-Publication Data
is available

ISBN 1-4051-1297-2

A catalogue record for this title is available from the British Library

Set in 9.5/11.5 Palatino
by DP Photosetting, Aylesbury, Bucks
Printed and bound in [country] using acid-free paper
by MPG Books Ltd, Bodmin, Cornwall

For further information on Blackwell Publishing, visit our website:
www.blackwellpublishing.com

To Mary
a constant inspiration

Contents

Preface

The Joint Contracts Tribunal Limited has taken a refreshingly bold step in producing a contract for major projects which is about one fifth the size of its other contracts suitable for such projects, e.g. JCT 98 With Contractor's Design, and which is written in a straightforward uncluttered style. More is said about this in Chapter 1.

Even so, this book is not a short one. Inevitably the new JCT Major Project Form 2003 is silent on many matters that typically arise on any construction project. In such circumstances reliance must be placed on implication and the background general law including contract and tort. In addition statute law is bound to be relevant.

The larger and more complex the project, the more potentially significant these matters become, hence this book not only explains the contract's provisions in detail and in their overall legal context, but also deals with, and offers solutions to, many common issues likely to arise in practice.

I am happy to acknowledge with gratitude the permission given by The Joint Contracts Tribunal Limited for the reproduction of the clauses of the contract.

I have many people to thank in making this book possible: the Pinsent Construction Team for ideas and constructive critical debate, with a special thanks to Paul Johnston for preparing the draft section on the Pricing Document, and Fran Crompton for her work on the practical implications of using the Third Party Rights Schedule in place of separate collateral warranties. Thanks also to Hemma Patel for her research and for all her work on those necessary and painstaking tasks including the tables of cases, statutes and clause references.

Finally, I have to thank my wife Mary for steadfastly typing the script while wondering why I should spend so much time writing a book when we could have been walking in the Lake District. Good point!

I have endeavoured to state the law as at 1 September 2003. Where the pronoun 'he' is used it should be taken to represent both he and she.

Neil F. Jones
September 2003

Abbreviations

ACE	Association of Consulting Engineers
CDM Regulations	Construction (Design and Management) Regulations 1994
CEDR	Centre for Effective Dispute Resolution
CIARB	Chartered Institute of Arbitrators
CIMAR	Construction Industry Model Arbitration Rules
CIS	Construction Industry Scheme
DOM/1	Standard Form of Subcontract for Domestic Subcontractors where main contractor under JCT 98
DOM/2	Standard Form of Subcontract for Domestic Subcontractors where main contractor under JCT 98 With Contractors Design
EDI	Electronic Document Interchange
ICE	Institution of Civil Engineers
IFC 98	1998 Edition of the JCT Intermediate Form of Building Contract
JCT	Joint Contracts Tribunal Ltd
JCT 63	Joint Contracts Tribunal Standard Form of Building Contract 1963 Edition
JCT 80	Joint Contracts Tribunal Standard Form of Building Contract 1980 Edition
JCT 98	Joint Contracts Tribunal Standard Form of Building Contract 1998 Edition
MPF	Joint Contracts Tribunal Major Project Form 2003 Edition
RIBA	Royal Institute of British Architects
RICS	Royal Institution of Chartered Surveyors
TCC	Technology and Construction Court
WCD 98	Joint Contracts Tribunal Standard Form of Building Contract 1998 With Contractors Design

Chapter 1
Introduction and background

Introduction and background to the Major Project Form

1-01 In publishing the JCT Major Project Form 2003 Edition (MPF), the Joint Contracts Tribunal Limited (JCT) has broken new ground. Firstly, it has produced a contract for major projects containing almost 80% fewer words than its other popular forms for such projects, namely JCT 98 With Contractor's Design (WCD 98) and JCT 98. Secondly, it has departed from its traditional approach in the production of new contracts, namely for them to be drafted by the secretariat with the assistance of a working party. The drafting of the MPF was outsourced to Brewer Consulting. The contribution made by Owen Fox and Geoff Brewer of Brewer Consulting to the drafting of the MPF on behalf of the JCT Steering Group responsible for this contract is significant. The document which has resulted from this exercise is to be warmly welcomed. It meets a market demand for a shorter, simpler form of contract which uses less legalistic language, is less procedural and is less burdened with detailed provisions in relation to legislative matters such as VAT, CIS and the CDM Regulations.

The MPF will undoubtedly be welcomed in the market place for major projects. It is a new, and some would say refreshing, approach from the JCT. However, any new contract, and more particularly one which departs from a recognised and well understood approach and style of drafting, will reveal teething problems and the MPF is no exception. Throughout this book attention is drawn to a number of issues where the contract may not clearly achieve what is probably its intention.

Essentially, the production of the MPF was driven by the British Property Federation on behalf of private client employers, and the Construction Confederation on behalf of the industry's larger contractors. It is intended to go at least some way towards reflecting current market practice. On any sizeable project, it is rare indeed, at any rate in the private sector, to find a JCT 98 or WCD 98 form being used without being subject to significant amendments produced, usually by lawyers, on behalf of the employer in order to transfer more of the risk to the contracting side. One particular problem with this is that the industry is faced with many different types of amendments aimed at essentially transferring similar risks. It is better by far therefore if at least some of this risk transfer can be achieved through common words which become part of the standard form itself.

The JCT has not suggested any minimum financial limit for the use of this contract. What is important is that the parties to it are fully familiar and experienced in the procurement, design and construction of significant projects. This could mean that where the parties are fully experienced the MPF will be suitable where the value is a few million pounds rather than only where the project is for tens of millions of pounds.

The reduction in length

1-02 The reduction in the length of the MPF when compared to, say, WCD 98, has been achieved by the removal of certain material found in other JCT standard forms, by the reduction or elimination of procedural matters, and by the drafting of other common provisions in a less fussy and more straightforward way. Examples of this approach can be seen in the following:

- There are no articles of agreement.
- There are no detailed insurance provisions, which are now treated as bespoke for each project.
- Sectional completion is dealt with very briefly.
- There are no fluctuations provisions.
- There are no detailed VAT provisions.
- There are no detailed provisions in relation to the CIS.
- There are no detailed provisions in relation to the CDM Regulations.
- The valuation provisions are significantly less detailed.
- There are no provisions dealing with payment for off-site materials or on-site materials prior to incorporation.
- The events giving rise to an adjustment to the completion date have been reduced and the procedures simplified.
- The payment provisions are simplified and instead much of the detail in relation to valuation and payment will be found in a pricing document.
- The determination provisions are simplified.
- The adjudication provisions incorporate the Statutory Adjudication Scheme in place of the JCT's own contractual adjudication procedure.
- There is no provision for arbitration in the MPF.

The above, together with the less legalistic drafting, have resulted in a contract of about 15 500 words, compared with approximately 85 000 words for WCD 98 taking into account its various supplementary provisions, supplemental provisions, appendices and annexures.

Who is this contract for?

1-03 The MPF is undoubtedly intended for employers who regularly undertake major projects, whether they be developers building for sale or lease or building owners who continue to own the buildings, such as hoteliers, supermarkets, etc.

The MPF is intended for contractors and sub-contractors who are familiar with large or complex projects and who in particular are fully familiar with the extent and nature of risks which may be allocated to them under construction contracts.

Under the MPF, contractors may be required to accept greater risks than are typically to be found in other JCT contracts. This is particularly noticeable in connection with design responsibility, ground conditions, meeting statutory requirements, responsibility for specialist contractors selected by the employer, and in respect of consultants whose appointments are novated by the employer to the contractor.

Having said this, in terms of the amendments commonly met in practice and made by or on behalf of employers in relation to WCD 98 and JCT 98, the MPF is still at the soft end of the scale so far as risk allocation is concerned. It is likely therefore that many employers using this new contract will still require some amendments to be made.

The MPF anticipates that the employer will produce requirements which will often involve the contractor in carrying out more or less design, although there is no reason why the requirements cannot be detailed and prescriptive, containing full design with the contractor then carrying out construction only. Nevertheless, in practice it is in the design and build field where this contract will undoubtedly make its mark. The intention is that the employer will produce requirements which will be readily understandable by the experienced contractor who will produce proposals demonstrating how the contractor will complete the design and construct the project.

The completion of the requirements should produce a cut-off point for the employer who thereafter should have little reason for providing more information to the contractor unless there are variations ('Changes') to the project instructed by the employer. Once the contract is entered into, the employer will provide the contractor with site access, will pay the contractor and will be involved in detail only in respect of comments upon the contractor's design documents.

The most detailed provisions to be found in the MPF concern the design submission procedure with regard to the contractor's design documents, enabling the employer to comment upon them in relation to their compliance with the requirements and proposals. In general terms, the employer is able to comment upon such documents without responsibility in that the contractor will remain responsible for ensuring that design documents are in accordance with the contract documents and will, if followed, result in a project built in accordance with the contract. Even though the contractor must take account of the employer's comments, he still has full responsibility for designing and constructing the project in accordance with the contract.

Some significant features

1-04 The contract has a number of significant features which will be listed here and dealt with in detail in subsequent chapters:

- Detailed design submission procedures (clause 6);
- The naming of specialist contractors uniquely or by list (clause 18);
- An option for the employer to novate the consultant's appointment to the contractor (clause 18);
- Wide ranging payment options in a pricing document including payment based upon:
  - value
  - stages or milestones
  - progress payments
  - any other method chosen by the parties;
- The valuation of changes which includes any loss and expense (clause 20);
- Interim payment advices and the final payment advice are to include as part of their calculation:
  - liquidated damages

  - ○ indemnities between the parties
  - ○ any other liabilities under the contract;

  An attempt has therefore been made in the MPF, so far as possible, to make the payments to the contractor all-inclusive (clause 22);
- 'Practical Completion' is defined (clause 39.2);
- A third party rights schedule has been incorporated to take advantage of section 1 of the Contracts (Rights of Third Parties) Act 1999 under which funders, purchasers and tenants obtain benefits which they can enforce directly against the contractor (clause 30). This is intended to remove the need for collateral warranties. It is a significant feature of this contract and will be discussed in detail when considering clause 30. Whether it will meet its objective of removing or reducing the practice of producing separate collateral warranties is debatable;
- There are provisions for acceleration (clause 13), bonus payments to the contractor for early completion (clause 14), and a cost savings scheme with shared savings in connection with suggestions for improvement put forward by the contractor (clause 19);
- The insurance requirements are minimal dealing essentially with regulating the insurances connected with the project. The types of policy, the risks or property covered, and the party responsible for obtaining the insurance are to be dealt with on a project specific basis (clause 27);
- There is express provision for the contractor to maintain a professional indemnity policy (clause 28);
- There are options in relation to the allocation of risk to the contractor where unexpected or unforeseen ground conditions are met (clause 8);
- There is no provision for retention from payments;
- There are no fluctuations provisions;
- There are no specific provisions for the payment of off-site goods or materials or on-site goods or materials prior to incorporation into the works;
- There are no provisional sums;
- There is no reference to a construction programme, unlike such contracts as the Engineering and Construction Contract;
- Adjudication, apart from the selection of the adjudicator, is dealt with under the Statutory Adjudication Scheme (clause 37);
- The possibility of mediation is expressly dealt with (clause 36);
- There is no provision for arbitration.

The contract documents

1-05 The contract documents comprise:

- Requirements from the employer (including where relevant the Employer's Form of Novation Agreement);
- Proposals from the contractor;
- A pricing document;
- The conditions;
- The appendix (to be completed on a project by project basis);
- The third party rights schedule.

The remainder of this chapter will briefly consider the requirements and proposals.

The remainder of this book will consider the conditions, together with the appendix and third party rights schedule. The pricing document is considered at the end of Chapter 5 dealing with valuation and payment.

The requirements

1-06 It is not intended in this book to deal in any detail with the requirements or proposals. They are mentioned in outline here. Parties entering into a contract under the MPF will be familiar with such documents.

The 'Requirements' are defined in clause 39.2 as:

> The documents identified in the Appendix that have been prepared by the Employer in order to set out its requirements for the Project and identify the boundaries of the Site.

In essence, the requirements set out what the employer requires from the contractor. This will necessarily deal with the end result required and then with the manner in which the desired outcome is to be achieved.

Considerable flexibility is permitted as to how the requirements are put together. They need to be well considered and competently prepared if the project is to be successful. This is especially so as under the MPF, it is anticipated that the requirements will be all that the contractor needs to carry out any further design and the construction of the project to the employer's requirements. It is not anticipated that further information will be supplied by the employer, at any rate once the requirements and proposals have been finalised and included as contract documents.

A choice has to be made on the employer's side between, on the one hand, being highly prescriptive in how the desired outcome is to be achieved giving little discretion to the contractor, and on the other, providing the contractor with not much more than carefully considered outcomes, giving the contractor the opportunity and discretion to provide a solution.

The requirements or an accompanying explanatory note from the employer should deal with how tenders are to be evaluated, especially where proposals can offer different design solutions.

Another matter of vital importance is for the employer to make it clear to the contractor the degree of design responsibility that is being retained by the employer as a consequence of any design which is within the requirements. The MPF anticipates this and in clause 5.1 provides that the contractor is not to be responsible for the contents of the requirements or the adequacy of the design contained in them. In practice, it will not be at all unusual for the employer either to amend this clause to make the contractor responsible for all of the design in connection with the project; or alternatively, to achieve the same result by ensuring that design carried out on behalf of the employer is in fact included as part of the contractor's proposals, making the contractor fully responsible for the design.

The MPF does not provide for the inclusion of provisional sums. No doubt the employer could specifically include them in the requirements. If so, they should be properly reflected in the proposals and in an attachment to the priced document. This should deal with how such provisional sums are to be expended and any relevant valuation rules that are to apply.

Some specific points, which will need to be addressed in the requirements, include:

- The appendix entries.
- The boundaries of the site.
- The present position in relation to planning consents and whether the payment of fees or charges will be met by the employer.
- Details of any restrictive covenants or easements affecting the land.
- Any performance specifications.
- Specific standards applicable to materials and workmanship and details of any tests and inspections required.
- Any requirements for the production of design documents, including any design submission programme.
- Any requirement for the division of the project into sections.
- Any restrictions there may be on access.
- Information in relation to others whom the employer will be authorising to work on the site at the same time as the contractor.
- The health and safety plan.
- Any specific requirements of the employer that must be met before practical completion can be achieved.
- Any pre-appointed consultants who are to be novated to the contractor, including full details of their original appointment to the employer, together with the model form of novation to be used.
- The identification of any designs to be prepared, or works to be undertaken by, named specialists, together with the name or a list of names for each category of work.
- Any requirements as to the format to be adopted by the contract sum analysis.
- Any requirements as to the pricing information to be provided by the contractor in connection with the pricing document.
- Information concerning the employer's VAT status and any procedures to be adopted.
- Confirmation of whether or not the employer is a 'Contractor' for the purposes of the Construction Industry Scheme.
- Preliminary details as to the insurance arrangements for the project.
- Details of any bonds and/or guarantees that will be required by the employer, including any that relate to advance payments (where applicable).
- Details of any amendments to the MPF being made by the employer.
- Details of any additional documentation which the employer may require of the contractor or, through him, of any consultants or sub-contractors, such as bonds, guarantees and warranties, forms of appointment etc.

The proposals

1-07 It is not intended to deal in any detail with the proposals. Contractors entering into the MPF will be fully aware of the nature of such documents.

The 'Proposals' are defined in clause 39.2 as:

> The documents identified in the Appendix that have been prepared by the Contractor in order to set out the manner in which it intends to satisfy the Requirements.

The proposals should respond to and meet any stipulations contained in the requirements. Where the proposals are at variance with the requirements it is vital that this is clearly stated so that, if acceptable, the requirements can be altered to reflect this. If the requirements are not altered then the MPF (see clause 4.4) entitles the employer to give an instruction as to which should be followed. That instruction is not to be treated as a change so the contractor will not be able to obtain any compensation or any adjustment to the completion date. Clearly therefore, if the contractor has put forward a less expensive solution, he could find himself faced with having to provide a solution in accordance with the requirements at the same lower cost to the employer as if the contractor's suggestion had been adopted.

The proposals will be set out in a form which the requirements stipulate or in such other suitable form as the contractor may choose. This could be by way of drawings and/or a specification or other documents.

Chapter Two
General obligations

Content

2-01 Clauses 1 to 8 inclusive fall under the general heading of 'General Obligations of the Contractor'. They cover the following topics:

- General obligations of the contractor
- Instructions
- Statutory requirements
- Conflict and discrepancy
- Standards of design, materials and workmanship
- Design submission procedure
- Copyright
- Ground conditions.

2-02 Most of these are areas of great importance in terms of both responsibility and risk allocation, dealing as they do with design, workmanship and materials, ground conditions, conflicts within or between documents and statutory requirements.

General obligations of the contractor – clause 1

2-03 Clause 1 deals in general terms with the contractor's obligations in relation to compliance with contract requirements in the execution of the project and his duties under the Construction (Design and Management) Regulations 1994 (the CDM Regulations).

'Contractor' is defined in clause 39.2 as:

> The party to the Contract named as such or any assignee to whom the Employer has consented in accordance with clause 29 (*Assignment*).

This is straightforward and requires no further comment.

Execution and completion

> 1.1 The Contractor shall execute and complete the Project in accordance with the Contract, including the completion of the design, the specification or selection of materials and the execution of the construction works.

2-04 The contractor is to execute and complete the project in accordance 'with the Contract'. The 'Contract' is defined in clause 39.2 as:

The Contract Conditions, the Appendix, the Third Party Rights Schedule, the Requirements, the Proposals and the Pricing Document.

2-05 Other JCT contracts avoid providing a definition of the contract. It is thought by some that it can defeat the intention of the parties, e.g. if there has been an important exchange of letters during the tender period which has not been reflected in alterations to the requirements or proposals and which are therefore not part of the contract when they were intended to be.

2-06 The contractor's obligation includes not only the execution of the construction work but also the completion of any design or specification or selection of materials where this is required by the requirements prepared by the employer or included by the contractor in his proposals.

2-07 In any building contract, the contractor's prime obligation is to execute and complete the contract works in accordance with the contract requirements. As with the MPF, this usually takes the form of an express term in the contract to this effect. However, even without such an express term, this obligation would be implied.

Obligation to complete

2-08 The contractor's obligation to complete can have quite extreme consequences unless the contract in some way limits or qualifies the obligation. In its unqualified form, the obligation will mean that if the contract works are damaged or destroyed, say by fire, even on the last day before completion, the contractor will generally be contractually responsible for reinstating or rebuilding them at his own cost and on time. Similarly, if the execution of the project is subject to a long suspension for a reason which is not a breach or act of prevention by the employer, the contractor will be obliged to complete on time and will be in breach and liable to pay damages unless the contract by its express terms limits or qualifies this general obligation. There are such contractual limitations in the MPF, for example, an adjustment to the completion date for delays due to *force majeure*, specified perils or terrorist activity under clause 12 and a qualified right for the contractor to terminate his own employment under clause 34.

2-09 Even apart from qualifications in the contract itself, there are certain legal excuses for a failure by the contractor to perform this prime obligation. Briefly they are as follows:

2-10 (1) FRUSTRATION

It is not possible in a book of this kind to consider in detail the law relating to the frustration of contracts. Suffice it to say that there must be some intervening event of a fundamental nature, which renders continued performance of the contract impossible or of a totally different nature to that which was envisaged when the contract was entered into. Only very rarely will a contractor be excused performance of his obligation to complete the contract works by reason of the contract becoming frustrated.

The mere fact that performance of the contract turns out to be more difficult or expensive than envisaged by the contractor at the outset will not be sufficient to frustrate the contract. Furthermore, if the contract itself expressly caters for the eventuality concerned, then it cannot be a frustrating event. A relatively modern example of a frustrating event in relation to a building contract occurred in the case of *Wong Lai Ying and Others* v. *Chinachem Investment Co Ltd* (1979), heard by the Judicial Committee of the Privy Council on appeal from the Court of Appeal of Hong Kong. It involved a landslip which took with it a block of flats of 13 storeys together with hundreds of tons of earth which landed on the site of partly completed buildings completely obliterating them. It was accepted for the purposes of argument that the landslip was an unforeseeable natural disaster. Following the landslip it was uncertain whether the partly completed contract could ever be completed and even if it could, it was uncertain when it could be completed. This was held to be a frustrating event.

In the case of *Davis Contractors Ltd* v. *Fareham UDC* (1956), a fixed price contract to build 78 houses in eight months was held not to be frustrated when, owing partly to severe and unforeseeable shortages of labour and materials, completion took 22 months. Lord Radcliffe in this case said:

> '... it is not hardship or inconvenience or material loss itself which calls the principle of frustration into play. There must be as well a change in the significance of the obligation that the thing undertaken would, if performed, be a different thing from that contracted for.'

In the case of *McAlpine Humberoak Ltd* v. *McDermott International Inc (No. 1)* (1992) the judge at first instance had held that a contract for providing seven steel pallets forming part of a weather deck structure for a tension leg platform had become frustrated. Judge John Davies QC had held that a contract, which was originally based on 22 drawings but which eventually became based on 161 drawings, had been transformed to such an extent that they were not 'changes' within the contract provision for extras and variations. The Court of Appeal overturned this decision, Lord Justice Lloyd saying:

> 'The revised drawings did not "transform" the contract into a different contract, or "distort its substance and identity". It remained a contract for the construction of ... pallets ...'

The contract had, properly construed, provided for such an eventuality even if exceptional in extent. Having said this, there must come a moment when, taking a typical contract provision for variations, the number or extent of variations reaches a point where it so changes the nature of the contract as to be outside the scope of the clause. In such a case the contractor would be contractually entitled to refuse to proceed, or to agree to do so only on renegotiating the contract.

It has been said that:

> 'Frustration is a doctrine only too often invoked by a party to a contract who finds performance difficult or unprofitable, but it is very rarely relied on with success. It is in fact a kind of last ditch.'
>
> (Lord Justice Harman in *Tsakiroglou & Co Ltd* v. *Noblee Thorl GMB* (1961))

A contract may become frustrated if the government prohibits or restricts the work contracted for (*Metropolitan Water Board* v. *Dick, Kerr & Co Ltd* (1918)).

In the event that the contract is frustrated, the financial position between the parties will generally be governed by sections 1 and 2 of the Law Reform (Frustrated Contracts) Act 1943, which gives the court discretion to award reasonable sums for work done or benefits conferred by the parties to the contract.

2-11 (2) EXCLUSION CLAUSES AND TERMS INTERFERING WITH COMMON LAW RIGHTS

The contract between the parties may seek to relieve the contractor from liability to perform under the contract or from liability for a tort connected with the contract. It may well be that in many instances, particularly where the parties are of equal bargaining power, such exclusions or restrictions do no more than apportion the risk between the parties which will, in turn, be reflected in the make-up of the tender sum. The more the contractor is able to exclude, limit or define a risk, the more likely it is that a lower tender sum will result. It is not possible in this book to deal with exclusion clauses in any detail and reference should be made to textbooks which deal with the topic, e.g. *Chitty on Contracts*, 28th edition, Chapter 14.

What can be said at this point is that, apart from statutory controls (referred to in the next paragraph), it is possible, provided the exclusion clause is appropriately worded in all the circumstances, to exclude liability even for fundamental breaches of contract. There is no rule of law that a fundamental breach committed by one party to a contract inevitably deprives that party of the right to rely on an exclusion clause: see *Photo Production Ltd* v. *Securicor Transport Ltd* (1980), in which the House of Lords disapproved the Court of Appeal's decision in *Harbutt's Plasticine Ltd* v. *Wayne Tank & Pump Co Ltd* (1970). The House of Lords unanimously rejected the view that a breach of contract by one party, accepted by the other as discharging him from further performance of his obligations under the contract, brought the contract to an end and, together with it, any exclusion clause. It all depends on the construction of the particular contract as to whether or not an exclusion clause is adequate to limit what would otherwise be the liability of the party at fault.

2-12 There is some statutory control of exclusion clauses in business and consumer contracts by virtue of the Unfair Contract Terms Act 1977 and in consumer contracts by the Unfair Terms in Consumer Contracts Regulations 1999. The 1977 Act and the 1999 Regulations provide some control over contract terms which exclude or restrict liability for breach of certain terms implied by statute or at common law into building contracts. They also control some contract terms, which purport to entitle one of the parties to render a contractual performance substantially different from that reasonably expected of him. Generally speaking, where a consumer transaction takes place, i.e. one of the parties is acting as a 'consumer' and the other contracting party is selling or providing the work in the course of a business, then the exclusion or restriction of liability in respect of such implied terms is rendered absolutely ineffective. On the other hand, if it is a business transaction, then the exclusion or restriction clause is likely to be effective but only in so far as it satisfies the requirement of reasonableness contained in the 1977 Act; the 1994 Regulations have no application to business transactions.

2-13 (3) BREACH OF CONTRACT BY EMPLOYER

Where the employer is guilty of a sufficiently serious breach of contract, the contractor is entitled to treat the employer's breach as a repudiation, which releases the contractor from any further obligation to perform the contract. It is not every breach that will have this effect. It must be of a very serious nature. If it is not, it will entitle the contractor to claim damages for breach of contract but will not entitle him to decline further performance of his contractual obligations.

2-14 (4) MISREPRESENTATION

A misrepresentation is an untrue statement of fact made by one contracting party to the other at, or before, the time of entering into the contract and which acts as an inducement to that other party to enter into the contract.

Where such a statement made by one party is relied on by the other party to his detriment, then that other party may, in appropriate circumstances, be able to obtain a rescission of the contract and also, depending upon the circumstances, to claim damages under the Misrepresentation Act 1967. If the statement concerned also becomes a term of the contract then the contractor will be entitled to sue for breach of contract. For a detailed discussion of misrepresentation the reader is referred to the appropriate textbooks covering this topic, e.g. *Chitty on Contracts,* 28th edition, Chapter 6.

2-15 (5) ACCORD AND SATISFACTION

There is nothing to stop the employer and contractor agreeing to excuse one another from any further performance of their contractual obligations upon agreed terms, e.g. the employer may be running short of funds and the contractor may have entered into a very unprofitable contract and they may well both be happy to terminate their obligations. The parties are of course always free to agree whatever terms they wish to end the contract before completion of the contract work. This is known as an accord and satisfaction.

In accordance with the contract

EXPRESS TERMS

2-16 First and foremost the contractor is required to 'execute and complete the project in accordance with the Contract' e.g. as to quality, the manner of execution and time. These express requirements appear throughout the contract as well as under the heading of 'General Obligations of the Contractor'. They will be commented on throughout this book when considering the relevant clauses.

IMPLIED TERMS

2-17 In the absence of an express term requiring the contractor to execute and complete the contract works in accordance with the contract, such a term would be implied. In addition, building contracts, as with all other contracts, are likely to have incorporated within them, to the extent that they are not inconsistent with the express terms, certain implied terms which may have a bearing on the contractor's obligation to execute and complete the project in accordance with the contract.

2-18 Certain terms are implied as matters of law, e.g. that in a contract for work and materials the goods will be of satisfactory quality; that to the extent that the contractor's skill and judgement is being relied on, for example where the contractor undertakes design or the selection of materials, the work will be reasonably fit for its purpose; and that reasonable skill and care will be used in relation to the workmanship employed. The express terms of the contract can have the effect of displacing an implied term if it is inconsistent with them. Clause 5 dealing with the standards of design, materials and workmanship is relevant in this respect (see paragraph 2-65 *et seq.*).

There can be many other implied terms as a matter of law in relation to building and other contracts. Examples are implied terms as to co-operation if this is required to enable the other party to carry out the works in a regular and orderly manner and similarly in relation to not hindering the other contracting party: see for example *Allridge (Builders) Ltd* v. *Grandactual Ltd* (1996). Such implied terms can, where the situation warrants it, extend to, for example, a contractor providing information to a sub-contractor in such manner and at such times as is reasonably necessary to enable the sub-contractor to fulfil his obligations under the sub-contract: see *J. & J. Fee Ltd* v. *The Express Lift Co* (1993) considering the DOM/2 Form of Domestic Sub-contract for use with the JCT 81 (now 98) WCD.

2-19 Yet again there are implied terms which are not implied as a matter of law but are implied as a matter of fact. Where the express terms of a contract do not cover a particular situation, the court may imply a term based on the imputed intentions of the parties gleaned from the actual circumstances of the case, where this is required to give the contract what is called 'business efficacy'. The extent and nature of such implied terms will clearly vary from case to case.

STATUTORY OBLIGATIONS

2-20 The contractor's obligation to carry out and complete the works is bound to be subject to a greater or lesser degree to statutory control, for example, in relation to the safety aspects of equipment and methods of working under the Health and Safety at Work etc. Act 1974. Furthermore, under the CDM Regulations made pursuant to section 15 of the 1974 Act, the contractor will have significant obligations in relation to health and safety. Also, the quality or standard of work to be achieved may be affected by statute, e.g. the Building Regulations under the Building Act 1984; the Defective Premises Act 1972.

Design, specification and selection of materials

2-21 Clause 1.1 makes it clear from the start that the contractor is responsible for completing the design and specification, and choosing suitable materials, to the extent that this has not already been done by the employer in the requirements. The nature of this obligation is discussed below under clause 5 when considering the design obligation in that clause (see paragraph 2-85 *et seq.*).

In addition, and importantly, subject to clause 5.1, the MPF has an optional facility for transferring to the contractor the responsibility for design services undertaken for the employer by his consultants before the MPF was entered into. This is achieved by selecting the option in the Appendix entry for clause 18 (see clauses 18.1 and 18.4). By clause 18.2, the contractor agrees to enter into a model novation agreement (no standard form of novation agreement is provided) with the pre-appointed consultants.

Contractor as planning supervisor and principal contractor

1.2 The Contractor is appointed as both Planning Supervisor and Principal Contractor, and the Contractor shall notify the Health and Safety Executive of its appointment in accordance with regulation 7(1) of the CDM Regulations. The Planning Supervisor previously appointed by the Employer (if any) is identified in the Appendix.

1.3 The Contractor warrants that it has the competence and will allocate the resources necessary to fulfil the roles of Planning Supervisor, Principal Contractor and designer in the manner referred to in the CDM Regulations.

The Construction (Design and Management) Regulations 1994 (the CDM Regulations)

2-22 All those concerned in the construction process will have statutory duties and obligations imposed on them under the CDM Regulations. This will include the employer and contractor. As such it is not necessary (though some consider it desirable) to include detailed provisions within a construction contract dealing with those statutory duties and obligations. This is generally the approach taken in the MPF which deals only with two aspects: the appointment of the contractor as planning supervisor and principal contractor (see clause 39.2 for definitions) and a warranty from the contractor that he has the competence and will allocate the resources necessary to fulfil those roles and the role of designer under the Regulations.

Planning supervisor and principal contractor

2-23 Clause 1.2 confirms that the contractor is appointed both as the planning supervisor and principal contractor. Presumably, if the appointment is not actually covered by a separate letter or document, this will effectively be the appointment once the contract conditions become live upon the contract being entered into. In most cases, employers will have their own letter or form of appointment and their own advance procedures for checking the contractor's competence and resources etc. It is advisable for the employer to ensure that there are separate and adequate terms of engagement between himself and the 'client' (if the employer is not the 'client' for the purposes of the CDM Regulations), and between himself and the contractor as planning supervisor and principal contractor, which includes appropriate indemnity clauses in the employer's favour should a failure by the 'client', or the contractor (as planning supervisor or principal contractor), to carry out their duties under the Regulations involve the employer in liability to third parties, particularly in the case of the contractor, where the liability may fall outside the indemnity clause 26, e.g. in respect of purely financial loss.

2-24 In practice, more often than not, employers who are using the MPF will have already found it necessary to appoint a planning supervisor before the contractor has been selected. Regulation 6(3) provides for a planning supervisor to be appointed by the client as soon as practicable after the client has such information about the project and the construction work involved in it as will enable him to comply with the requirements imposed upon him by Regulation 8(1) (satisfying himself as to the proposed planning supervisor's competence) and Regulation 9(1) (satisfying himself as to the intended allocation of adequate resources by the planning supervisor). If the employer already has a planning supervisor in place who will be replaced by the contractor, such person is to be identified in the appendix.

2-25 The contractor, as the appointed planning supervisor, is required under clause 1.2 to notify the Health and Safety Executive of his appointment in accordance with Regulation 7(1). This Regulation states that the planning supervisor is to ensure that notice of the project in respect of which he is appointed is given to the Executive. By Regulation 7(2) the notice is to be in writing or such other manner as the Executive may approve. By Regulation 7(3) the notice must provide the particulars specified in Schedule 1 to the Regulations to the extent that they are known or can reasonably be ascertained at the time of appointment; otherwise the information is to be provided as soon as practicable thereafter. In any event, the particulars are required before the start of construction work. Schedule 1 requires details of such matters as the location of the site; name and address of client; type of project; name and address of the planning supervisor and principal contractor with a declaration signed by each that they have been appointed; planned date for the start of the construction phase and its duration; estimated maximum number of people that work on the site; and the planned number of contractors on the site. The Executive provides a suitable form for this purpose (Form 10). It should be sent to the area office covering the site. A copy of the notified particulars must be displayed on site (Regulation 16(1)(d)).

Contractor's warranty of competence

2-26 By clause 1.3 the contractor warrants that he has the competence and will allocate the resources necessary to fulfil the roles of planning supervisor, principal contractor and designer in the manner referred to in the CDM Regulations. While this warranty is of significant contractual benefit to the employer in the event that any incompetence or failure to allocate necessary resources will be a clear breach of warranty, making the contractor liable to the employer in connection with any foreseeable losses flowing from the breach, it does not of course discharge the employer's own duty, as a 'client' under the Regulations. By Regulation 8, no client shall appoint a planning supervisor, designer or contractor unless he is reasonably satisfied as to their competence to fulfil their respective duties under the Regulations. Similarly, by Regulation 9, the client must be reasonably satisfied that the planning supervisor, designer and contractor have allocated or will allocate adequate resources for the purpose of performing their functions under the Regulations.

Instructions – clause 2

2-27 Clause 2 deals with the issue by the employer to the contractor of instructions in connection with the design, execution and completion of the project. Clause 2 also deals with the matter of consequential payment or adjustments to the completion date. Finally, if the contractor fails to comply with an instruction the employer is entitled to engage others at the contractor's cost.

Compliance with instructions

2.1 The Contractor shall comply with all written instructions issued by the Employer in connection with the design, execution and completion of the Project, except to the extent that the terms of the Contract restrict the Employer's right to issue any particular instruction.

2-28 The contractor is to comply with all written instructions issued by the employer in connection with the design, execution and completion of the project, except to the extent that the terms of the contract restrict the right to issue any particular instruction.

Oral instructions

2-29 If the instruction is not in writing, the contractor is not obliged to comply with it. However, it appears that an oral instruction is still an instruction. By its wording, this clause does not appear to make oral instructions invalid, it simply relieves the contractor from the obligation to comply with them. There is no provision, as there is for instance in WCD 98 (clause 4.3), for an oral instruction to be made effective by subsequent confirmation from the employer, or the contractor without dissent. What then, if the contractor does comply with an oral instruction? Although an instruction is a means of communication and clause 38.1 provides that all communications required to be made by one party to the other in accordance with the contract shall be in writing or in a selected electronic or other agreed format, it could be argued that an instruction is not *required* to be made at all, though the better view must be that if the employer is issuing an instruction as permitted under the contract then the communication of that instruction is required to be made in writing, etc. On this basis, in giving an oral instruction, the employer would be stepping outside the contract, technically being in breach of it. However, it is still an instruction which amounts to a change as defined in clause 39.2, and there seems no reason why the contractor cannot, under clause 20 which deals with changes, notify the employer under clause 20.1 that he considers that the instruction has given rise to a change.

The matter is not absolutely clear and it would have been preferable for clause 2.1 to provide that instructions must be in writing in order to have any validity at all. This would then have taken the matter of payment for compliance with an oral instruction outside the terms of the contract altogether and the contractor would have had to seek recompense outside it, e.g. in terms of a restitutionary claim (in which case the recompense to the contractor would be based not on the cost of carrying out the work but on the value of it to the employer which can of course, depending upon the circumstances, be significantly advantageous to one party or the other – see for example, *Costain Civil Engineering Limited and Tarmac Construction Limited* v. *Zanen Dredging & Contracting Company Limited* (1996).

Scope of instructions

2-30 Many standard forms of building contract in providing for instructions expressly state that they must relate to matters which the instructing party is expressly empowered by the contract conditions to issue. For instance, clause 4.1.1 of WCD 98 requires the contractor to:

> 'comply with all instructions issued to him by the Employer in regard to any matter in respect of which the Employer is expressly empowered by the Conditions to issue instructions;'

Clause 2.1 is very different, providing that the contractor is to comply with all written instructions in connection with the design, execution and completion of the project 'except to the extent that the terms of the Contract restrict the Employer's right to issue any particular instruction'. An example of this is to be found in clause 13 under which the employer and contractor can agree that the contractor will achieve an early completion date. Clause 13.3 expressly provides that saving any such agreement, 'the Employer may not instruct the Contractor to achieve Practical

Completion on a date earlier than the Completion Date'. Further examples are to be found in clause 4.4, providing that no instructions shall require the contractor to act otherwise than in accordance with statutory requirements and clause 19.5 under which, by implication, the employer may not instruct a change in relation to improvements suggested by the employer otherwise than under that clause.

2-31 However, this is not as far reaching as it may appear. Firstly, most instructions, though by no means all, in connection with the design, execution and completion of a project will fall within the definition of a change (see clause 39.2), and will attract payment and an appropriate adjustment to the completion date by virtue of clause 20. However, it is to be noted that the definition of change excludes the situation where the need for the instruction is as a result of any negligence or default on the part of the contractor. Secondly, there must be a natural limit on the scope of instructions. It would not, for example, be possible for the employer to issue an instruction which changed the terms of the contract unless there was a much more specific contractual provision enabling this to be done, which is not the case under the MPF. So, for example, the employer could not by an instruction unilaterally and without just cause (e.g. the contractor's failure to fulfil his statutory duties or obligations) terminate the contractor's appointment as planning supervisor or principal contractor under the CDM Regulations.

2-32 Instructions can cover a wide variety of matters. The most obvious is of course in connection with changes as defined. In addition, instructions can be given which do not amount to a change:

- For dealing with discrepant provisions under clause 4.3 and 4.4;
- Under clause 4.5 where statutory requirements are altered after the base date, having been announced beforehand;
- Relating to opening up for inspection or testing for which provision is made in the contract or where the opening up etc. reveals non-compliant work (clause 16.1);
- To remove work which is not in accordance with the contract (clauses 16.2.1 and 16.3), or alternatively an instruction that such work may remain subject to appropriate compensation (clauses 16.2.2 and 16.3);
- Consequential upon discovering work which does not comply with the contract (clauses 16.2.3 and 16.3);
- As to further opening up or testing (clauses 16.2.4 and 16.3);
- To rectify defects (clause 17).

Compliance by the contractor with written instructions

2-33 The contractor is to comply with all written instructions. Unlike WCD 98 (clause 4.1.1), the obligation is not to comply 'forthwith'. The instruction must therefore be complied with within a reasonable time having regard to both the interests of the employer and contractor and having regard to the urgency, if any, of the situation.

Again, unlike WCD 98 (clause 4.2), there is no machinery by means of which the contractor can request the employer to specify the provision of the conditions, which empowers the issue of an instruction. The absence of such a provision is of course a consequence of the fact that under the MPF it is only if the terms of the contract expressly restrict the employer's right to issue an instruction that it can be challenged.

Instructions to be treated as not giving rise to a change

> 2.2 Where the Contract provides that an instruction is not to be treated as giving rise to a Change, the Contractor shall not be entitled to any additional payment or adjustment to the Completion Date as a consequence of complying with the instruction and the issue of the instruction shall not relieve the Contractor of any of its obligations under the Contract.

2-34 In a number of instances the contract provides that an instruction is not to be treated as giving rise to a change, in which case the contractor receives no additional payment and no adjustment to the completion date. In addition, the issue of such an instruction does not relieve the contractor of any of his obligations under the contract. Instructions falling into this category include those listed above (see paragraph 2-32).

2-35 The only instructions that entitle the contractor to any additional payment or any adjustment to the completion date, are those which fall within the definition of a change under clause 39.2. Changes are considered in detail later (see paragraph 5-02 *et seq.*). Briefly, a change is:

- Any alteration in the requirements or proposals which alters design quality or quantity;
- Any alteration of any restriction or obligation in the requirements or proposals as to the manner in which the contractor is to execute the project;
- Any matter which the contractor requires to be treated as giving rise to a change.

Provided in each case that if it has been prompted as a result of any negligence or default on the part of the contractor, then it is excluded from the definition.

Contractor failing to comply with instruction

> 2.3 Where the Contractor fails to comply with an instruction, the Employer may engage others to give effect to the instruction provided it has first given seven days' notice in writing to the Contractor of its intention to do so. The Contractor shall be liable to pay the Employer's additional costs of engaging others, after taking into account any sums that would have been payable to the Contractor under the terms of the Contract had the Contractor complied with the instruction.

Employer's right to engage others if the contractor fails to comply with an instruction

2-36 If the contractor fails to comply with an instruction the employer can, having first given 7 days' notice in writing to the contractor, engage others to implement that instruction. Although this clause refers to the failure to comply with any instruction giving rise to the employer's right to engage others, it must, in view of the opening words of clause 2.1 which require the contractor to comply with all written instructions, exclude the contractor's failure to comply with an oral instruction. It may have been appropriate to have referred expressly to written instructions at the beginning of clause 2.3.

2-37 As stated earlier, the absence of any requirement for the contractor to comply forthwith or immediately with a written instruction means that the contractor must comply within a reasonable time in all the circumstances. This may make it difficult for the employer to be certain as to the exact point at which there has been a failure to comply. Bearing this in mind, it is surprising that the 7-day notice to the contractor of the employer's intention to engage others is not expressed in such a way as to give the contractor the opportunity to comply with the instruction within that 7-day period. The equivalent provision in WCD 98 (clause 4.1.2) provides that the employer may employ others only if the contractor does not comply with the instruction within 7 days after receipt of a written notice from the employer requiring compliance. In other words, it is truly a warning notice. This is both neater and more sensible than clause 2.3 of the MPF. Having regard to the fact that under clause 4.1.1 of WCD 98 the contractor is to comply 'forthwith', clause 4.2 enables the employer to give the written notice to the contractor requiring compliance at any time. The result under the MPF is that, contractually at least, at the point at which the contractor first becomes aware that the employer considers he has failed to comply with an instruction, namely upon the giving of the 7-day notice, the contractor has no opportunity to put matters right. In practice, if the contractor, upon receipt of such a notice actually then complies with the instruction in the sense of diligently commencing compliance without having finished complying, it may put the employer in a difficult position. Strictly speaking, once the employer has given notice that he intends to engage others, the contractor appears to have no contractual right to begin complying or to continue complying with the instruction but in practice he may well do so.

2-38 Placing work with another contractor under this provision is an extreme remedy which can cause serious practical problems on site and hinder the smooth running of the contract. In practical terms, unless the content of the instruction relates to some discrete element of work, which does not involve significant integration or interfacing with other work, it is unlikely to be a satisfactory option for the employer. In practice it is most unlikely to be employed without first giving the contractor every opportunity to comply with the instruction. Further, it is likely in many instances to be symptomatic of a greater problem which may mean that termination of the contractor's employment is in the end a cleaner, more decisive option, particularly where, as is highly likely, the relationship between the parties has broken down.

Financial consequences

2-39 Clause 2.3 provides that the contractor is to be liable for the employer's additional costs of engaging others, after taking account of any sums that would have been payable to the contractor had he complied with the instruction. The additional costs are those which are over and above any amount which the employer would have paid to the contractor had the contractor complied with the instruction. In the vast majority of cases, the instruction will be in relation to a change so the amount which the employer would have paid the contractor for implementing the change will be deductible from the total costs which the employer pays to the other contractor or other consultant (for example an architect or structural engineer where it is design work which is the subject of the change instruction).

2-40 The additional cost for which the contractor is liable to pay the employer is taken into account in determining the sum due to the contractor (see clauses 22.5.4 and 22.6.2). As it is the amount stated in the employer's payment advice which determines the sum due for payment to the contractor, the inclusion in the calculation of this sum will not amount to the withholding of a sum due under the contract and will not therefore fall under the provisions of section 111 of the Housing Grants, Construction and Regeneration Act 1996 which requires the employer to give advance notice of any intention to withhold payment. If for any reason the employer has not included this sum in the calculation reflected in a payment advice, then, by clause 23.2, an effective notice of withholding must be given to the contractor no later than 7 days before the final date for payment of the sum from which the withholding is to be made (namely 14 days after receipt by the employer of a VAT invoice from the contractor in respect of the sum from which the withholding is to be made - clause 22.4), identifying the ground or grounds and the amount attributable to each ground.

Statutory requirements - clause 3

2-41 Clause 3 requires the contractor to make applications or give notices required by 'Statutory Requirements' as defined in clause 39.2. It also requires the contractor to comply generally with them. In addition, copies of all applications and notices together with approvals, rejections and other communications received in connection with the statutory requirements are to be provided by the contractor to the employer. Finally, unless the employer's requirements state that fees and charges are the employer's responsibility, the contractor is to pay them.

Compliance with statutory requirements

3.1 The Contractor shall make any applications, give any notices required by and comply with the Statutory Requirements. The Contractor shall provide the Employer with copies of all applications made and notices given, and pass to the Employer all approvals, rejections or other communications received in connection with the Statutory Requirements.

2-42 Clause 3.1 imposes a general requirement upon the contractor to comply with the statutory requirements. These are defined in clause 39.2 as:

In relation to the Project:

- any Act of Parliament, any instrument, rule or order made under any Act of Parliament.
- any regulation or byelaw of any local authority or of any statutory undertaker which has jurisdiction with regard to the Project or with whose systems those of the Project are or will be connected.
- any directive of the European Community having the force of law.

2-43 There is of course in any event a legal obligation on the contractor to comply so it does not need to be expressly stated as such. However, clause 3.1 has the advantage of making any failure to comply a breach of an express term of the contract as opposed to what would certainly be an implied term. The inclusion of an express provision is sometimes seen as an advantage when linked to an indemnity provision such as that contained in clause 26.1.

2-44 Many JCT contracts, such as WCD 98 and JCT 98 provide detailed provisions in relation to relevant legislation including VAT; the operation of the CIS; the CDM Regulations; the Building Regulations; the Defective Premises Act 1972 and so on. This is not so with the MPF. It is a contract for use with employers and contractors who are fully familiar with the construction industry and who are or should be familiar with relevant legislation. Detailed provisions relating to such statutory matters tend to be long-winded and inherently complex.

2-45 If the contractor fails to comply with the statutory requirements in relation to the project, causing loss to the employer, this will of course be generally recoverable as damage for breach of contract and may give rise in appropriate situations to a claim for an indemnity by the employer under clause 26. In addition it may be a 'Material Breach' (see clause 39.2) giving the employer the right to terminate the contractor's employment under clause 32.

Conflict between statutory requirements and employer's requirements

2-46 The contractor is obliged to comply with the employer's requirements as well as with the statutory requirements. What if they are inconsistent and the employer's requirements do not comply with the statutory requirements (see clause 4.4)? The contractor must still comply with the statutory requirements. Indeed the contractor is bound to follow the statutory requirements rather than the employer's requirements which would, it is submitted, be invalid and unenforceable. Accordingly there could be no claim against the contractor arising out of such technical non-compliance. Clause 4.4 provides that in such a situation the employer is to instruct which of the discrepant provisions it wishes the contractor to adopt. It goes on to provide that the instruction will not be treated as giving rise to a change. It then provides, importantly and necessarily, that no instruction shall require the contractor to act otherwise than in accordance with the statutory requirements. In effect this requires the employer to issue an instruction for the contractor to comply with the statutory requirements. It does mean however that in complying with statutory requirements, which are at odds with the employer's requirements, the contractor will not receive compensation. It is clearly vitally important therefore that the contractor thoroughly checks the employer's requirements at tender stage to ensure that they do comply with the statutory requirements. Any discrepancy should preferably be brought to the employer's attention during the tender period, or as a minimum should be dealt with expressly in the proposals, stating how the contractor intends to deal with the matter so as to comply with the statutory requirements.

Applications made and notices given by the contractor in connection with statutory requirements

2-47 Clause 3.1 requires the contractor not only to provide the employer with copies of applications and notices given pursuant to statutory requirements but also to pass to the employer all approvals or rejections received and any other communications received in connection with the statutory requirements. This should ensure that the employer is kept fully informed as to the position in relation to such matters.

Fees and charges

> 3.2 Unless the Requirements state that specific fees and charges have been or are to be paid by the Employer, the Contractor shall pay all fees or charges payable in connection with the Statutory Requirements.

2-48 The contractor is to pay all fees or charges in connection with the statutory requirements unless the employer's requirements state that specific fees and charges have already been or will be paid by the employer. Where the contractor is responsible for paying fees and charges, he will also be responsible for paying any increases in them. The contractor will also be responsible for any newly imposed fees or charges, though if fees or charges are the result of the contractor complying with a change instruction, he will be able to recover these as part of any accepted quotation or any loss or expense forming part of the calculation of a 'fair valuation' (clause 20).

Civil liability for breach of 'Statutory Requirements'

2-49 It is not every breach of an obligation imposed by statute that will give rise to a civil claim for damages by someone who has been injured or suffered loss or damage by reason of the non-compliance. The injured party must show firstly that the injury suffered was within the ambit of the statute; secondly that the statutory duty imposes a liability to civil action; thirdly that the statutory duty was not fulfilled; and fourthly that the breach of the statutory duty caused the injury.

Where the statute is silent as to a remedy, there is a presumption that the injured party will have a right of action for breach of statutory duty. For a detailed discussion of this topic the reader is referred to leading textbooks, e.g. *Clerk and Lindsell on Torts*, 18th edition.

CDM Regulations

2-50 So far as the Construction (Design and Management) Regulations 1994 (the CDM Regulations) are concerned, made pursuant to section 15 of the Health and Safety at Work etc. Act 1974, the position is as follows. The Act deals with civil liability in section 47. It provides in section 47(2) that breach of duty imposed by health and safety regulations 'shall, so far as it causes damage, be actionable except in so far as the Regulations provide otherwise'. Turning to the CDM Regulations, Regulation 21 provides that:

> 'Breach of a duty imposed by these Regulations, other than those imposed by Regulation 10 and Regulation 16(1)(c), shall not confer a right of action in any civil proceedings.'

Regulation 10 requires that every client shall ensure so far as is reasonably practicable that the construction phase of any project does not start unless a health and safety plan has been prepared; Regulation 16(1)(c) provides that the principal contractor shall take reasonable steps to ensure that only authorised persons are allowed into any premises where construction work is being carried out. Accordingly it follows that so far as civil liability for breach of statutory duty is concerned, it is only contravention of those two provisions which confers a right of action in any civil proceedings.

In relation to contraventions of Regulation 10 it will be necessary for any injured party to demonstrate that the damage to his or her health or safety was as a result of the construction phase starting without a health and safety plan having been prepared.

In relation to Regulation 16(1)(c), it will be necessary for the injured party to demonstrate that the damage to health or safety was caused as a result of the principal contractor not having taken reasonable steps to ensure that only authorised personnel were allowed into premises where construction work was being carried out.

In relation to both liabilities, causation is likely to be a particularly troublesome issue in pursuing any claim.

CONTRACTUAL IMPLICATIONS OF FAILURE TO COMPLY

A breach of the CDM Regulations is a 'Material Breach' (see clause 39.2) that may entitle the employer to terminate the contractor's employment under clause 32. It may also bring into play the indemnity provisions of clause 26 if the employer is faced with any expense, liability, loss, claim or proceeding under statute or common law involving personal injury or damage to property.

Building Regulations

2-51 An issue of particular importance is whether a building contractor owes an independent statutory duty to the employer to comply with the Building Regulations 2000 (as amended by Building (Amendment) Regulations (2002)). Section 38 of the Building Act 1984 provides that, subject to the provisions of that section, breach of a duty imposed by the Building Regulations, shall, so far as it causes damage, be actionable except in so far as the Regulations provide otherwise. There is provision for the regulations to provide prescribed defences.

2-52 It is probable that until section 38 of the Building Act 1984 is brought fully into force, there is no liability for breach of statutory duty. However the situation is still not absolutely clear. See *Keating on Building Contracts,* 7th edition, paragraphs 15–38, where the relevant cases are summarised in footnotes.

Turning to local authorities, the question of whether a duty of care in the tort of negligence is owed at all in favour of occupiers and users has not been finally determined: see *Murphy* v. *Brentwood District Council* (1990). Even if such a duty is owed, the position is likely to be the same as it is for any contractor who owes a duty of care, namely it will be confined to actual personal injury or physical damage to separate property. It does not generally extend to liability in respect of the cost of averting damage to health or safety of person or property.

The employment by building owners of 'approved inspectors' under the 1984 Act on contractual terms is likely to enhance significantly the building owner's position by providing non-occupying building owners with a contractual remedy where none existed before.

Conflict and discrepancy – clause 4

Introduction

2-53 Clause 4 deals with discrepancies within and between the requirements and the proposals and between either of those documents and the statutory requirements. The requirements and proposals are two of the most important contract documents. The documents making up the requirements are to be identified in the appendix.

They are prepared by the employer in order to set out his requirements for the project and also to identify the boundaries of the site. The documents making up the proposals are also to be identified in the appendix and will have been prepared by the contractor in order to set out the manner in which he intends to satisfy the requirements. Ideally there will be no discrepancy either internally within either of these documents or between them. Similarly ideally, there should be no discrepancy between either of these documents and the statutory requirements .

However, in practice discrepancies do arise. It is sensible therefore for the contract conditions to provide a mechanism for choosing which of the discrepant provisions is to prevail. Any consequential financial issues also need to be covered. Clause 4 provides this code. In addition it provides for the situation where statutory requirements change after the base date (the date is to be identified in the appendix – it will usually be a period of days, e.g. 10, prior to the date for submission of tenders).

Notification of discrepancies

4.1 If either party identifies any discrepancy within or between the Requirements, the Proposals and/or the Statutory Requirements, including any discrepancy that arises as a consequence of an alteration to the Statutory Requirements, it shall immediately notify the other party accordingly.

2-54 If either party identifies a discrepancy within either the requirements or the proposals or between them, or between either of them and the statutory requirements, that party is to notify the other party immediately. That notification, being a communication, must be made in writing or by any alternative procedure, e.g. such electronic means as may have been agreed by the parties in writing (see clause 38.1). Any discrepancy between the requirements or the proposals and the statutory requirements includes discrepancies that arise out of any alterations to the statutory requirements. The notification must be made immediately.

2-55 There is no express contractual duty on the contractor to search for discrepancies within the requirements though the contractor has a general duty when submitting his intended proposals to demonstrate the manner in which he intends to satisfy the proposed requirements. If at tender stage the contractor comes across discrepancies within the proposed requirements, he will generally of course point this out as well as indicating how he proposes to deal with the discrepancy in terms of his intended proposals. However as the contract only comes into existence after a tender has been accepted, there can be no duty on the tendering contractor to notify discrepancies during the tender period. Very often in practice of course, the proposed requirements will be both extensive and technical. The contractor will have little time during the tender period to trawl through detailed technical requirements. Further, he may not at that stage have the resources, particularly if he is to take on the employer's designers if successful in obtaining the contract. In any event, even if resources are available it would be uneconomic to expend these as part of the tendering costs.

Discrepancies within the requirements

4.2 Where a discrepancy is identified within the Requirements, the Contractor shall notify the Employer which of the discrepant provisions it intends to adopt and proceed accordingly.

If the Employer wishes the Contractor to proceed otherwise, it shall so instruct the Contractor and that instruction will be treated as giving rise to a Change.

2-56 If either party has notified the other that there is a discrepancy within the requirements, the contractor must notify the employer which of the discrepant provisions he intends to adopt. Often the contractor will already have proceeded with his design, which reflects his interpretation of the requirements, and will simply be confirming the position he has already adopted. However this may not always be so where the contractor has still to complete some design and/or selection of goods or materials. If the employer would prefer a different result from that which the contractor's choice of discrepant provisions indicates, an instruction can be issued to produce such different result. This will be treated as giving rise to a change and accordingly will be valued as appropriate under clause 20 and an appropriate adjustment made to the completion date under clause 12.

The boundaries of the site

2-57 The requirements, as defined in clause 39.2, include documents which not only set out the employer's requirements for the project but also identify the boundaries of the site (see also the definition of 'The Site' – clause 39.2). If there is a discrepancy between the identification of the site and the rest of the requirements, the practical situation may be that the identification of the site, presuming it is accurate, must prevail. Depending upon how the contractor has or intends to deal with the situation, it may of course lead to an instruction, which will be treated as a change.

Can the contractor change his mind having already chosen in his proposals between discrepant provisions in the requirements?

2-58 Clause 4.2 requires the contractor to notify the employer which of the discrepant provisions in the requirements he intends to adopt. Of course the contractor may in his proposals have already followed a particular route which has effectively made that choice, even where he was unaware of the discrepancy. In such a situation, WCD 98 states that the contractor's proposals will prevail (clause 2.4.1). It would have been sensible to have a similar provision for the MPF. Without such a provision, can the contractor, having become aware of the discrepancy, choose to do something different to what is contained in his proposals? While the proposals are the contractor's statement as to the manner in which he intends to satisfy the requirements, clause 4.2 expressly provides that he can choose between discrepant provisions within the requirements. Literally therefore, having discovered the discrepancy, the contractor could elect to choose the less expensive option and this will not be treated as a change. Can the contractor save money in this way? Almost certainly not. Firstly, he cannot unilaterally alter his proposals – see clause 1.1 and the definition of 'Contract'; secondly, even if he could select the less expensive option having included the more expensive option in his proposals, the employer could issue an instruction to proceed in accordance with the original proposals. Even if, and it is arguable, this instruction does fall to be treated as giving rise to a change under clause 4.2, the valuation of it under clause 20, particularly 20.6 based on a fair valuation, would result in nothing being added to the contract sum. The pricing document will already have included for this. The same reasoning would mean that the contractor would not suffer any delay and there would be no adjustment to the completion date pursuant to clause 12.

Discrepancies within the proposals

4.3 Where a discrepancy is identified within the Proposals, the Employer shall instruct the Contractor which of the discrepant provisions it wishes the Contractor to adopt and that instruction will not be treated as giving rise to a Change.

2-59 Where there are discrepancies identified within the proposals, it is for the employer to instruct the contractor which of the discrepant provisions is to be adopted by the contractor. That instruction will not be treated as giving rise to a change. So, if the contractor's proposals are internally inconsistent, the employer can choose which is to apply without that being treated as a change and giving rise to additional cost or any adjustment to the completion date. This is the case whatever the contractor may have allowed for in the pricing document. It is therefore an area of significant risk for the contractor and it is of prime importance that he checks and double-checks his proposals for any inconsistencies.

Discrepancies between the requirements, proposals and statutory requirements

4.4 Where a discrepancy is identified between the Requirements, the Proposals and/or the Statutory Requirements the Employer shall instruct which of the discrepant provisions it wishes the Contractor to adopt and that instruction will not, subject to clause 4.5, be treated as giving rise to a Change. No instruction shall require the Contractor to act otherwise than in accordance with the Statutory Requirements.

2-60 If a discrepancy is identified either between the requirements and the proposals or between either of those documents and the statutory requirements, the employer instructs which of the discrepant provisions the contractor is to adopt. That instruction is not treated as giving rise to a change unless the discrepancy arises because of an alteration in the statutory requirements, which had not been announced before the base date (clause 4.5). However, it is expressly provided that no instruction of the employer shall require the contractor to act in any way otherwise than in accordance with the statutory requirements. This issue has already been discussed in connection with the contractor's obligation to comply with 'Statutory Requirements' under clause 3.1 (see paragraph 2-42).

2-61 This clause makes it clear that if there is an inconsistency between the requirements and the proposals, the responsibility for this will rest firmly with the contractor. It is his responsibility to ensure that the proposals meet the requirements. If the proposals do not meet the requirements the employer can choose to follow either document without that decision, giving rise to a change. This again therefore is an area of high risk for the contractor who needs to be diligent in ensuring that the proposals do not conflict with the requirements. For instance, the contractor may have included in his proposals a particular design solution which is less costly than some others. If this solution is found to be inconsistent with the requirements, the employer can insist on the requirements being met so that the contractor will have to adopt a more expensive solution for the same price as his less expensive one.

2-62 WCD 98 does not address this situation, at any rate not head on. It has no provision to deal with such discrepancy. All it does provide is a declaration by the employer in the third recital that having examined the contractor's proposals he is satisfied that they appear to meet the employer's requirements. This is seen by some employers as a dangerous statement to make and the recital is often deleted in practice, usually together with an added condition similar to clause 4.4 of the MPF.

Alterations to statutory requirements after the base date

> 4.5 If the Statutory Requirements alter after the Base Date in a manner that necessitates an amendment to either the Requirements or the Proposals, the Employer shall instruct the necessary amendments. That instruction shall be treated as giving rise to a Change where the alteration to the Statutory Requirements had not been announced at the Base Date. In any other case the instruction shall not be treated as giving rise to a Change.

2-63 The statutory requirements (see earlier paragraph 2-42) may alter after the base date (stated in the appendix to the MPF) and this may require an amendment to either the requirements or the proposals. The employer must instruct the necessary amendment so that there is compliance with the altered statutory requirements. That instruction is to be treated as a change unless the alteration to the statutory requirements had been announced at the base date. In other words, even though the alteration to the statutory requirements may take effect after the base date, if the alteration has already been announced by then, that alteration will not be treated as a change. The contractor is therefore expected to keep abreast of imminent alterations to the statutory requirements and any failure to do so will be at his peril. This is yet another risk area for the contractor especially as there is generally a short tender period, giving little opportunity to check such matters thoroughly.

2-64 The result of this clause together with clause 4.4 is that if the employer, no doubt through his consultants, has produced requirements which are inconsistent with statutory requirements whether in force or announced but not in force at the base date, the contractor, whose proposals comply with requirements which later require amendment to meet the statutory requirements, will take the risks associated with this. Some may regard that as harsh. The employer's consultants are likely to be far more familiar with recent or imminent changes to statutory requirements and should, in fulfilling their duty to exercise reasonable skill and care to the employer, have known about and reflected this in the preparation of the requirements. Where the consultant's appointments have been novated to the contractor under clause 18 using the Model Form of Novation Agreement which is not part of the MPF documentation, there may, depending upon its terms, be a liability on the part of the consultants in favour of the contractor.

Standards of design, materials and workmanship – clause 5

2-65 Clause 5 covers the contractor's warranties in relation to matters of design, materials and workmanship. Before considering clause 5 it is appropriate to provide in outline a general statement of the law.

Background law – design – workmanship and materials

What is design?

2-66 The precise meaning of design will depend on the context in which it is used. It can range from the depiction of a preliminary concept or idea by means of sketches, plans or drawings as a means of communicating that concept or idea, to the making of a decision as to whether the retaining screws in a skirting board should be set at 60 cm or 90 cm centres in order secure it adequately to the wall.

2-67 Within the building industry generally, the feature determining whether something is a design matter or issue will generally depend upon whether there is a choice to be made. Contractually, the person making that choice will generally have some responsibility for it, subject always to express terms of the relevant contract. The nature of the liability of a contracting party for design for which he is responsible will be governed by the express terms of the contract. Subject to what the express terms provide, the law will imply a term. The nature of this implied term will depend upon whether the contract concerned is one for the provision not only of a design service but also a contractual responsibility to physically construct the design on the one hand, or limited to the provision of a service only, i.e. the creation of the design or the exercise of design functions, on the other.

(1) PROVISION OF A DESIGN SERVICE COUPLED WITH THE OBLIGATION TO PHYSICALLY CONSTRUCT

2-68 It goes without saying that in determining the legal obligations owed by one party to a contract to another, the first thing to do is to look at the express terms of the contract. Subject however to such express terms, the law will imply a number of terms in contracts for work and materials. These terms are similar in nature to those implied in contracts for the sale of goods. An authoritative case in this connection is the well-known case of *Hancock* v. *Brazier (Anerley)* (1966).

In this case a house was built in which hardcore containing sulphates was used. This expanded and cracked the floor. This happened through no fault of the builder who was unaware that the hardcore contained such sulphates. Nevertheless it was held that the builder had an obligation in respect of the hardcore not merely that it should be selected with skill and judgement but that it should be fit and proper and suitable for its purpose. That liability was absolute, i.e. independent of fault. In this case the court stated that in the provision of a dwelling there was a threefold obligation upon the builder (subject to what the express terms of the contract may state):

(1) To do work in a good and workmanlike manner;
(2) To supply good and proper materials;
(3) To provide a house reasonably fit for human habitation.

This last implied term of reasonable fitness for purpose covers design by the contractor and has been endorsed in many decisions, e.g. *Greaves (Contractors) Limited* v. *Baynham Meikle & Partners* (1975) in which Lord Denning said:

> '... as between the building owners and the contractors, it is plain that the owners made known to the contractors the purpose for which the building was required, so as to show that they relied on the contractors' skill and judgment. It was, therefore, the duty of the contractors to see that the finished work was reasonably

> fit for the purpose for which they knew it was required. It was not merely an obligation to use reasonable care ... in this case Greaves undertook an obligation towards Duckham that the warehouse should be reasonably fit for the purpose for which, they knew, it was required, that is, as a store in which to keep and move barrels of oil.'

And in the case of *IBA* v. *EMI and BICC* (1980), per Lord Scarman:

> 'The extent of the obligation is, of course, to be determined as a matter of construction of the contract. But, in the absence of a clear, contractual indication to the contrary, I see no reason why one who in the course of his business contracts to design, supply and erect a television aerial mast is not under an obligation to ensure that it is reasonably fit for the purpose for which he knows it is intended to be used... The critical question of fact is whether he for whom the mast was designed relied upon the skill of the supplier to design and supply a mast fit for the known purpose for which it was required.'

This fitness for purpose implied term will not operate unless it is reasonable for the employer to rely on the contractor's skill and judgment. So, where the employer seeks to specify precisely the materials which are believed to be suitable for his requirements, the implied term will have no room in which to operate. See for example the employer's specification of hardcore requirements in the case of *Rotherham Metropolitan Borough Council* v. *Frank Haslam Milan & Co Ltd and M. J. Gleeson (Northern) Ltd* (1996) where the main contractor was held not liable for providing hardcore which, while it met the specification, was nevertheless (unknown to the parties) unsuitable for use in confined spaces due to its propensity to expand on hydration.

Statute

These implied terms imposing contractual liabilities on the producer of a physical product such as a building, are now embodied in statutory form (see sections 4 (as amended by the Sale and Supply of Goods Act 1994) and 13 of the Supply of Goods and Services Act 1982). The ability of a contracting party to exclude or restrict these implied terms is limited by the provisions of the Unfair Contract Terms Act 1977 (see section 7) and, where a 'consumer' is involved, the Unfair Terms in Consumer Contracts Regulations 1999, derived from Directive 93/13EEC of 5 April 1993 on Unfair Terms in Consumer Contracts.

Conclusion

2-69 In contracts for work and materials where the contractor carries out the design work and the client relies upon his skill and judgment in that regard, then, subject to the express terms of the contract, the contractor has an absolute obligation to provide a finished product which is reasonably fit for its purpose. This is independent of fault. It is an absolute obligation whether the contractor designs in-house or brings in an outside consultant to do the design work on his behalf. This may be a very heavy burden falling upon the contractor because the exercise of all reasonable skill and care in the design is no excuse and no defence if it in fact fails.

(2) PROVISION OF A DESIGN SERVICE ONLY WITH NO OBLIGATION TO CONSTRUCT

2-70 If the designer is providing the design without entering into a contractual obligation to construct the design, e.g. architects, structural engineers etc. engaged by the employer, his obligation, subject to what the terms of his engagement may provide, will be to carry out that service using reasonable skill and care. This is the obligation applied generally to professionals carrying out a professional service, e.g. surgeons, lawyers, accountants or architects. So, if an operation carried out by a surgeon fails in its intended purpose, i.e. the curing of the patient, or if the case upon which the lawyer is engaged is lost, or if the architect's design does not achieve the result which, to the architect's knowledge, the employer intended, that in itself will not create any liability on the part of the service provider. Where the obligation is to use reasonable skill and care, the emphasis is not upon looking at the result (the output) but judging the quality of the exercise of professional care in seeking to achieve the result (the input). It is only if there has been a breach of that duty to exercise reasonable skill and care that a cause of action will be available.

The basis of this test appears to be that in many spheres of professional activity involving the provision of a service, it is just not possible to judge performance against results, e.g. a talented surgeon can perform careful skilful operations which nevertheless fail to achieve their objective and do not cure the patient; a lawyer can present a case to its very best advantage but the case may still be lost. Instead therefore of relying upon the success or failure of the end result, in the case of professionals the test is the exercise of reasonable skill and care in the performance of the professional task.

The law may not be particularly logical in this area for while in some instances to judge performance by the result is just not reasonable, in other areas it may well be possible, e.g. the designer of a building or a structure, such as a dam or bridge, which has a specific purpose to fulfil. If the design fails then why should the designer not be liable even if he has exercised reasonable skill and care? In some professional spheres, including those involved in construction, the law may yet develop along these lines. In *Greaves (Contractors) Limited* v. *Baynham Meikle & Partners* (1975) Lord Denning said:

> 'The law does not usually imply a warranty that he (the professional man) will achieve the desired result, but only a term that he will use reasonable care and skill. The surgeon does not warrant that he will cure the patient. Nor does the solicitor warrant that he will win the case. But, when a dentist agrees to make a set of false teeth for a patient, there is an implied warranty that they will fit his gums, see *Samuel* v. *Davis* (1943) 1 KB 526.
>
> What then is the position when an architect or an engineer is employed to design a house or a bridge? Is he under an implied warranty that, if the work is carried out to his design, it will be reasonably fit for the purpose? or is he only under a duty to use reasonable care and skill? this question may require to be answered some day as a matter of law.'

However, the test generally accepted is that of the achievement by the professional of a standard of care ordinarily to be expected from a professional man. The leading statement of this test is still that of Mr Justice McNair in *Bolam* v. *Friern Hospital Management Committee* (1957). He said:

> 'The test is the standard of the ordinary skilled man exercising and professing to have that special skill. It is well established law that it is sufficient if he exercises the ordinary skill of an ordinary competent man exercising that particular art...
>
> I myself would prefer to put it this way, that he is not guilty of negligence if he

> has acted in accordance with a practice accepted as proper by a responsible body of medical men skilled in that particular art... Putting it the other way round, a man is not negligent, if he is acting in accordance with such a practice, merely because there is a body of opinion who would take the contrary view.'

However, it should not be thought from this statement of the law that it is necessarily a defence to demonstrate that other professionals in the same sphere (or even all professionals in the same sphere) adopt a similar practice. It could be that the profession as a whole has fallen short of proper standards of skill and care. In the case of *Board of Governors of the Bethlem Royal Hospital* v. *Sidaway* (1985) it was made clear that the practice or conduct called into question may be judged not only against the generally accepted practice but also in accordance with a practice *rightly* accepted as proper by a body of skilled and experienced men. In other words, the judiciary reserves to itself the right to condemn certain practices or routines even if generally accepted as proper within the profession in question.

In specific instances, the court may decide that the *Bolam* test is not applicable – see for example a case involving architects, *J. D. Williams & Co Ltd* v. *Michael Hyde & Associates Ltd* (2001) in which the Court of Appeal had to consider the application of the *Bolam* test which had previously been applied to architects in *Nye Saunders* v. *Alan E. Bristow* (1987). The court found that the defendants' argument that the *Bolam* test should apply failed. The test did not apply in this case. Two of the judges held that there were some instances in which the judge was entitled to judge for himself what the appropriate standard of care was, rather than to defer to the standard of care set by a responsible body of opinion within the profession; the third judge held simply that the judge was entitled to find that the exercise of judgment involved did not in itself require any special architectural skills; that the judge did not have to get 'under the skin of a different profession', but was entitled to judge for himself what warning light shone from the disclaimer in this case.

Lord Justice Sedgley said that the principle was that professional negligence meant falling below a proper standard of competence. One application of the principal found its expression in the test, while the other comprehended a single forensically determined standard. For him, the reason for rejecting the test was not that the court did not need expert help at all, but rather that this was a case in which the judge could determine the issue before him. He went on to say that in general the test was typically appropriate where the negligence was said to lie in a conscious choice of available courses made by a trained professional, and typically inappropriate where the negligence was said to lie in an oversight.

Statute

2-71 The implied terms in relation to contracts for services are now to be found in statutory form in the Supply of Goods and Services Act 1982 Part 2.

A typical building contract is within the statutory definitions both of a contract to provide work and materials and a contract to provide a service.

Conclusion

2-72 The exercise of reasonable skill and care relates to the quality of the performance (the input) rather than an assessment of the result following the performance (the output). Negligence imports some element of fault or culpability or shortcoming. The test of fitness for purpose is independent of fault. It does not concern itself with the skill or care involved in the performance of the task. It is simply a judgment on the result, i.e. does it work?

FITNESS FOR PURPOSE VERSUS SKILL AND CARE

2-73 The employer is clearly in legal terms better off with the former obligation being owed to him. The contractor who provides the design would be in a better position if the test was based upon the latter.

2-74 Compared to the 'traditional' procurement method where the employer designs, a contractor design and build contract can lead to what must appear to be an injustice or at any rate extremely hard law on a contractor. The parties' positions are affected as follows:

(1) The employer's remedies are stepped up from a remedy against the designer (providing a service only) for failure to exercise reasonable skill and care, to a remedy against the contractor (providing a design service and also physically constructing the building) for any failure of the building to be reasonably fit for its intended purpose. Where therefore the design fails through no want of skill and care on the part of the designing contractor, the employer has a remedy which he does not have under the traditional procurement route;
(2) The contractor consequently carries a new liability greater than that which applies to a professional designer who provides a design only service. If held liable the contractor may, in turn, seek a remedy against his designer where success however will depend upon a failure of the designer to exercise reasonable skill and care. Where therefore the failure in design is not attributable to a failure to exercise reasonable skill and care, the contractor is liable to the employer on a fitness for purpose warranty but in turn cannot recover from the non-negligent designer whose design has failed.

Very often the failure in design will be of such a nature as to result not only in the building not being reasonably fit for its intended purpose but also in a want of reasonable skill and care, in which case the difference between the two tests will not be significant. However, there could be occasions, especially the 'frontiers of knowledge' type cases, where the defective design is not the result of a failure to exercise reasonable skill and care and in such cases liability stops at the contractor's door.

The contractor may have employed a designer in just the same way as the employer would have done and yet the employer has greater rights against the contractor for defective design than he has against the designer.

Contractor's solution

2-75 The contractor's answer is to either:

(1) Reduce liability to the employer in relation to defective design to equate with that of a professional man, e.g. as in WCD 98 clause 2.5.1; or
(2) Increase the liability of the designer from reasonable skill and care to fitness for purpose. This can sometimes happen unwittingly – see *Greaves (Contractors) Limited* v. *Baynham Meikle & Partners* (1975). Lord Justice Goff in the case of *Holland Hannen & Cubitts (Northern) Ltd* v. *Welsh Health Technical Services Organisation* (1985) (Court of Appeal) expressed the view that where the pro-

fessional's service resulted in a physical product, then it sometimes appeared as if his obligation approached that of fitness for purpose.

In practical terms, the most likely way to resolve this problem is for the contractor to reduce liability to the employer to equate with that of a professional man in relation to defective design. Firstly, the contractor may have difficulty obtaining adequate insurance to cover a fitness for purpose warranty. Secondly, consultants will resist accepting an obligation based on fitness for purpose as they will certainly experience difficulty in obtaining adequate insurance cover in respect of such a liability.

DESIGN AND TORTIOUS DUTIES

2-76 It should not be forgotten that in some circumstances a contractor who designs and constructs may be found to owe an independent duty of care in the tort of negligence to the employer. Firstly, there will undoubtedly be an independent duty to take reasonable care so as not to cause personal injury to the employer or physical damage to the employer's separate property (that is other than the project works). So far as liability in negligence for purely economic loss is concerned, it has been held that, as with consultants providing a service only, an independent duty to exercise reasonable skill and care exists quite apart from the contractor's contract – see for example, *Storey* v. *Charles Church Developments Limited* (1997), which was however not followed in *Samuel Payne and Others* v. *John Setchell Ltd* (2001).

Very often it does not matter whether the action is brought for breach of contract or for breach of an independent duty to exercise reasonable skill and care. However, it may be that because the cause of action in negligence arises when damage is suffered, whereas in actions for breach of contract the cause of action arises on the breach of the contract, the statutory limitation period in respect of the breach of contract expires earlier than that for an action in negligence and in these circumstances the ability to frame a claim in negligence will be of significant benefit to the employer.

CONTRACTOR'S DUTY TO WARN

2-77 Even to the extent that a contractor does not take on design duties, as an experienced contractor there is still a possibility that he may have a duty to warn of obvious inadequacies in the design of the employer – see for example *Equitable Debenture Assets Corporation Limited* v. *Moss* (1984); *Brunswick Construction* v. *Nowlan* (1974). However, where the employer is experienced or engages consultants or specialists as is likely under the MPF, the chances of such a duty being owed are likely to be quite remote. The duty is likely to be easier to establish where the inadequacies of the design create a dangerous situation – see *Plant Construction Plc* v. *Clive Adams Associates* (2000).

Materials and workmanship

2-78 As stated when discussing design (see paragraph 2-67 above), in determining the contractor's responsibilities in connection with materials and workmanship, first and foremost regard must be to the express terms of the contract concerned. However, as also stated earlier (see paragraph 2-67 *et seq.*), subject to what the express terms of the contract say and provided the express terms are not inconsistent with them, certain terms are implied as matters of law, e.g. that in a contract for work and materials the goods will be of satisfactory quality; and that reasonable skill and care will be used in relation to the workmanship employed.

These implied terms were the product of the common law. They are now, however, embodied in statute law - see section 4 (as amended by the Sale and Supply of Goods Act 1994) and 13 of the Supply of Goods and Services Act 1982.

We turn now to consider the relevant clauses of the MPF in relation to standards of design, workmanship and materials.

The contractor's responsibility for the requirements

5.1 The Contractor shall not be responsible for the contents of the Requirements or the adequacy of the design contained within the Requirements.

2-79 This is a brief and strong statement that the contractor has no responsibility for what the employer has put into his requirements in terms of their content or the adequacy of any design. However it cannot be taken completely at its face value.

Design in conflict with statutory requirements

2-80 Clause 4.4 (dealt with earlier - see paragraph 2-60) provides that if the employer's requirements are inconsistent with the statutory requirements, the employer instructs which provision is to prevail. That instruction will not be treated as giving rise to a change. This means that the contractor does take responsibility for the cost of bringing the employer's requirements into line with statutory requirements where they are at odds. The contractor's warranty under clause 5.2.1, to comply with statutory requirements being made subject to clause 5.1 (contractor not responsible for the contents of the requirements or the adequacy of the design within the requirements), does not assist the contractor here.

Duty to warn

2-81 A further point to consider is whether clause 5.1 is sufficient to remove any duty of the contractor to warn the employer of obvious inadequacies within the employer's requirements or the design contained in them. As was noted earlier (see paragraph 2-77), there may be circumstances in which despite the employer being responsible for design, the contractor nevertheless has a duty, probably as part of his duty in connection with carrying out the work in a good and workmanlike manner, to point out obvious errors. The question is whether that duty is inconsistent with clause 5.1. It should be pointed out that whether the duty to warn is incorporated in any event will depend to some extent upon whether the employer is experienced and knowledgeable in such matters and also may depend upon the extent to which the employer has adequate expertise available. If in any given situation the duty to warn would be implied, it is quite possible that clause 5.1 is not sufficiently clear to exclude its operation.

Contractor's obligation to use skill and care to check employer's design

2-82 In the case of *Co-operative Insurance Society Limited* v. *Henry Boot Scotland Limited* (2002), it was held that a contractor, under a JCT 80 Private Edition with Quantities form of contract incorporating a Contractor's Design Portion Supplement, who took forward the employer's design for contiguous bored pile walls had, by accepting an obligation to complete the design, assumed as part of that overall obligation, an obligation to satisfy himself that the design prepared by or on behalf of the employer had been prepared with reasonable skill and care. Judge Richard Seymour said:

> 'In my judgment the obligation of Boot ... was to complete the design, that is to say, to develop the conceptual design [of the employer] into a completed design capable of being constructed. That process of completing the design must, it seems to me, involve examining the design at the point at which responsibility is taken over, assessing the assumptions upon which it is based and forming an opinion whether those assumptions are appropriate. Ultimately, in my view, somebody who undertakes, on terms such as those of the Contract ... an obligation to complete a design begun by someone else agrees that the result, however much of the design work has been done before the process of completion commenced, will have been prepared with reasonable skill and care. The concept of "completion" of a design of necessity, in my judgment, involves a need to understand the principles underlying the work done thus far and to form a view as to its sufficiency'.

2-83 The contractor's arguments to the contrary were not helped by the absence of separate employer requirements, contractor's proposals (mentioned only as sections of the bills of quantities) and contract sum analysis; or by the inclusion of different portions of design within a single supplement together with specific employer amendments making the contractor responsible for the integration and compatibility of the various elements of the works as a whole and design co-ordination.

However, the contract in the *Co-operative* case did not contain a clause such as 5.1 of the MPF, and the MPF makes no mention of co-ordination or integration. It is likely therefore that the result would be different under the MPF form. It should not be forgotten, however, that the contractor is likely to need to look at the pre-existing design in order to complete it and a duty to warn might arise (see paragraph 2-81).

Employer amendment?

2-84 Finally, it must be said that many employers who find themselves in a strong bargaining position will not find clause 5.1 acceptable and will no doubt, as they do in respect of the employer's requirements under WCD 98, transfer responsibility for the whole of the requirements to the contractor upon his appointment. This may be achieved by ensuring that whatever the employer requires and wishes to stipulate for is in fact kept out of the requirements and instead is included only in the proposals from the contractor; alternatively the employer may amend the MPF to expressly place responsibility for the requirements onto the contractor. If either approach is adopted, it will be important for the contractor to ensure that there is a novation agreement which provides him with an adequate remedy against the consultant should a consultant's negligent error in the preparation of the proposals or requirements result in breach of the MPF on the part of the contractor.

Contractor's design warranties

> 5.2 Subject to clause 5.1 the Contractor warrants that the design of the Project will:
> .1 comply with the Statutory Requirements,
> .2 satisfy any performance specification contained within the Requirements, and
> .3 use materials selected in accordance with 'Good practice in the selection of construction materials', prepared by Ove Arup & Partners and sponsored by the British Council for Offices and the British Property Federation, as current at the Base Date.

2-85 We will consider these three elements in turn, all of which are subject to clause 5.1. Clause 5.3 goes on to deal with the contractor's warranty to exercise skill and care. Clause 5.4 deals with the contractor's obligations in connection with the provision of goods and materials, and clause 5.5 with standards of workmanship.

Warranty to comply with statutory requirements

2-86 This is contained in clause 5.2.1. It is made subject to clause 5.1, which provides that the contractor is not responsible for the contents of the employer's requirements or the adequacy of the design within the requirements. The relationship between clause 5.1 and 5.2.1 was commented on above (paragraph 2-80). The obligation to comply with statutory requirements is absolute and even compliance by the contractor with the employer's requirements, if this results in a failure to comply with statutory requirements, will be no defence to the contractor who is in breach of the statutory requirements. The statutory requirements most commonly relevant to construction work have already been considered (see paragraph 2-50 *et seq.*).

Warranty to satisfy any performance specification in the requirements

2-87 The warranty to satisfy any performance specification contained in the requirements is to be found in clause 5.2.2. It is absolute in nature. Any remedy of the employer for the contractor's failure to achieve any such performance specification is not dependent upon establishing that the contractor has failed to exercise reasonable skill and care. Liability here is independent of fault. In practice many aspects of the requirements are likely to be expressed in terms of performance criteria. Harrison and Keeble in *Performance Specifications for Whole Buildings: Report on BRE Studies 1974–1982*, published by British Research Establishment, Garston (1983), define performance specification (p. 52) as 'a detailed description of performance requirements, which may refer to tests'.

2-88 A warranty to meet a performance specification is an example of an obligation to provide whatever may be the subject of the performance specification in such a manner that it will be fit for its intended purpose. If the contractor has used a specialist contractor or supplier to achieve the performance requirement, he needs to ensure that the relevant sub-contract or supply contract mirrors this obligation both in terms of a fitness for purpose obligation and the absence of any financial limitation clause. If the contractor has used consultants to carry out a design service only, almost certainly their obligation will be restricted to the use of reasonable skill and care which will leave the contractor exposed to a fitness for purpose claim from the employer which cannot be stepped down. The contractor should, if it is obtainable, ensure that insurance cover is available to cover this risk.

Warranty to use materials selected in accordance with good practice

2-89 By clause 5.2.3 the contractor warrants that he will select any materials he uses in accordance with 'Good practice in the selection of construction materials', prepared by Ove Arup and Partners and sponsored by the British Council for Offices and the British Property Federation current at the base date. This is a sensible provision which is also commonly found in collateral warranties. An alternative, and much less satisfactory option, is to include a specific list of prohibited materials. Many such lists which are encountered in practice, would, if they were adhered to strictly, prevent the project ever being built.

The reference to the edition current at the base date strikes what is probably a reasonable balance of risk. Some employers may well seek to amend this provision by replacing the reference to base date with the time of use, imposing on the contractor the risk of a material which is regarded as satisfactory at the base date, subsequently being declared unsuitable in a later revision of the guide.

Contractor to use skill and care

> 5.3 The Contractor warrants that it will exercise in the performance of its obligations in relation to the design of the Project the skill and care to be expected of a professional designer appropriately qualified and competent in the discipline to which such design relates and experienced in carrying out work of a similar scope, nature and size to the Project. The Contractor does not warrant that the Project, when constructed in accordance with its designs, will be suitable for any particular purpose [b].
>
> [b] See Guidance Note which sets out alternative model clauses

2-90 Clause 5.3 is the contractor's warranty to exercise skill and care in the design of the project.

Comparison with the implied term of professional skill and care

2-91 The implication of terms in relation to design only and design and construct contracts were discussed earlier (see paragraph 2-66 *et seq*.). It will be recalled that in the absence of express terms of the contract to the contrary, a contract for design and construction will contain an implied term requiring the design and build contractor to construct a project which is reasonably fit for its intended purpose. On the other hand, in a contract for the provision of design services only, with no obligation to build, the implied term is different. In this case the professional designer will be required to exercise his professional skills with the reasonable skill and care to be expected of the ordinary man exercising and professing to have that special professional skill (see the *Bolam* test referred to earlier – paragraph 2-70). For the reasons given earlier (see paragraph 2-75), contractors will generally seek to move away from the implied term of reasonable fitness for purpose which guarantees a result (the output), and instead replace that obligation with one similar to that of a professional designer to use reasonable skill and care in the carrying out of the design (the input).

2-92 It is important to realise that clause 5.3 does not simply require the contractor to use reasonable skill and care. It goes on to say much more. The skill and care required is that to be expected of a professional designer appropriately qualified and competent in the discipline to which such design relates, and experienced in carrying out work of a similar scope, nature and size to the project. These qualifying words are of vital significance.

Without such express words, the debate would be open as to exactly what was to be expected from the ordinary man exercising and professing to have the special skills required to carry out design. Without these further qualifying words, would the contractor be judged against the standard of the average professional designer such as the average architect or average structural engineer on the one hand, or against a specialist within the particular sphere of design activity, e.g. an architect specialising in art gallery design, on the other? The law has not as yet given a definitive answer to this. Supposing a professional person is engaged and paid at a higher than average fee because of his or her specialist knowledge. Should the skill and care to be expected of that person be greater than that of someone with less specialist knowledge or of lesser experience? This can apply across the professions. For instance, in a case concerning a solicitor, *Duchess of Argyll* v. *Beuselinck* (1972) Mr Justice Megarry said:

> 'No doubt the inexperienced solicitor is liable if he fails to attain the standard of a reasonably competent solicitor. But if the client employs a solicitor of high standard and great experience, will an action for negligence fail if it appears that the solicitor did not exercise the care and skill to be expected of him, though he did not fall below the standard of a reasonably competent solicitor? If the client engages an expert, and doubtless expects to pay commensurate fees, is he not entitled to expect something more than the standard of the reasonably competent? . . . If, as is usual the retainer contains no express term as to the solicitor's duty of care, and the matter rests upon an implied term, what is that term in the case of a solicitor of long experience or specialist skill? Is it that he would put at his client's disposal the care and skill of an average solicitor, or the care and skill that he has? . . . I wish to make it clear that I have not overlooked the point, which one day may require further consideration.'

The same issue was considered further in the case of *George Wimpey & Co Limited* v. *D. V. Poole and Others* (1984) where Mr Justice Webster made the point that the more stringent test would be appropriate where the issue is related to the actual knowledge of a specialist or experienced professional where, having that knowledge, he acts with a want of reasonable skill and care. In such a case the question whether he has fallen short of the appropriate standard cannot be answered by reference to a lesser degree of knowledge than he actually has on the grounds that an ordinarily competent practitioner would have had that lesser degree of knowledge. Nevertheless, the judge emphasised that the *Bolam* test applies where the professional man causes damage because he actually lacks some knowledge or awareness. The *Bolam* test then establishes the degree of knowledge or awareness which he should have. See also *Gloucestershire Health Authority* v. *M. A. Torpy & Partners* (1997).

It is much more sensible and satisfactory from the employer's point of view to make the appropriate test more certain and this is achieved by the wording of clause 5.3. The result is that whatever the actual discipline, expertise or experience of the designer, the skill and care which the contract requires and the contractor warrants is set against the standard of the appropriately qualified competent professional designer who is experienced in carrying out work of similar scope, nature and size.

2-93 The express reference to the *professional* designer is clearly appropriate as otherwise there would be a debate as to what was the reasonable skill and care to be expected of a design and build contractor in a situation where there is of course no profession as such of design and construct contractors by which to measure ordinary competence, let alone specialist skills.

No warranty of fitness for purpose

2-94 Clause 5.3 goes on to provide expressly that the contractor does not warrant that the project, when completed, will be suitable for any particular purpose. This negatives any fitness for purpose obligation with respect to the project as a whole. It is sensible to expressly state this to avoid the argument that expressly requiring the contractor to use skill and care in the design (input) is insufficient to exclude the implied term that the project will be reasonably fit for its intended purpose (output). Without such express words, it would at least be arguable that the contractor would have a dual obligation both to use reasonable skill and care in the design and to produce a project which was reasonably suitable for its intended use.

2-95 This express absence of a fitness for purpose warranty is in respect of the project as a whole. In relation to materials or goods incorporated in the project, clause 5.4 provides that, where the contractor selects them, they are to be 'reasonably fit for their intended purpose'. This is a significant limitation to the application of the express exclusion of the warranty of fitness for purpose contained in clause 5.3. Indeed it is fundamentally inconsistent with it. If the project is unsuitable for its purpose because of materials or goods being unfit for their intended purposes, the contractor will be in breach of clause 5.4.

2-96 The express exclusion in clause 5.3 must be targeted at what might be called the pure design element, e.g. the calculations to achieve the required floor areas for a building (even here if the requirement is specified in performance terms it carries a fitness for purpose obligation under clause 5.2.2 – see paragraph 2-87). In many situations a design incorporates a choice of suitable materials or goods. For example, the contractor's structural engineer may without negligence miscalculate the loads to be imposed on piled foundations, resulting in the selection of an inadequate pile. The pile chosen will, in one sense at least, not be reasonably fit for its intended purpose, i.e. it will not sustain the actual load imposed by the building and its intended contents. However, in another sense it will be suitable for its intended use, which is to sustain the load shown in the structural engineer's design. The contractor would not therefore be liable.

2-97 Consistency would have required the contractor to have a contractual obligation limited to the use of skill and care in seeking to achieve any performance specification and in the selection of suitable materials and goods. As it stands, the MPF in clauses 5.2.2, 5.2.3 and 5.4 takes away a considerable amount of what it purports to give the contractor in clause 5.3.

The relationship between the failure to exercise skill and care and the obligation to achieve practical completion

2-98 In effect, the contractor is only liable for failures in the design if they amount to professional negligence. There is no warranty that the project, when constructed in accordance with the contractor's design, will be suitable for any particular purpose.

So what is the position if during the course of construction it becomes apparent that there is a design defect which is not the result of a failure to exercise skill and care but which will, unless the design is changed, prevent the project reaching practical completion at all. Suppose, for example, a large two-storey wholesale warehouse complex is being built using a structural steel frame and pre-cast concrete planks bound together with ready-mixed concrete to make the solid floor between the two storeys. Suppose further that there is a code of practice covering such a form of construction and that one of the matters for consideration is the question of vibration. This is critical as the upper floor of the warehouse is intended to be used by fork lift trucks. The structural engineers engaged by the contractor to carry out the design consider this matter by reference to the code of practice. They interpret it in a certain way which is reasonable but wrong because (unappreciated by them) the code of practice is ambiguous and can be read in two different ways. The result of this non-negligent mistake is that the design of the floor is inadequate to withstand the vibration caused by stacker trucks running across it. This is realised during construction when cracking appears in this deck when construction plant, which has a similar effect to stacker trucks, operates across it. Until the floor problem is resolved and rectified, it is impossible to continue with the construction in accordance with the design as the lack of structural integrity of the floor following the cracking means that the loading of the walls on top of the floor and the roof cannot be safely imposed.

2-99 If the contractor is not liable for this non-negligent design error, does this relieve the contractor of the obligation to complete? Alternatively, does it mean that the employer must issue an instruction under clause 2.1 in connection with the design of the project, and if so does it amount to a 'Change' under clause 20, entitling the contractor to payment and an adjustment to the completion date? If the employer does not issue an instruction, can the contractor regard himself as effectively discharged from further performance on the basis that as a result of a matter for which he is not contractually responsible, the design is incapable of achieving practical completion of the project?

2-100 A number of clauses are relevant in addition to clause 5.3. Clause 1.1 requires the contractor to execute and complete the project in accordance with the contract, which includes his proposals (which would include the design of the floor). Under clause 9.2 the contractor is to achieve practical completion. Practical completion is defined in clause 39.2; it 'takes place when the Project is complete for all practical purposes'. Practical completion of the project is also defined in clause 39.2, which provides that it occurs 'upon Practical Completion'.

From this it can be deduced that the contractor has a contractual obligation to achieve practical completion in accordance with his proposals. Is he warranting that his proposals are, if followed, capable of achieving practical completion? In other words, does he guarantee that this is possible so that if there is a non-negligent design error which makes this impossible, the contractor is in breach of contract? Or can the contractor argue that what he is warranting is that he will use skill and care in endeavouring to ensure that the proposals, if followed, will result in the project achieving practical completion?

The problem in having design obligations aligned to those of a professional is that the professional has no contractual link with the intended end result. The design and build contractor has an obligation not only to design but also to physically construct the building and to achieve practical completion, that is, to hand it over to the employer in a state in which it is 'complete for all practical purposes'; in the above example, that is as a warehouse in which to store and move barrels of oil, and which can therefore, at least at that time, be used for the purpose for which it was intended. If the building can be handed over and taken into use for its intended purpose, the subsequent discovery of a non-negligent design defect which renders it unfit for continued use, will not result in any liability for the contractor as he will have met his obligation to achieve practical completion and will not have failed to use appropriate skill and care.

In the example given, unless the design is changed, the project can never be handed over in a state in which it can be used at all. One way of looking at it is to say that in effect the contractor is promising that the use of skill and care in the design will enable him to achieve practical completion; in other words, that he will by the use of skill and care in the design be able to hand over a completed project which, at that point in time at least, can be used for its intended purpose. Equally, it could be contended that the contractor, becoming aware part-way through the design and construction process that his defective design will prevent the project reaching practical completion, must redesign (and rebuild as necessary) in order to fulfil his obligation at that point in time to design using the appropriate skill and care.

On the other hand, the contractor would point out that clause 5.3 expressly states that the contractor does not warrant that the project will be suitable for any particular purpose. However, it is to be noted that the precise wording is that the contractor does not warrant that the project 'when constructed in accordance with its designs, will be suitable'. In other words it assumes that the project can be constructed in accordance with his design and only then, that is in effect at practical completion, is it possible to judge whether or not it is suitable for the employer's purpose. Put another way, the fitness for purpose obligation may not actually be excluded until practical completion is reached.

On this basis the contractor would be responsible for the failure. The employer would not be obliged to issue a change instruction requiring the contractor to alter his proposals. The MPF prescribes that certain matters are to be treated as being a change. However, none of these would apply (see paragraph 5-15). The contractor would however need to redesign.

It is worth noting also that the definition of 'Change' in clause 39.2, while it includes 'Any alteration in the ... Proposals that gives rise to an alteration in the design' is subject to the proviso that 'the alteration or matter referred to ... is not required as the result of any negligence or default on the part of the Contractor'. Here, there has been no negligence on the part of the contractor but arguably the contractor is in default with the result that there is no change and the contractor will not be paid in connection with the revised design and its implementation. On the basis that the contractor is in default, he will of course not be recompensed for any rebuilding or rectification works required as a result of the design error. However, the issue of whether the contractor is in default in such circumstances does beg the question whether the non-negligent design defect amounts to a default at all.

The position is similar if the contractor discovers such a non-negligent design error before any relevant construction has taken place. Again, he will need to redesign and bear the costs of the alterations to his proposals.

Quite apart from all of this, in this particular case the contractor would find himself liable, whether the failure occurs before or after completion, for breach of the warranty under clause 5.2.2 if the requirements were in the form of a performance specification.

In passing it is worth noting that a respectable argument of a similar kind can be raised in connection with WCD 98 (see clauses 2.1, 2.5.1, 8.4, 8.5, 12.1, 16.1 and 23.1.1 of that contract). It is made stronger by the fact that the definition of a change in clause 12.1 is limited to changes in the employer's requirements (which would remain unchanged in the example used above) and does not extend, as does the MPF form, to encompass alterations to the contractor's proposals).

If this interpretation of the MPF is correct, it may leave the contractor bearing a risk without the protection of insurance cover.

Employers wishing to amend the MPF to impose a fitness for purpose obligation

2-101 The MPF Conditions do not provide an option for the contractor to give a warranty of fitness for purpose in connection with his design of the project. However, a model *clause* (not alternative model *clauses* as footnote [b] states) is provided in the Guidance Note (page 7) for such a purpose. It provides as follows:

> 'The Contractor warrants that the Project, when completed, shall be suitable for the purpose stated in the Requirements.'

This warranty of fitness for purpose is related to the suitability of the project for the purpose stated in the requirements. It is important for both employer and contractor that this purpose or the purposes are clearly stated. It can be expected that the employer will refer to a considerable range of possible uses for the finished project without restricting it to any intended immediate use. It will of course be for the employer to decide whether, if a fitness for purpose obligation is to be included, he wishes to adopt the model clauses or alternatively produce his own clauses.

Kinds and standards of materials and goods

> 5.4 The Contractor shall use in the execution of the Project materials and goods of the kinds and standards described in the Contract or, if no such kinds or standards are described, materials and goods that are reasonably fit for their intended purpose. All materials and goods used for the Project shall be of satisfactory quality. Where materials or goods of the kinds and standards described in the Contract are not procurable the Contractor shall propose for the acceptance of the Employer an alternative that is wherever possible of an equivalent or better kind or standard, such acceptance not to be unreasonably delayed or withheld. The use of any alternative shall not be treated as giving rise to a Change unless the alternative accepted by the employer is of a lesser kind or standard to that described in the Contract, in which case the provisions of clause 20 (Changes) shall apply as though the Employer had instructed a Change.

2-102 This clause covers the basic obligation of the contractor to execute the project using materials and goods of the kinds and standards described in the contract; this will generally mean as described in the requirements or else in the proposals. Where no kinds or standards are described, the materials and goods selected by the contractor are to be reasonably fit for their intended purpose. All materials or goods used on the project are to be of satisfactory quality. The clause goes on to deal with the position where materials or goods of the required kinds or standards are not available, in which case the contractor has to propose an alternative for acceptance by the employer, which wherever possible is to be of an equivalent or better standard. This does not result in a change unless the accepted alternative is of a lesser kind or standard.

2-103 The contract documents will usually specify the materials or goods. They may also specify the standard of workmanship. Where they do not so specify, there will be an implied term that the materials or goods will be of satisfactory quality and that the workmanship will be carried out with reasonable skill and care (see sections 4 (as amended by the Sale and Supply of Goods Act 1994) and 13 of the Supply of Goods and Services Act 1982). Even where the contract documents do specify, there will still be room for the statutory implied term to apply to the extent that it is not inconsistent with the specified requirements.

2-104 Where the specification as to quality leaves room for debate, it is a matter for argument as to whether the price contained in the contract for the materials or work is a relevant factor. In reliance on the case of *Cotton* v. *Wallis* (1955) it is possible to argue that the price may be taken into account. However, the first point of reference is the description in the contract documents and if this assists in determining the quality or standards then it would be wrong to take the price or rate for the work into account. If, however, the contract documents leave the matter genuinely in doubt, this can bring into consideration the price or rate. It was confirmed in the case of *Brown and Brown* v. *Gilbert-Scott and Another* (1992) that an architect could have regard to the price in determining matters of standards and quality. He could not be expected to ignore the fact that the works may have been 'built down' to a price. Even so, the line has to be drawn somewhere between work which exhibits a low quality or standard and that which can be regarded as defective when compared with the contract specification.

The Supply of Goods and Services Act 1982 (as amended by the Sale and Supply of Goods Act 1994), which applies to building contracts, states in relation to the implied condition that goods supplied under such a contract are to be of satisfactory quality, that they are of satisfactory quality if they meet the standard that a reasonable person would regard as satisfactory, taking account of any description of the goods, the price (if relevant) and all the other relevant circumstances (see section 4(2A)). This provision to some extent, therefore, begs the question. If the price is a relevant factor, i.e. where it is clearly intended to have a direct bearing upon quality or standards, then it will be taken into account.

2-105 The Unfair Contract Terms Act 1977 places restrictions on the ability of a contracting party to exclude or limit the operation of the implied terms as to quality. Where the contract falls within the definition of 'business transaction' any such exclusion or restriction must satisfy 'the requirement of reasonableness', see sections 7 and 11 and schedule 2 to the Act.

By far the majority of contracts let under the MPF will form part of a business transaction. In most contracts if the contract documents specifically require a lesser standard than that which would be implied under the Supply of Goods and Services Act 1982, this will be the requirement of the employer, and there is no exclusion or restriction upon which the Unfair Contract Terms Act 1977 can operate.

It is also possible that a standard form such as MPF could be covered by the Unfair Contract Terms Act 1977 as being the employer's written standard terms of business. See, for example, a case concerning the British International Freight Association Standard Trading Conditions: *Overland Shoes Ltd* v. *Schenkers Ltd* (1998) in which it was held that the section 3 requirement of reasonableness could apply to any contractual term seeking to exclude or restrict the employer's liability in respect of any breach by the employer of a contract term. Such terms are likely to be few and far between in MPF but a possible example could be in relation to the binding effect of the final payment advice (clause 22.7).

Materials and goods of the kinds and standards described in the contract

2-106 The description of the kinds or standards of the materials and goods is likely to be found in the requirements, failing which in the proposals. They will often of course be in the form of a specification, or sometimes noted on drawings. Some information on kinds or standards may be located in the pricing document.

2-107 The description of kinds or standards may in some cases be specific in relation to particular materials or goods, e.g. use of a particular manufacturer's product or equipment stating its item or product number; in other cases there may a general description, e.g. that all materials and goods shall be of the best kind and highest standard. From the contractor's point of view, such a description is likely to cause considerable concern. It prevents what would otherwise be, in the absence of such an express general provision, an obligation on the contractor under clause 5.4, to provide materials and goods that are reasonably fit for their intended purpose. It may place upon the contractor an obligation to do more than meet a reasonable fitness for intended purpose obligation, with some greater obligation which is difficult to assess. In the absence of a specific description, the employer's rights are adequately protected both by the express requirement for the contractor to provide materials and goods reasonably fit for their intended purpose and in any event by the contractually implied term to the same effect (see earlier paragraph 2-78). Faced with such a provision in the requirements, the contractor may wish to specifically describe what is being provided in his proposals, though this would not in itself be capable of defining the kind or standard needed to meet such a requirement. Further, it would be open to the employer to argue that there is a discrepancy between the requirements and the proposals and to instruct the contractor to comply with the description in the requirements (see clause 4.4).

Reasonable fitness for purpose and skill and care

2-108 Clause 5.3 provides the contractor's warranty to exercise skill and care in relation to the design of the project. Design generally includes the selection of suitable materials or goods. However, clause 5.4 provides that in making that selection, the materials and goods are to be reasonably fit for their intended purpose. This is of course an absolute warranty not dependent upon the use of the skill and care in making the selection. It is a significant inroad into the skill and care warranty in clause 5.3 (this has been discussed earlier – see paragraph 2-95).

Materials and goods to be of satisfactory quality

2-109 Not only must the materials or goods be of the kind and standards described in the contract or alternatively of the kind or standard such that they are reasonably fit for their intended purpose, they must also be of satisfactory quality. This mirrors the implied term as to satisfactory quality (Supply of Goods and Services Act 1982 – see paragraph 2-78). This means that even if materials or goods described in the contract or otherwise selected by the contractor are suitable for their intended use, e.g. a specific high grade roof tile is selected which is ideally suited for the particular application, such as Somerset 13 tiles, if nevertheless the particular batch provided contains a defect, they will fail to meet the requirement that they are to be of satisfactory quality. (See *Young & Marten Ltd* v. *McManus Childs* (1969).) The contractor will therefore be responsible for this defect in quality even if the particular materials or goods concerned are as described in the requirements.

Materials or goods not procurable

2-110 If the materials or goods of the kinds or standards described in the contract are not procurable, the contractor must propose an alternative which, wherever possible, is to be either of an equivalent or better kind or standard.

The obligation to provide an alternative which is of an equivalent or better kind or standard is only marginally qualified by the reference to 'wherever possible'. So provided it is possible to obtain an equivalent, it must be obtained unless something even better is available which the contractor might wish to propose, e.g. it may be less expensive. What does 'wherever possible' mean? Even though it is absolute in terms, regard must be had to the overall context. If the only equivalent or better alternative is disproportionately expensive, the contractor might be entitled to propose an available lesser kind or standard (which would be treated as a change) if the lowering of the kind or standard is not significant. If an equivalent is not available but it is possible to obtain a better kind or standard then the contractor must propose this, presumably even if the additional cost is significantly greater. The use of equivalent or better is expressly stated as not giving rise to a change. This can be seen to be a significant risk area for the contractor.

2-111 Clause 5.4 provides for the execution of materials or goods of kinds and standards described in the contract and goes on to deal with the situation if no such kinds or standards are procurable. It also states that all materials and goods shall be of satisfactory quality. It does not, however, expressly provide that the materials and goods are to be of the quality described in the contract. What if goods of the same kind and standards as described in the contract are available but they are of lesser quality than the quality described in the contract, nevertheless being still of satisfactory quality? Presumably the reference to 'standards' must mean something different to the use of the word 'quality' in clause 5.4. In such a case is the contractor bound to achieve that stated contractual degree of quality? No doubt he is under an obligation to execute the project in accordance with the contract (clause 1.1) which includes the requirements (clause 39.2).

The position could have been made clearer if clause 5.4 required materials or goods for the project to be of the quality described in the contract and that if no such quality was described that they should be of satisfactory quality.

2-112 Another problem is that if the required quality is not procurable there is no option to propose an alternative quality even though it may still be satisfactory.

EMPLOYER'S ACCEPTANCE OF ALTERNATIVE PROPOSAL

2-113 Clause 5.4 requires that the employer's acceptance of the contractor's proposal of an equal or better alternative is not to be unreasonably delayed or withheld. Even if the contractor's proposed alternative is of a lesser kind or standard, the employer's agreement is still not to be unreasonably delayed or withheld if an equal or better kind or standard is unobtainable, or possibly even if obtainable but only on a totally unreasonable basis (see previous paragraph).

EMPLOYER WITHHOLDING ACCEPTANCE

2-114 There may be a number of grounds upon which the employer can reasonably withhold acceptance of the contractor's alternative proposal. An example might be where the contractor has a choice of materials or goods of an equivalent kind or standard and proposes a particular alternative which is not the most appropriate choice from the employer's point of view, e.g. where the employer's property portfolio is being rationalised so as to limit the number of different lift manufacturers used. Difficult issues could arise where the employer's objection is based upon aesthetic rather than functional considerations. It is likely to introduce a debate about the extent to which the particular selection of materials or goods should be based upon function and/or aesthetic quality.

Depending upon the meaning to be given to the statement that the use of any alternative shall not be treated as giving rise to a change, another ground for withholding acceptance may be that the alternative proposal, if accepted, would lead to a consequential need to alter other aspects of the project which might give rise to a change for which the contractor could seek payment.

ALTERNATIVE MATERIALS OR GOODS NOT TO BE TREATED AS GIVING RISE TO A CHANGE UNLESS OF A LESSER KIND OR STANDARD TO THAT DESCRIBED

2-115 Clause 5.4 provides that the use of materials or goods of an alternative but equivalent or better kind or standard does not give rise to a change. This means that the contractor will not be able to seek payment under clause 20. The fact that clause 5.4 expressly provides that the use of any such alternative shall not be treated as giving rise to a change suggests that without such express words, it might be capable of being treated as a change. Change is defined in clause 39.2 to include any alteration in the requirements or proposals giving rise to alterations in design quality or quantity. However, part of the definition of change also includes a proviso to the effect that the alteration must not be required as a result of any negligence or default on the part of the contractor. The express provision that if the materials or goods of the required kinds and standards are not procurable, the contractor's choice of an equivalent or better kind or standard is not to be treated as a change suggests that the contractor's inability to procure is not a default. The problem is that clause 5.4 is not written in these terms. It provides that the contractor is to provide materials and goods of the kinds and standards described in the contract. It then goes on to deal with the situation where these are not procurable. That does not in itself necessarily lead to the conclusion that it is not still a breach of contract on the part of the contractor. If it is a breach, then provided the reference to 'default' in the proviso contained within the definition of change, includes such a breach of contract, the proviso would apply and prevent a change arising in any event.

This point becomes relevant in relation to the possible knock-on effects to the project if such an alternative proposal is agreed. While in itself it does not give rise to a change, it could result in consequential alterations to other aspects of the design, quality or quantity required in accordance with the contract. The contractor may seek to argue that the wording of clause 5.4 does not embrace such consequential alterations. Two issues arise. Firstly, the same question arises as in the previous paragraph as to whether the consequential alteration is the result of any negligence or default on the part of the contractor so as to prevent it becoming a change. Secondly, if there is any chance that the proviso does not apply with the effect that the consequential alterations could become a change, for which the contractor would be compensated, could the employer reasonably withhold his acceptance of the proposed alteration? Probably yes. Employers should be aware of this possibility when considering any proposed alternative.

ACCEPTANCE OF A LESSER KIND OR STANDARD IS A CHANGE

2-116 If the employer does accept the contractor's alternative proposal to provide materials or goods of a lesser kind or standard than that described in the contract, this would be treated as a change and valued in accordance with clause 20. The clear intention here must be that it is to be treated as a change in order for the employer to obtain the benefit of a reduction in the contract sum to reflect the lesser kind or standard. No doubt the amount of the reduction can be agreed between the parties, using the provisions in clause 20.3 to 20.5 for the contractor to provide a quotation prior to the employer instructing a change. Failing agreement, a fair valuation will be required under clause 20.6 and in either event regard is required to be had to the matters identified in clauses 20.6.1 to 20.6.4.

If the intention is that clause 20 should apply in order to achieve a reduction in the contract sum, the most relevant clause would be clause 20.6.3 which requires regard to be had in reaching a fair valuation to the prices and principles in the pricing document so far as applicable. The other matters – the nature and timing of the change (clause 20.6.1), the effect of the change on other parts of the project (clause 20.6.2) and any loss and expense (clause 20.6.4) – seem for the most part inapplicable.

It should be noted that clause 5.4 does not specifically provide that the acceptance of the employer of a lesser kind of standard should not also be treated as a change where the overall effect is to nevertheless add to the cost of the project. Presumably therefore it is still necessary to use the provisions of clause 20 to determine the matter. Whether this is achieved by the quotation procedure in clauses 20.3 to 20.5 or a fair valuation made by the employer under clause 20.6, regard is to be had to the same matters set out in clauses 20.6.1 to 20.6.4. On its face this would appear to entitle the contractor to additional costs, and possibly even to recovery of loss and expense unless the contractor's failure to procure is to be regarded as a contributory element under the proviso to clause 20.6.4. Can the employer argue that agreeing to accept materials or goods of a lesser kind or standard means that in principle a fair valuation should not compensate the contractor who is providing a project of a lower standard than that described in the contract? Probably not. The contractor is entitled to a fair valuation having regard to the matters set out in clauses 20.6. That is what clause 5.4 expressly provides in such circumstances. Having determined that a fair valuation is to take place, regard must be had to the stated matters. This is not to say that they must inevitably be applied, though they must not be disregarded. It is the amount of the valuation having regard to those matters, which is to be fair, not whether it is fair to have a valuation at all. This must mean that a fair valuation should seek to genuinely assess the effect upon costs of the nature and timing of the change (clause 20.6.1) and the effect of the change on other parts of the project (clause 20.6.2), as well as a reduction based upon the prices and principles set out in the pricing document where applicable (clause 20.6.3), and possibly also loss and expense (clause 20.6.4).

If this interpretation of clause 5.4 is correct, it is important that the employer appreciates the situation before deciding whether to accept the contractor's proposal. The contractor's position in relation to this matter should be clarified and clearly recorded. If an acceptance by the employer would have the effect of increasing the overall cost to the employer, it may well be reasonable to withhold such acceptance.

ADJUSTMENTS TO THE COMPLETION DATE

2-117 As the acceptance by the employer of the contractor's proposal to provide materials or goods of a lesser kind or standard is to be treated as a change, the contractor will be entitled to a fair and reasonable adjustment to the completion date (see clauses 12.1.5 and 12.4). It could well be that this was unintended. Here, it will clearly be arguable that the reference to fair and reasonable is not just linked to a fair and reasonable assessment of actual delay but carries with it consideration of whether it is in principle fair to adjust the completion date by extending it. Even so, it would be sensible for the employer to seek clarification of the contractor's position before accepting the contractor's alternative proposal.

Standards of workmanship

5.5 All workmanship shall be of the standards described in the Contract or, if no such standards are described, shall be executed in a good and workmanlike manner.

2-118 This clause deals with workmanship, which is to be of the standard described in the contract. It goes on to provide, rather confusingly, that if no standards are described then workmanship shall be executed in a good and workmanlike manner.

2-119 The obligation to achieve standards of workmanship as described in the contract is a typical provision to be found in construction contracts. It is normally targeted at the requirement to use a sufficient degree of skill to achieve the standard of workmanship required by the contract. However, the standard or quality of the workmanship is not the only relevant factor. It is important also that in carrying out the work, the contractor undertakes that it be executed in a good and workmanlike manner. This further provision is targeted at ensuring that not only are the appropriate standards of workmanship reached in terms of the quality of the project, but that the contractor also has a responsibility for carrying out the work in such a way as to not cause collateral problems unassociated with the quality of the project. An example would be where, while the contractor achieved adequate load bearing in employing a certain piling design, nevertheless in installing the piles the amount of vibration adversely affected the soil conditions and caused damage to adjoining properties (*Greater Nottingham Co-operative Society Ltd* v. *Cementation Piling & Foundations Ltd* (1989)). Other examples would be carrying out the work so as to avoid trespassing with an overhanging tower crane. There should accordingly be two requirements, one to achieve the appropriate standards of workmanship, and the other to execute the work in a good and workmanlike manner so as not to cause collateral risks of injury or damage. It is for this reason that it appears inappropriate to state that if no standards of workmanship are described, workmanship shall be executed in a good and workmanlike manner. It is the work that should be executed in a good and workmanlike manner, but the workmanship which should be of the required standard. Comparison may be made with WCD 98 which, in clause 8.1.2, deals with standards of workmanship and provides that to the extent that no standards are specifically described, the workmanship shall be of a standard appropriate to the works. It goes on to provide in clause 8.1.3 that all work shall be carried out in a proper and workmanlike manner. They are separate obligations.

Design submission procedure – clause 6

Introduction

2-120 Clause 6 provides for a design submission procedure. In respect of many areas of the MPF there is much less in the way of detailed procedures than with other JCT contracts, e.g. in relation to extensions of time, loss and expense, contractor's quotations in respect of changes, the valuation of changes, off-site goods and materials, insurances, the use of named specialists. However, when it comes to the contractor complying with the requirements and proposals, this clause provides for a specific procedure allowing the employer to comment in relation to what are called 'Design Documents' (drawings, specifications, details, schedules of levels, setting out dimensions and the like). Anything therefore which falls into this category and which is not already part of the proposals, but which is required for the purpose of explaining or amplifying the requirements or the proposals in order that the contractor can execute the project, will have to be submitted to the employer who is given the opportunity to comment.

This is a major issue in practice. The employer or his retained professional advisors will generally want firstly to check by this means whether the requirements and proposals are being adhered to and faithfully interpreted, and secondly will in any event want the opportunity to see exactly what is to be done in case they want the opportunity to make a change. WCD 98 does not provide for such a mechanism although procedures are regularly to be found set out in the employer's requirements. Such procedures clearly have implications in relation to the contractor's construction programme and also the resources expended in complying with such procedures. The MPF brings this out into the open with the contract conditions themselves providing a timetable which the contractor can and should build into his programme for the project. The procedure and timetable are shown in Fig. 2.1.

The purpose of providing for the employer to comment on design documents is to identify firstly whether the employer has formed the view that they are not in accordance with the contract, and secondly, to identify why he considers that they are not in accordance with the contract.

This clause also covers other points of considerable importance in practice, including the contractual status and effect of the employer's and contractor's participation in the design submission procedure.

Contractor to prepare the design documents

6.1 The Contractor shall prepare the Design Documents.

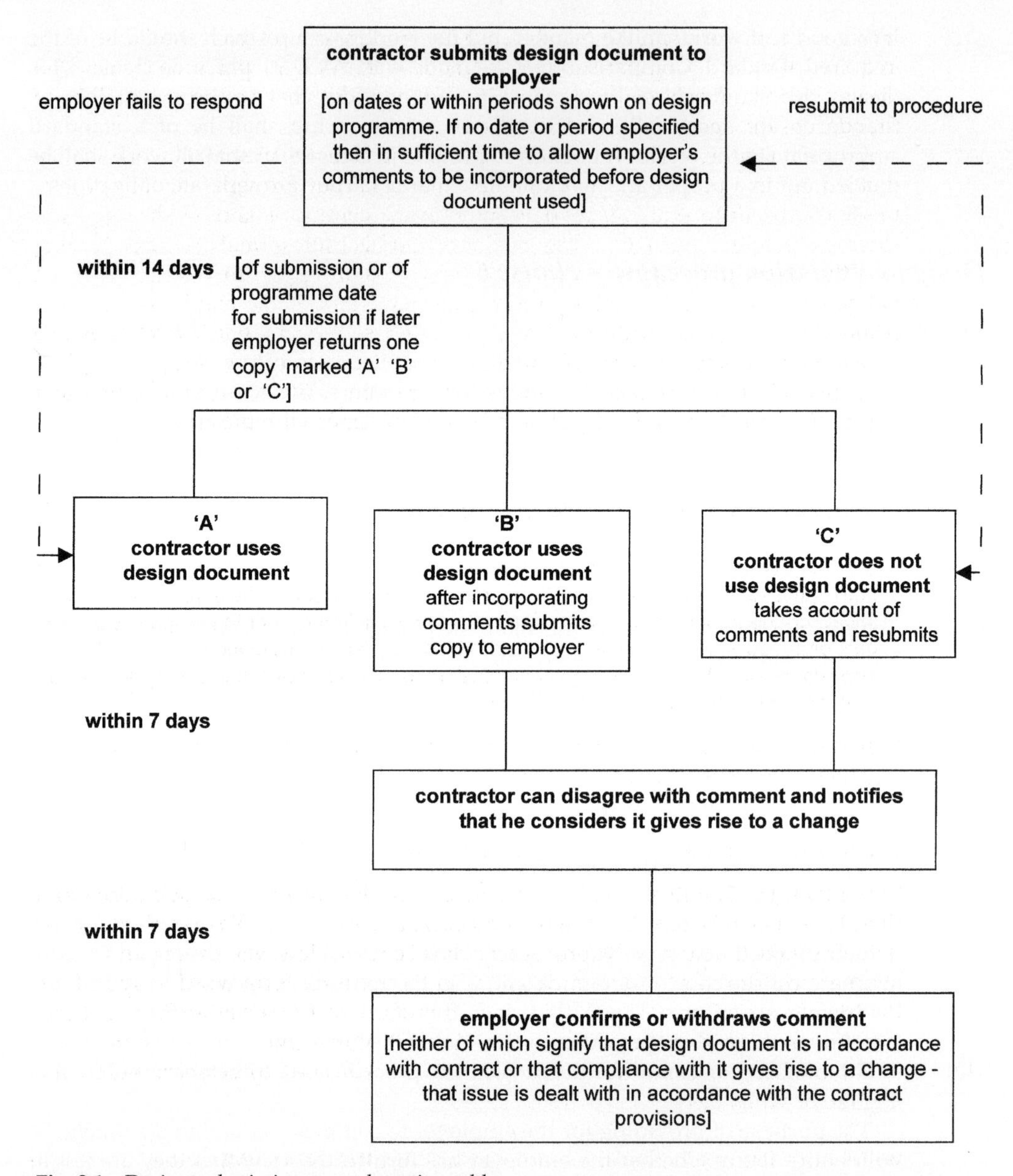

Fig. 2.1 Design submission procedure timetable.

Design documents

These are defined in clause 39.2 as:

> Drawings, specifications, details, schedules of levels, setting out dimensions and the like which are required to be prepared by the Contractor for the purposes of explaining and amplifying the Requirements and/or Proposals, which are necessary to enable the Contractor to execute the Project or which are required by any provision in the Requirements.

2-121 It follows from the contractor's obligation to complete the design as well as execute and complete the project (see clause 1.1) that the design, particularly detailed design, which is not already contained in the requirements or proposals, will be in the form of the design documents as defined.

2-122 For the major projects for which the MPF is intended, the inevitable result will be that a mass of such documents are required to be put through this design submission procedure. This places a burden on the contractor and also a very considerable potential burden on the employer and his professional advisors. Having said this, from the employer's point of view, it is an opportunity rather than an obligation. As discussed below (see paragraph 2-136), if the employer does not respond in accordance with the timetable, the design document is treated as one which the contractor is to comply with. As also discussed below (see paragraph 2-157) this will not diminish the contractor's responsibility for ensuring that any design document and the project itself is in accordance with the contract.

Content and timing of submission of design documents

> 6.2 The Contractor shall submit the Design Documents to the Employer in the quantities and format identified in the Appendix on the dates or on or before the expiry of the periods shown on the design programme contained in the Requirements or the Proposals or, if no date or period is shown, in sufficient time to allow any comments made by the Employer in accordance with clause 6 to be incorporated prior to the Design Document being required for procurement and/or execution of the Project.

Clause 6.2 deals with the submission of the design documents in a prescribed form together with the timing of the submission.

Quantities and format of design documents to be submitted

2-123 The quantities (number of copies) of the design documents to be submitted and their format is to be identified in the appendix to the contract. The required format will often take the form of the provision of hard copies. However, the appendix may alternatively provide for documents to be in an electronic form which may include the use of a stated model or protocol, e.g. that of the EDI Association Standard EDI Agreement or perhaps the European Model EDI Agreement.

2-124 The cost of providing the design documents will need to be included by the contractor when costing his proposals.

WHEN DOES THE CONTRACTOR HAVE TO SUBMIT THE DESIGN DOCUMENTS?

2-125 The contractor submits the design documents to the employer on or before the date, or expiry of the periods, shown on any design programme which is contained in the requirements or the proposals. If no date or period is shown, the contractor's obligation is to provide them in sufficient time to allow the employer to comment in accordance with the time-scales and for those comments to be incorporated into the design document prior to its use in the procurement or execution of the project.

THE DESIGN PROGRAMME

2-126 There is no definition of design programme. It can basically take any form which reflects planned design activity in terms of periods or times. It could be that for certain matters the employer would want to provide for the timing of the submission in the requirements. Otherwise it will be based on any design programme contained in the proposals. The timing of the submission may be linked by the programme to specific calendar dates or more likely to a formula which can be translated into a calendar date, e.g. by reference to a week number of the contract period.

2-127 The main purpose in having a design programme is to enable the employer and his professionals to plan their resources to meet the likely flow of design documents which they wish to have the opportunity to review. Additionally, if the design programme is included in the requirements, it will enable the employer and the professional team to ensure that design documentation is produced at the right time and in the right order so far as they are concerned.

2-128 Read literally, clause 6.2 requires the contractor to comply with any dates or periods shown on the design programme, even if they have become inappropriate, e.g. if the contractor decides to revise the sequence of construction work, which means that some design is to be advanced and some deferred. A similar situation could arise where the employer instructs a change that requires additional design, which has the effect of deferring the production of otherwise scheduled design documents. The completion date could be adjusted to a later date under clause 12, both because of a change and for other reasons which could also have a knock-on effect on the production of design documents. The clause provides no answers to these matters; there is no provision for revising the design programme. There is no mention of the contractor revising the design programme where dates or periods have become inapplicable. As it is for the contractor to undertake the completion of the design and execution of the project, it would have been sensible for the contractor at least to have been able to amend any design programme, with the consent of the employer not to be unreasonably delayed or withheld.

2-129 The contractor's failure to provide a design document by some date or within some period which has ceased to be applicable is likely, if a breach at all, to be more often than not a technical breach in respect of which the employer would have some difficulty in establishing loss, particularly if nevertheless the contractor provided the design document in sufficient time to allow for the employer's comments to be incorporated before the document was to be used.

2-130 This clause envisages that the client's comments will be 'incorporated' in the design documents (clause 6.6). The comment itself is a means by which the employer can identify why he considers that the design document is not in accordance with the contract. Clearly a literal incorporation of the comment cannot have been envisaged. It might have been more appropriate to talk about the comments being reflected (rather than incorporated) in the design document before being used for the project.

Employer's response

6.3 The Employer shall within 14 days of receipt of any Design Document, or before the expiry of 14 days from the date or the expiry of the period for submission of the same shown on the design programme, whichever is the later, return one copy of the Design Document to the Contractor marked 'A Action', 'B Action' or 'C Action' provided that the Employer shall only mark a Design Document 'B Action' or 'C Action' where it considers that the Design Document is not in accordance with the Contract.

2-131 This clause provides for the time limit within which the employer can comment following receipt of any design document and provides for the employer to mark the document with an 'A', 'B' or 'C' Action together with the significance of such a comment in terms of its contractual status and effect.

Time limit

2-132 The employer has 14 days to mark the document with an 'A', 'B' or 'C' Action from whichever is the later of the following:

- Receipt of the design document;
- The date or expiry of the period for submission shown on the design programme.

2-133 This means that if the contractor provides the design document earlier than shown on the design programme, the 14-day time limit does not begin until the date or expiry of the period for submission shown on the design programme. Where the employer has retained professionals to check the design documents for compliance with the contract or to otherwise consider them, they may wish to be able to plan their work, particularly if large volumes of design documents are anticipated. Having said this, it is extremely hard on a contractor who has his hands tied in this manner. It takes away the flexibility the contractor might otherwise have to resequence his design work, possibly to catch up on or limit the delay associated with the occurrence of some event for which the contractor is not entitled to an adjustment to the completion date or even where he is so entitled but is simply trying to fulfil his obligation to use reasonable endeavours to prevent or reduce delays (see clause 9.3). As already mentioned (see paragraph 2-128), it might have been sensible to allow the contractor to reprogramme the production of design documents subject to the reasonable consent of the employer.

2-134 The contractor's ability to earn a bonus (see clause 14) may also be jeopardised.

The employer's comment

The employer can choose to do nothing, in which case the design document is treated as being marked 'A Action' (see clause 6.4). Alternatively, the employer can mark the document with one of the following:

- A Action;
- B Action;
- C Action.

The means by which the employer communicates the comment is by returning one copy of the design document to the contractor.

The action required of the contractor following receipt of the marked design document is dealt with in clause 6.6 (see paragraph 2-137 *et seq.*).

Employer of the view that the design document is not in accordance with the contract

2-135 If the employer considers that the design document reflects design which is in accordance with the contract, the document can only be marked 'A Action'. The 'B Action' or 'C Action' can only be used where the employer considers that the design document is not in accordance with the contract. As we shall see when considering clause 6.5 (see paragraph 2-136), the employer must include a comment identifying why he has reached this view.

Employer not responding to a submitted design document

> 6.4 If the Employer does not respond to a Design Document in accordance with the procedures in clause 6.3, the Employer shall be regarded as having marked that Design Document 'A Action'.

2-136 This can give rise to some interesting and difficult situations. If the employer marks the design document with a 'B Action' or a 'C Action', which indicates that he considers it is not in accordance with the contract, but is late in returning the copy to the contractor so marked, the contract treats it as having been marked as an 'A Action' which requires the contractor to execute the project in strict accordance with the design document (clause 6.6.1). It may be that the 14-day time limit in clause 6.3 will be regarded as merely procedural and not as rendering the employer's comments invalid. But if so, what if the design document is not returned for say, 6 weeks, during which time the contractor has used it for the project relying on the effect of clause 6.4? The position is unclear. The contractor is faced, belatedly, with a design document which identifies why the employer considers that it is not in accordance with the contract, and indeed, if it is marked with 'C Action' the contractor is not to execute the project in accordance with the design document (clause 6.6.3) and yet it is also to be regarded as having been marked 'A Action', which requires that the contractor shall execute the project in strict accordance with it. The contract requires that the contractor shall proceed in accordance with the design document and at the same time forbids him to do so. Literally, as the contractor has not received the 'C Action' in accordance with clause 6.3, it does not bite under clause 6.6.3 and clause 6.4 prevails.

It can be anticipated that the scenario outlined above will frequently happen. No doubt the employer will have recourse to the wording of clause 6.10 (see paragraph 2-157) to the effect that compliance by the contractor with the design submittal procedure will not diminish the contractor's responsibility for ensuring that the design document is in accordance with the contract. Even so there is bound to be argument about it.

Employer's explanation of why it considers that the design document is non-compliant

> 6.5 Where the Employer marks a Design Document as 'B Action' or 'C Action', it shall also identify by means of a written comment why it considers that the Design Document is not in accordance with the Contract.

Where the employer has marked the design document either 'B Action' or 'C Action', he is thereby stating that he considers that the design document, by which he must mean the design content of the document, does not comply with the contract. Having so marked it, the employer must then by way of a written comment identify why he considers that the design document does not comply. The comment could be contained on the design document itself or alternatively in a covering note or letter. As a matter of record, it would be sensible, even if a separate note or letter is provided, to both cross-refer to it by writing on the design document itself, and to physically attach the note or letter to the document. The resulting action required of the contractor is covered by clause 6.6 and raises some difficult questions. Particularly because of these difficulties, the employer or his advisors should take the greatest care in formulating any comment.

Action required from the contractor

6.6 When the Employer returns any Design Document under clause 6.3, the Contractor shall take the following action in relation to such Design Document:

.1 if it is marked 'A Action', the Contractor shall execute the Project in strict accordance with such Design Document;

.2 if it is marked 'B Action', the Contractor shall execute the Project in accordance with such Design Document, provided that the Employer's comments are incorporated into such Design Document and a further copy of it is promptly submitted to the Employer; and

.3 if it is marked 'C Action', the Contractor shall take account of the Employer's comments on such Design Document and shall forthwith resubmit it to the Employer for comment in accordance with the provisions of clause 6.2. The contractor shall not execute the Project in accordance with any Design Document marked 'C Action'.

This is a critical clause which sets out the action required of the contractor following an 'A Action', 'B Action' or 'C Action' marked on the design document and the employer's written comments in relation to the latter two.

Design document marked 'A Action'

2-137 Clause 6.6.1 provides that if the design document is returned marked 'A Action' the contractor is to execute the project in strict accordance with it.

2-138 This clause might have said that a design document marked 'A Action' was the equivalent of the employer having no comment to make. However it goes much further than this by requiring the contractor to proceed strictly in accordance with the design document, arguably even if it is not actually in accordance with the contract. Employers are sensibly very wary of any action they take in connection with design documents being treated as in some way their approval of the design documents, with the potential risk that even if the design documents are not in accordance with the contract the employers will nevertheless have sanctioned them and will find it difficult subsequently to complain. Even so, bearing in mind that 'B Action' and 'C Action' can only be marked on a design document where the employer considers that it is not in accordance with the contract, the implication with an 'A Action' mark is that the employer considers that it is in accordance with the contract. If in fact it is not, the 'A Action' mark still tells the contractor that he must execute the project in strict accordance with it. This is not just an approval from the employer; it is a contractual obligation. What is its effect?

There is an overall obligation on the contractor to execute and complete the project in accordance with the contract and this includes completing the design (clause 1.1). There is a provision in clause 6.6 requiring any design document bearing the 'A Action' mark to be used by the contractor. Clause 6.10 (dealt with later – see paragraph 2-157) states that complying with the design submittal procedure does not diminish the contractor's responsibility for ensuring both that any design document he prepares is in accordance with the contract and that the project, when completed, is in accordance with the contract. The overall intention is probably to enable the employer not only, in effect, to 'approve' a design document but also to instruct its use with no comeback for the employer if it turns out that the design reflected in it is not in accordance with the contract. The employer seeks the best of both worlds: the exercise of power and control with no responsibility. This is probably the overall effect, though the emphatic nature of the wording in clause 6.6.1 causes some hesitation in reaching this conclusion.

It should be noted that the strict requirement in clause 6.6.1 applies also where the employer does not respond to the design document under the procedure (see clause 6.4). Here, it is even more surprising that the strict wording of clause 6.6.1 applies. The contractor's general obligation to ensure that the design is in accordance with the contract should be adequate in such circumstances without a requirement that the project is to be executed in strict accordance with the design document.

Clearly, even if a design document is marked or treated as having been marked 'A Action', if the contractor forms the view that in fact it does not comply with the contract, despite the strict wording of clause 6.6.1, he should submit an amended design document under the same procedure, preferably with an explanation of what is being done.

2-139 This submission procedure does have the virtue that the contractor cannot, in submitting a non-complying design document which the employer inadvertently fails to detect, obtain approval by default.

Design document marked 'B Action'

2-140 Clause 6.6.2 provides that if the design document is marked with a 'B Action', the contractor is still to execute the project in accordance with it, provided that the employer's comments identifying why he considers that the design document is not in accordance with the contract are incorporated into the design document with a further copy being promptly submitted to the employer.

2-141 This method of proceeding might be appropriate where the non-compliance is relatively insignificant in terms of its effects on the project, or where the construction programme makes it critical that the contractor should at least begin to use the design document in the execution of the project as intended so that the employer's comment does not create any delay.

2-142 The contractor is to promptly submit a revised design document incorporating the employer's comments. Does this mean that it will then go through the same procedure until it gets marked 'A Action'? This would seem sensible. However, unlike clause 6.6.3 dealing with design documents marked 'C Action', clause 6.6.2 does not expressly say that the submitted design document is to be dealt with in accordance with the procedure. Indeed, whereas clause 6.6.3 dealing with documents marked 'C Action' requires the contractor to 'resubmit' it, under clause 6.6.2 the contractor is to merely 'submit' a copy of the design document. In addition, the implicit approval of the contractor using design documents marked 'B Action' in clause 6.7 suggests that they do not go back through the procedure. It seems unlikely therefore that it is intended that the procedure should apply in such a case.

2-143 The requirement that the employer's comment (which is a statement as to why the employer believes that the design document is not in accordance with the contract) is to be 'incorporated' into the design document is an inelegant way of expressing what must be the intention, namely that the contractor will revise the design document to address the employer's comments.

2-144 The most worrying aspect of clause 6.6.2 from a contractor's point of view is that he is obliged, subject to the effects of clause 6.8 (see paragraph 2-152), to reflect the employer's comment in the design document. The employer may be wrong. Indeed, the employer or his design professionals may be making a clear mistake. If the employer does not change his view under the clause 6.8 procedure (see paragraph 2-153) the contractor will have to continue with what amounts to an amended employer design. The contractor may be able to demonstrate that this amounts to a change in order to obtain a fair valuation, but this will not help the contractor in the event that the amended design is defective. Remember that clause 6.10 (dealt with later – see paragraph 2-157) provides that complying with the design submittal procedure is not to diminish the contractor's responsibility for ensuring both that any design document he prepares is in accordance with the contract and that the project, when completed, is in accordance with the contract. If the employer's comment is in relation to what might be termed a matter of 'pure' design and the contractor's liability is based on skill and care he will not be liable, at any rate post practical completion (see paragraph 2-98 *et seq.*), unless perhaps, if the employer's mistake is obvious and the contractor has not gone through the clause 6.8 procedure, the contractor is in breach of a duty to warn (see paragraph 2-81).

However, the comment may be such as to have an impact in connection with a performance specification or the choice of materials or goods for which the contractor carries a fitness for purpose obligation which is independent of fault on his part (see paragraph 2-87 and paragraph 2-102 *et seq.*).

This situation is most likely to arise where the contractor has failed to appreciate that the employer's comment is the result of a mistake on the employer's part. If of course the contractor is aware of the mistake he will raise it with the employer; if the employer does not budge it will be open to the contractor to take the matter further, e.g. by way of obtaining an adjudicator's decision.

Design document marked 'C Action'

2-145 Clause 6.6.3 provides that if the design document is marked 'C Action', the contractor is to take account of the employer's comments by producing a revised design document which must be resubmitted to the employer for comment under the procedure.

2-146 The contractor is not to execute the project in accordance with any design document marked 'C Action'. In other words, the contractor must wait until the procedure has been repeated and he has received a response in terms of either 'A Action' or 'B Action'.

2-147 Though clause 6.6.3 provides that the revised design document must be submitted for comment in accordance with clause 6.2, clearly the time-scales contained in clause 6.2 in relation to the timing of the contractor's submission will be inapplicable.

2-148 The points made under the previous heading in relation to the contractor's contractual position, where the contractor takes account of mistaken employer's comments which results in the project not being executed and completed in accordance with the contract, are also relevant here (see paragraph 2-144).

Employer not liable to pay unless work executed pursuant to design documents marked 'A' or 'B'

6.7 The Employer shall not be liable to pay for any work executed otherwise than in accordance with Design Documents marked 'A Action' or 'B Action' in accordance with clauses 6.3 or 6.4.

2-149 Clause 6.7 provides that the employer is not liable to pay for any work forming part of the project in connection with which the contractor is required to prepare design documents, unless that work has been carried out in accordance with the design documents marked 'A Action' or 'B Action' in accordance with clauses 6.3 or 6.4.

2-150 It is not clear if this clause is meant to have one of two meanings, or possibly both of two meanings. Firstly, it could be intended to mean that until a design document has been through the design submission procedure and has been cleared for use, any work carried out by the contractor reflecting that design document need not be paid for, even though the work, as designed, is in accordance with the contract so that any relevant design document will be marked 'A Action'. Secondly, it could just mean that if, after a design document has been marked 'A Action' or 'B Action', the contractor executes the work in a manner which is not in accordance with that design document, the employer will not be liable to pay for it.

Whether the clause has one or both of the meanings suggested above, if the employer accepts the benefit of the executed work which complies with the contract, while refusing to pay anything for it in reliance on this clause, it will raise the issue of whether the clause could amount to a penalty in law. There is every chance that it does.

2-151 This clause is of course a very strong incentive for the contractor to comply with the design submission procedure in respect of every conceivable design document and to execute the work strictly in accordance with it. It is worth remembering that design documents do not just include drawings and specifications but also include details, schedules of levels, setting out dimensions and similar documentation which the contractor is required to prepare in order to explain and amplify the requirements or proposals, and which are necessary to enable the contractor to execute the project or which are required by any specific provision in the requirements.

Contractor's challenge

6.8 If the Contractor disagrees with a comment and considers that the Design Document is in accordance with the Contract it shall, within 7 days of receipt of the comment, notify the Employer that it considers compliance with the comment would give rise to a Change. Such notification shall be accompanied by a statement from the Contractor setting out the reasons why it considers that compliance with such comment would give rise to a Change. Upon receipt of such a notification the Employer shall within 7 days either confirm or withdraw the comment and, where the comment is confirmed, the Contractor shall amend its Design Document accordingly. The confirmation or withdrawal of a comment in accordance with clause 6.8 does not signify acceptance by the Employer that the Design Document, or amended Design Document, is in accordance with the Contract or that compliance with its comment would give rise to a Change.

Clause 6.8 provides a procedure and timetable should the contractor wish to disagree with the employer's comments identifying why the employer considers that the design document is not in accordance with the contract. It also deals with the contractual significance of having gone through the procedure.

The challenge

2-152 If the contractor is of the view that the design document is in accordance with the contract and so he disagrees with the employer's comment, he has 7 days from receipt of the comment within which to notify the employer that he considers that compliance with the comment would give rise to a change. That notification is to include a statement setting out the reasons why he considers that compliance would give rise to a change. As any failure of the parties to reach agreement could lead to the matter proceeding to adjudication, it is important for the employer to carefully consider his comments, and equally for the contractor to carefully consider his response, as this will identify any disagreement and therefore the dispute between the parties.

Having received the notification, the employer has 7 days in which to either confirm the comment, in which case the contractor will be obliged to amend the design document to reflect the comment; or alternatively withdraw the comment, in which case the contractor can proceed on the basis of the original design document submitted.

Contractual significance of the contractor's challenge

2-153 This challenge procedure can add up to 14 days to the overall process. If the employer withdraws his comment the contractor may well argue that the project has in consequence been disrupted or delayed. In these circumstances the contractor will be entitled to an appropriate adjustment to the completion date and to recover loss and expense on the basis of the employer's act of prevention (see clauses 12.1.8 and 21.2.1).

However, clause 6.8 provides that any withdrawal of a comment does not signify acceptance by the employer that the design document is in accordance with the contract. In other words, it is not to be regarded as an admission.

Clause 6.8 also provides that any confirmation of the comment, which means that the contractor has to reflect this in an amended design document, does not signify acceptance by the employer that the design document as amended to reflect the comment is in accordance with the contract. Again therefore, it is not an admission. Nevertheless it will clearly be evidence that at that point in time the employer is of the view that if the design document reflects his comments it will then comply with the contract.

2-154 Clause 6.8 further states that any confirmation or withdrawal of a comment does not signify acceptance by the employer that compliance with the design document or an amended design document would give rise to a change. It is difficult to see how this could be the effect of a confirmation or withdrawal in any event. If the comment is attached to a design document marked with a 'C Action', the contractor is not to comply with it anyway. If it is attached to a design document with a 'B Action', the employer's confirmation of the comment has the effect of requiring the contractor to incorporate the comment in the design document and then to comply with it. The employer is simply restating in effect that he is of the opinion that incorporation of the comment is required in order for the contractor to comply with the contract, which can hardly be an acceptance that it gives rise to a change. If the employer withdraws the comment, in effect confirming that the design document is in accordance with the contract, that cannot amount to an acceptance by the employer that compliance with the design document would give rise to a change. The only exception may be if the effect of making a challenge to a document marked 'B Action' does not in fact mean that there is a holding period in respect of its use (see next paragraph). This might mean that the contractor proceeds with the design document as amended in accordance with the comment. Any subsequent withdrawal of the comment by the employer will not then amount to the acceptance by the employer that the amended design document has produced a change.

Clause 6.6.2 provides that a design document marked with a 'B Action' requires the contractor to execute the project in accordance with that document, having incorporated the employer's comment into it. Presumably, the requirement to execute the project in this situation is subject to the challenge procedure in clause 6.8, which provides that it is only upon confirmation of the comment that the contractor is to amend the design document. The contract appears therefore to envisage a holding period if the contractor wishes to disagree with the comment.

If a comment is confirmed so that the contractor must amend the design document and then comply with it, he may nevertheless take the matter further, e.g. adjudication would appear to be a logical and appropriate step to take. The result of the adjudicator's decision, if in the contractor's favour, is likely to mean that the amendment made to reflect the employer's comment will then be treated as a change, as it will have amounted to an alteration in the requirements or proposals giving rise to an alteration in design quality or quantity.

Employer failing to respond to contractor's challenge

2-155 The position where the employer fails to respond to the contractor's notification of a disagreement is unfortunately not covered. An appropriate default position would have been that the employer's original comment was confirmed. Presumably, the contractor is required to proceed on that assumption.

Significance of contractor not employing the challenge procedure

6.9 Where the Contractor does not notify the Employer in accordance with clause 6.8, any comment by the Employer will not be treated as giving rise to a Change.

2-156 Clause 6.9 is highly significant. It provides that if the contractor does not notify the employer in accordance with clause 6.8 by engaging the challenge procedure, the employer's comment is not to be treated as giving rise to a change. This means that in failing to make the challenge in accordance with clause 6.8, the contractor is barred from later contending that the comments, once reflected in the design document, require something different or over and above what is contained in the requirements or proposals or in what the contractor is to do to comply with the contract. This makes the 7-day deadline from the receipt of the employer's comments in which to respond, with a statement of reasons why he considers that reflecting the comments in the design document would amount to a change, of paramount importance. Failure to follow this procedure will prevent the contractor from ever raising the point. The contractor needs to be fully aware of the implications of the contract and ensure that a system is in place to produce a rapid and effective response. Having regard to the 7-day time limit, it may be sensible for the contractor, wherever possible, to refrain from submitting design documents in the period running up to a holiday, particularly the Christmas to New Year construction industry close-down.

Contractual effect of compliance with the design submittal procedure

> 6.10 Compliance with the design submittal procedure in clause 6 and/or with any comments from the Employer under it shall not diminish the Contractor's responsibility for ensuring both that any Design Document it prepares is in accordance with the Contract and that the Project, when completed, is in accordance with the Contract.

2-157 Clause 6.10 provides that compliance by the contractor with the design submittal procedure or with any comment from the employer under it is not to detract from the contractor's responsibility for ensuring that design documents and the project itself, when completed, are in accordance with the contract. The fact that the contractor goes through this process, which may or may not result in the employer expressing views as to whether the design document is in accordance with the contract, does not in any way dilute or qualify the contractor's fundamental contractual obligation to prepare design documents in accordance with the contract and to ensure that the project is also in accordance with the contract. This clause is aimed at preventing the employer's comment, or indeed the absence of any comment, from in any way entitling the contractor to rely upon this in determining whether any design document is or is not in accordance with the contract. While the employer may be seeking the best of both worlds in the sense that he wishes to be able to check design documents and even give an indication as to why they may not be prepared in accordance with the contract, on the one hand, and on the other, provides contractual provisions to ensure that no responsibility attaches to that exercise, this is only a reflection of modern day practice. See also on this paragraph 2-138.

Copyright – clause 7

2-158 Clause 7 deals with the matter of copyright in the 'Design Documents' prepared by the contractor. Copyright is a complex area of the law that is not appropriate for discussion in a book of this nature. Reference can be made to standard works on the subject such as *Copinger and Skone James on Copyright* (14th edition) and *Intellectual Property in Europe*, Guy Tritton (2nd edition), both published by Sweet & Maxwell. Accordingly, this book limits itself to a very brief background before considering clause 7 itself.

Brief background

2-159 The relevant statute dealing with copyright is the Copyright, Designs and Patents Act 1988. Previously, so far as statute is concerned, the law was governed by the Copyright Act of 1911 and the Copyright Act of 1956.

2-160 There is a special type of copyright in the building itself as opposed to any plans and drawings relating to it. This was effectively created by the 1911 Act and applied to architectural works of art (and see *Meikle* v. *Maufe* (1941)). It now also includes models for buildings and any fixed structures or parts of a building or fixed structure. This form of copyright protects against the copying of a building either by the construction of a subsequent building or by drawings etc. being made of it. For this type of copyright there has to exist an element of artistic quality. This copyright will be the property of the designer and not of the builder unless he is also the designer.

2-161 Turning to the copyright in the plans and drawings themselves, the 1988 Act applies to drawings, diagrams and plans. It is originality which is protected, which does not mean novelty but means the exercise of independent skill and labour rather than artistic quality. It will therefore apply to plans and drawings for a building produced typically by structural engineers as much as it does to an architect's plans.

2-162 Although the design may be in copyright, it is always possible that a licence can be granted either expressly or implicitly for someone else to make use of the design by using plans and drawings.

2-163 In relation to a design and build contract, copyright will, unless the contract expressly provides otherwise, remain with the contractor if he is the author of the design. Should the contractor employ independent professionals to carry out the design, it will be readily inferred that the contractor will have a licence from the designer for the use. However, it will depend upon what the terms of engagement provide in this regard, including whether any licence is to be exclusive to the contractor or otherwise.

2-164 The 1988 Act introduced certain 'moral right' provisions covering the right to identification of the author of the copyright in certain circumstances and also a right of objection to derogatory treatment, i.e. a distortion or mutilation of the copyright work. The right to identification exists, provided it is asserted as required by the Act, if a model of a building is exhibited in public, or a visual image of it is broadcast or filmed in public, or photographs showing it are issued to the public. In the case of a building, the author also has a right of physical identification on the building.

2-165 The effect of owning the copyright means that no one else, including the employer, will be able to reproduce the plans or repeat the design in a further building without the designer's consent, express or implied.

2-166 In terms of infringement of copyright, there must not only be an element of actual copying of a substantial part of the design, but also a sufficiently close and immediately perceived resemblance between the two pieces of work. The reconstruction of the original building itself, e.g. following a fire, is not a breach of copyright.

2-167 The main remedies for breach of copyright are damages, which historically have not been particularly high. In appropriate cases an injunction to restrain the breach may be obtained, although this is not common. If the defendant was unaware that the copyright existed and had no reason to believe that it did, a claimant would not be entitled to damages. A starting point for the assessment of damages is the sum that might otherwise have been charged for a licence to use the copyright for the purpose for which it was in fact used.

2-168 Design and build contracts generally have express clauses dealing with the issue of copyright. Some also deal with royalty and patent rights. The MPF deals only with copyright.

Copyright under the MPF

> 7.1 The copyright in all Design Documents prepared by the Contractor in accordance with the Contract shall remain vested in the Contractor but the Contractor grants to the Employer an irrevocable, royalty-free non-exclusive licence to copy and use the Design Documents and to reproduce the designs and content of them for any purpose relating to the Project including, without limitation, the design, execution, completion, maintenance, letting, sale, promotion, advertisement, reinstatement, refurbishment and repair of the Project. Such licence shall enable the Employer to copy and use the Design Documents for the purposes of an extension to the Project, but such use shall not include a licence to reproduce the designs contained in them for any extension to the Project. The Contractor agrees that the Employer may grant sub-licences to other persons to use and to reproduce the Design Documents and the designs and content of them for any purposes relating to the Project.
>
> 7.2 To the extent that the Contractor does not have ownership of the copyright in any Design Document the Contractor shall procure from the copyright holder licences with full title guarantee and shall grant a sublicence to the Employer in respect of that Design Document in the same terms as are set out in clause 7.1.
>
> 7.3 The Contractor shall not be liable for any use made of the Design Documents that is outside of the scope of the licence granted by clause 7.

2-169 Clause 7.1 provides that copyright in all 'Design Documents' prepared by the contractor is vested in the contractor. The employer nevertheless is granted an irrevocable, royalty-free and non-exclusive licence to copy and use the documents and to reproduce the designs and content of them for any purpose relating to the project, including without limitation, the design, execution, completion, maintenance, letting, sale, promotion, advertisement, reinstatement, refurbishment and repair of the project.

It is expressly provided that this licence enables the employer to copy and use the design documents in connection with an extension to the project but does not permit a licence to reproduce the designs contained in them for the design of the extension. This means that the licence does not allow an extension to be built which mirrors the design of the building itself. It simply enables the employer to copy and use the design documents to enable him to extend the building, with the extension itself being based on a different design. Of course express permission may be sought and obtained from the copyright holder to incorporate the design in the design of the extension itself.

2-170 By this clause, the contractor also agrees that the employer can grant sub-licences to others to use and reproduce the design documents and the designs and contents of them for any purpose relating to the project.

Some points to note

'Design Documents'

These are defined in clause 39.2 as:

> Drawings, specifications, details, schedules of levels, setting out dimensions and the like which are required to be prepared by the Contractor for the purposes of explaining and amplifying the Requirements and/or Proposals, which are necessary to enable the Contractor to execute the Project or which are required by any provision in the Requirements.

As defined, design documents would probably not cover models.

Purposes for which the design can be used

2-171 Clause 7.1 refers to 'any purpose relating to the Project' and then goes on to include certain specific purposes but these are stated to be 'without limitation'. Even so, while the specifically listed purposes will not be an exhaustive list, any further purposes must, it is suggested, be of a similar kind. It would not therefore be possible to use the designs in further buildings. This is made all the more clear by the specific treatment and limitation of the right to use the designs in connection with any extension to the project.

Position where contractor's employment is terminated

2-172 The contractor's employment may be terminated on stated grounds by the employer (clause 32), by the contractor himself (clause 33) or by either if certain circumstances arise (clause 34). If the termination arises under clause 32 or 34, the contractor is required to provide the employer with all design documents prepared in accordance with the project. If the termination takes place under clause 33, there is no such provision.

It should be noted that the express termination provisions in clauses 32 to 34 are without prejudice to any other rights or remedies the parties may possess (see clause 31.1). It may be that rather than a termination taking place in accordance with the express provisions of the contract, the employer chooses to accept a serious breach of contract by the contractor as a repudiation, so bringing the contract to an end at common law. It is important for the employer to realise that if the repudiation is accepted and the contract brought to an end by this route rather than by operating the express termination provisions of the contract, the express provision requiring the contractor to provide the design documents prepared in connection with the project will not apply. Some employers may be tempted to amend clause 7 and the termination clauses to overcome this problem and also that of the position in relation to the handing over of design documents where clause 33 applies (see the previous paragraph). A general requirement for the employer to call for design documents to be handed over at any time is likely to be inserted into clause 7 by employers.

Breaches of copyright

2-173 Clause 7 does not deal with the situation where the contractor purports to license the employer to copy and use the design documents when in fact the contractor does not have the copyright and does not have a licence from the copyright holder (see clause 7.2). In this situation the employer may find himself facing a claim for breach of copyright. In such a case the employer could in turn claim against the contractor for breach of clause 7. But in practice, unlike the MPF, many similar clauses in other contracts also contain an indemnity provision whereby the contractor indemnifies the employer against any claims, losses or expenses, etc. Express indemnities can have advantages over simple claims for breach of contract (see paragraph 6-06).

The position of purchasers, tenants and funders

2-174 It should be noted that where a funder has the benefit of third party rights (see clause 30 and clause F12 of the third party rights schedule), provided the employer has paid the contractor as required by the contract (or the funder has done so), the funder is granted the same rights as the employer has been granted by clause 7. Similar rights are given to a purchaser or tenant (see clause PT7 of the third party rights schedule).

Patent rights

2-175 Nothing is said in clause 7 on the question of patented articles, processes or inventions. These may be required to be used by the contractor by virtue of the requirements. Alternatively, the contractor may choose to use them. In either case, if there has been an infringement of patent rights there could be claims and cross-claims. Some contracts specifically provide indemnities or cross indemnities in this situation. However, the MPF does not do so. This is no doubt because in practice such occurrences are rare.

Contractor not liable for misuse by employer

2-176 Clause 7.3 provides that the contractor is not to be liable for any use made of the design documents outside the scope of the licence granted by clause 7. No doubt it will be difficult to hold the contractor liable for use outside the scope of the licence in any event. However, this makes the position clear. Implicitly it does of course make the contractor liable for any use made of the design documents which is within the scope of the licence granted by clause 7. So, for example, if there is negligent design contained in a design document which is used in connection with reinstatement, refurbishment or repair, the contractor could find himself liable for any foreseeable losses which may result.

Design documents becoming out of date

2-177 Some contracts expressly deal with the position where a design document is used for its licensed purposes but which, at the time of use, has become out of date or superseded. It may, for example, be that a change in building regulations would result in the use of the design document being in breach of the regulations. Is the contractor to be liable? The answer is probably not, as the employer, funder or purchaser, as the case may be, will be deemed to be aware of the current law. Even so it is not unusual in practice to find an express exclusion of liability to cover this situation.

Ground conditions – clause 8

Background

2-178 Clause 8 tackles the very important question of problems in the ground. This needs to be considered against the general common law background relating to responsibility for the condition of the site. Under the general common law, an employer does not impliedly warrant the condition of the site. He does not guarantee that the site is such that his design will work – see *Appleby* v. *Myers* (1867). Unexpected difficulties or expense due to unforeseen adverse site conditions are no excuse for failure to perform and no ground for extra payment – see *Bottoms* v. *York Corporation* (1892). This is of course subject to the express terms of the contract made between the parties and the contract documents. In particular the contract documents may give rise to a warranty or representation of some kind in relation to ground conditions. In the unlikely event that the employer has prepared requirements which include a bill of quantities prepared in accordance with a particular standard method of measurement, such as the Standard Method of Measurement of Building Works, 7th Edition (SMM7), published by the Royal Institution of Chartered Surveyors and the Construction Confederation, the contract as a whole will have to be examined to decide the effect of the bill of quantities. SMM7 requires that particulars be given of soil and ground conditions such as water level, trial pits or boreholes and over or underground services. If that information is not available, a description of the ground and strata, as assumed, shall be stated (see section D20 of SMM 7). In such a case, to a greater or lesser extent, a representation of site conditions may be given either from site investigation results in the bills, or alternatively the contractor will be required to use assumptions. If the actual information given turns out to be inaccurate or misleading, it is conceivable that the contractor may have a claim for misrepresentation. The precise extent of any such representation given in this way is likely to be difficult to assess, e.g. to what extent does information obtained from a trial pit or a bore hole amount to a representation by the employer that the contractor can freely assume in pricing that such information is typical of the whole of the site? Furthermore, such information as is given is in practice likely to be heavily qualified (see paragraph 2-180 and *Co-operative Insurance Society Limited* v. *Henry Boot Scotland Limited* (2002)). The issues surrounding misrepresentation are discussed in more detail below (paragraph 2-185).

2-179 Further, in other specific situations, the employer may expressly state in the contract documents the assumptions as to ground conditions on which the contractor is asked to design the works, and this may well amount to a term of the contract and a warranty on the part of the employer as to site conditions: see *Bacal Construction (Midlands) Ltd* v. *Northampton Development Corporation* (1975).

SITE INVESTIGATION REPORTS

2-180 It may well be that the employer has commissioned or otherwise obtained site investigation reports, which deal with ground conditions and also possibly artificial obstructions in the ground. These are rarely included as a contract document but may well be referred to somewhere in the contract documents, such as the requirements. Such reports often contain within them statements to the effect that they are provided specifically for the employer and that anyone else, e.g. a contractor, who may choose to make use of them does so entirely at his own risk. Additionally or alternatively, a clause may be inserted into the contract documents which refers to the reports and expressly provides that they do not amount to a warranty or representation. A statement may also be included requiring the contractor to himself interpret and form his own opinions from the site investigation reports in terms of facts, e.g. the contents of boreholes or trial pits.

MISREPRESENTATION

2-181 Reliance by contractors on site investigation reports prepared by or on behalf of employers and disclosed to contractors can also raise questions as to whether in law an actionable misrepresentation may have been made.

2-182 A misrepresentation is basically an untrue statement of fact made by one contracting party to the other, at or before the time of contracting, which is one of the causes that induces the contract. It must be distinguished from an honest opinion or belief which is not itself a representation of fact. Bearing in mind the uncertainties in the science or art of soil and rock mechanics, it is not unreasonable to suggest that the contractor should not invariably rely upon the opinions expressed by the author of a site investigation report, without carrying out some elementary checks whether by inspections or examinations or acquiring geological information.

Misrepresentations may be:

- innocent;
- fraudulent;
- negligent.

Innocent misrepresentation

2-183 An innocent misrepresentation is simply a misrepresentation which is made with due care but which in fact turns out to be untrue and which causes the contractor loss after he has placed reliance upon it.

The basic remedy for innocent misrepresentation is rescission of the contract and restitution. In some circumstances, by virtue of section 2(2) of the Misrepresentation Act 1967, the court has a discretion to award damages for innocent misrepresentation.

The ability to rescind can be lost when the person entitled to rescind either leaves it too late so that third party rights have been affected, or he takes any step consistent with his treating the contract as continuing, in which case he is held to have affirmed the contract.

Fraudulent misrepresentation

2-184 The leading case on this is *Derry* v. *Peek* (1889) in which it was said that: 'fraud is proved when it is shown that a false representation has been made:

(a) knowingly;
(b) without belief in its truth, or
(c) recklessly, carelessly whether it be true or false.'

It is fortunately rare for actions for fraudulent misrepresentation to arise in the construction industry. However, it is as well for employers and their advisors to bear in mind that they can be guilty of fraudulent misrepresentation if they make statements as to site conditions which are reckless, not caring whether they are true or false. An example of this is to be found in the case of *Pearson* v. *Dublin Corporation* (1907). The contractor undertook to carry out a high-level outfall sewer and outfall works for a lump sum. It was an elaborate scheme and an essential feature was the use of a wall shown on the maps and drawings supplied by the employer's engineers. The wall was shown as extending to a depth of 9 ft below the ordnance datum line. In fact it rarely extended as much as 3 ft. The wall therefore proved useless for the proposed purpose and the contractor was put to extra expense. The engineers had carried out an accurate survey and even had doubt themselves whether the wall existed or if it was in accordance with the description stated. They had, said the court, 'rashly and without enquiry represented 9 feet of wall when no wall existed'.

Dublin Corporation were held liable for fraudulent misrepresentation. This was the result despite a provision in the contract that the contractor was to satisfy himself as to the dimensions of all existing works and that the Corporation did not hold itself responsible for the accuracy of statements about existing works. It was held that while such an exclusion clause may have afforded a defence to a claim for innocent misrepresentation, it afforded no defence to an action for fraudulent misrepresentation. As Lord Loreburn said, 'no one can escape liability for his own fraudulent statements by inserting in a contract a clause that the other party shall not rely upon them.'

The facts which can give rise to a fraudulent misrepresentation can also give rise to an action by the contractor for the tort of deceit against the employer or his advisor.

Negligent misrepresentation

2-185 By virtue of section 2(1) of the Misrepresentation Act 1967, a person entering into a contract as a result of a misrepresentation having been made by the other party to the contract, and who suffers loss, can claim damages from that other party unless that other party proves that he had reasonable grounds to believe and did believe up to the time the contract was made that the facts represented were true. Unlike the case with fraudulent misrepresentation, dishonesty is not an essential element in the making of the representation. All that is required is that the facts which were represented as true were in fact untrue and that the maker had no reasonable grounds for believing them to be true. The test as to reasonable grounds for belief is objective. It is not therefore for the claimant to prove negligence but for the defendant to prove reasonable grounds for belief.

2-186 Quite separately from actionable misrepresentation which acts as an inducement to a party to enter into a contract, the maker of a negligent misstatement or the giver of negligent advice may find himself liable under an independent implied duty of care to exercise due skill and care in circumstances where the maker knows or ought to have known that reliance was being placed on such skill and judgment by the recipient.

Negligent misstatement or negligent advice

2-187 Before 1964 it was generally thought that there was no claim as a result of a negligent misstatement which caused economic loss rather than personal injury or physical damage to property, unless there was a breach of contract or a breach of some fiduciary duty. However, in the case of *Hedley Byrne & Co Limited* v. *Heller* (1964), the House of Lords held that such a claim could arise where there existed what was called a special relationship between the parties. A claim for damages for financial loss may be successful on the basis of an implied duty of care when one party seeks information from another who has special skills and knowledge, which the second party knew or ought to have known that the first party was relying upon. See also *Henderson* v. *Merrett Syndicates Limited* (1995). For example, an engineer in making statements, before or during a project, to the contractor, which he knows will be or are likely to be acted on by the contractor, may owe a duty to take reasonable care in making such statements. This could include information as to the site or ground conditions. Liability is likely to be confined to cases where the statement or advice is given to a specific person for a specific purpose of which the maker of the statement is aware, in the knowledge that the recipient is likely to act on it and upon which the recipient has relied and acted to his detriment. See *Esso* v. *Mardon* (1976).

2-188 Clause 8.1 of the MPF, as we shall see below, gives the contractor no relief from the rigours of the common law position where unforeseen ground conditions or man-made obstructions make the work more expensive or time consuming. However, where clause 8.2 is selected through the appendix and to the extent that an amendment to the requirements or proposals is necessitated, an option is provided as to risk allocation, which provides the contractor with some relief.

Ground conditions at contractor's risk

> 8.1 If the Contractor encounters ground conditions or man-made obstructions in the ground that necessitate an amendment to the Requirements and/or Proposals it shall notify the Employer of the amendments it proposes for the agreement of the Employer, such agreement not to be unreasonably delayed or withheld. Unless clause 8.2 applies, such amendment shall not be treated as giving rise to a Change.

2-189 This clause covers the situation where the contractor encounters ground conditions or man-made obstructions necessitating an amendment to the requirements or proposals. Unless clause 8.2 has been selected, this will not be treated as a change. Not only therefore does the contractor take the risk of the work being unexpectedly difficult and therefore expensive to execute and of any delays caused, he also meets the cost of any changes, e.g. to design, which may become necessary.

2-190 If the contract in one of its documents actually gives a warranty as to ground conditions (see paragraph 2-179) upon which the design in the requirements and/or the proposals is based, is clause 8.1 sufficient to prevent a contractor, who has relied on it in preparing his proposals and putting together a price for the work, from successfully claiming damages for breach of warranty, misrepresentation or negligent misstatement? There is no reason to suppose that it excludes such a claim. While under the contract it is not to be treated as a change and therefore nothing is added to the contract sum, this would not prevent a claim. To exclude such a claim the contract would have to be very clear in stating this. Clause 8.1 does not even begin to do that. The fact that a contract document, e.g. the requirements, makes reference to another document, such as a soil investigation report which is supplied to the contractor, will not make that report a contract document so that its contents cannot form the basis of a warranty, though exceptionally they may give rise to a representation – see *Co-operative Insurance Society Limited* v. *Henry Boot Scotland Limited* (2002) in which, on the facts, no representation was made.

The employer may well ensure that any liability for site information contained in the requirements or referred to in them is excluded. Even if, as between the employer and the contractor, the employer excludes liability for the contents of any report, it may be that a relevant consultant is a pre-appointed consultant under clause 18, in which case the contractor will be interested in ensuring that terms of the model form of novation between him and the consultant referred to in clause 18.2 provide a right of recourse if the consultant has been negligent.

Ground conditions or man-made obstructions in the ground

2-191 These words are different to those found in some other contracts which deal with ground conditions. For example, in clause 12 of the ICE Design and Construct Conditions, reference is made to 'physical conditions'. This has been held to include not only ground conditions but also the unforeseen instability of a 300 tonne crane on a jack-up barge causing the barge to become unstable, collapsing and causing extensive damage – see *Humber Oil Terminals Trustees Limited* v. *Harbour & General Works (Stevin) Limited* (1991). The wording used in clause 8.1 is therefore more restrictive.

Action by contractor having encountered ground conditions or man-made obstructions

2-192 If the contractor does encounter ground conditions or man-made obstructions in the ground which do necessitate an amendment to the requirements or proposals, the contractor is required to put forward to the employer proposals for dealing with the situation. The employer must not unreasonably delay or withhold agreement. Bearing in mind that unless clause 8.2 applies, any such amendment is not to be treated as a change, so that the contractor meets the extra costs and does not get an adjustment to the completion date, he will of course be keen to do the absolute minimum in satisfying this obligation. Clearly it is only amendments which are necessary which need to be included in the contractor's proposal. It would not include matters which may be merely desirable from the employer's point of view. Further, even if the ground conditions or man-made obstructions encountered ought reasonably to have been foreseen by the contractor when calculating the contract sum, this should not have a significant bearing on whether agreement to the contractor's proposals should be withheld when contrasted with the situation where the difficulties encountered were unforeseen and unforeseeable. In either situation, had they been known about, the chances are that the proposals would have been different and the contract sum would have reflected this.

Contractor's proposal treated as giving rise to a change

8.2 When the Appendix states that clause 8.2 is to apply, any amendment agreed by the Employer under clause 8.1 shall be treated as giving rise to a Change to the extent that the ground conditions or man-made obstructions in the ground could not reasonably have been foreseen by an experienced and competent contractor, on the Base Date, having regard to any information concerning the Site that the Contractor had or ought reasonably to have obtained.

2-193 Where an appendix entry states that clause 8.2 is to apply, then any agreed amendment under clause 8.1 is to be treated as giving rise to a change but only to the extent that the ground conditions or man-made obstructions could not reasonably have been foreseen by an experienced and competent contractor at the base date having regard to any information concerning the site that the contractor had or ought reasonably to have obtained. The 'Base Date' is defined in clause 39.2 as 'The date identified in the Appendix' and will be a date some time before the tender is submitted. In effect, it is aimed at the time when the contractor is preparing his proposals and calculating the contract sum. If the appendix is not completed in relation to this item, then clause 8.2 will not apply.

2-194 The first point to appreciate is that this does not of course compensate the contractor, or provide a ground for an adjustment to the completion date, if he encounters ground conditions or man-made obstructions which result in extra expense or delay without requiring an amendment to the requirements or proposals. So, for example, at any rate where neither the requirements nor the proposals contained relevant quantities, encountering unexpected quantities of rock making excavation more difficult but not necessitating any amendment to the requirements or proposals would not be covered even if clause 8.2 is selected in the appendix.

Ground conditions or man-made obstructions that could not reasonably have been foreseen by an experienced and competent contractor

2-195 The test of whether what is encountered will be treated as a change, enabling the contractor to receive compensation and an adjustment to the completion date where the requirements or proposals require amendment, depends upon the ground conditions or man-made obstructions not being reasonably foreseeable by an experienced and competent contractor. Clearly therefore, if the conditions encountered are unforeseen only because the contractor is inexperienced or, even if experienced because he has been incompetent in not appreciating them, then this will not be treated as a change and the contractor will have to bear the costs associated with the amendment and will also be deprived of any adjustment to the completion date as a result of any delays suffered.

Information concerning the site that the contractor had or ought reasonably to have obtained

2-196 In judging whether the conditions encountered could not reasonably have been foreseen, regard is to be had to any information concerning the site that the contractor either has or ought reasonably to have obtained. Such information may be provided by the employer, for example, in the form of a site or soil investigation report. Even if this is provided on the express basis that the contractor relies upon it at his own risk, nevertheless he will still be expected to have been mindful of it when anticipating the sort of ground conditions or man-made obstructions in the ground, which may occur. The extent and nature of information provided by the employer will be important. The more detailed and authoritative it is, the less perhaps that the contractor needs to undertake in relation to site investigation. Likewise, the less information which is provided, the more the contractor should consider carrying out at least some basic site investigation work, e.g. trial pits where feasible. However great or little the information provided by the employer, the contractor should in any event seek to obtain any readily available public information regarding the site and its history, e.g. local library records as to the history of the area and specifically the site if available.

2-197 It is not only information that the contractor actually had that is relevant, but also that which he ought to have obtained. So the issue arises as to the extent to which the contractor should inspect the site and also the extent to which he should carry out site investigations. Some other contracts, e.g. clause 11(2) of the ICE Design and Construct Conditions of Contract, state that the contractor shall be deemed to have inspected and examined the site and its surroundings etc. so as to be aware of the nature of the ground and subsoil but only so far as is practicable and reasonable before the award of the contract. There is no such provision in the MPF. In practice, tendering contractors often have very little time to carry out anything other than a superficial site inspection. Certainly the possibility of them being able to carry out detailed site investigations is unrealistic both in terms of time and tendering costs. Accordingly, information which the contractor 'ought reasonably to have obtained' will have to take account of these factors.

The fact that the contractor takes the risk of unforeseen ground conditions or man-made obstructions which do not result in an amendment to the requirements or proposals should in any event encourage the contractor to take this issue very seriously.

Chapter 3
Time

Content

Clauses 9 to 14 inclusive fall under the general heading of 'Time'. They cover the following topics:

- Commencement and completion
- Damages for delay
- Taking over parts of the project
- Extension of time
- Acceleration
- Bonus (for early completion).

Commencement and completion – clause 9

Clause 9 deals with the start on site of the project and its completion.

Access

> 9.1 The Employer shall give the Contractor access to the Site on the date stated in the Appendix and shall give to the Contractor access to such part or parts of the Site at such times and for such periods as may be reasonably necessary to enable the Contractor to execute and complete the Project in accordance with the Contract. Access to the site shall be subject to any restrictions set out in the Requirements. The Contractor shall not be entitled to exclusive possession of the Site.

3-01 The employer is to give the contractor access to the site on the date stated in the appendix. However, the insertion of a specified date may well be impracticable from the employer's point of view, bearing in mind such matters as the tendering procedures involved or last minute difficulties in obtaining possession of land. For one reason or another there may be difficulty in accepting a tender in sufficient time to give possession of the site to the contractor by a stipulated date. Often, therefore, the date will be stated in terms of some days or weeks from the date of acceptance of the contractor's tender. The insertion of such words as 'on a date to be agreed' at tender stage is fraught with danger for the employer. If such a phrase is used, it is vital that before the contractor's tender is accepted, the date for giving access is agreed and inserted in the appendix. If this is not done at the time of acceptance of the contractor's tender and no date can subsequently be agreed, then it is likely that access must be given within a reasonable time and this could arguably invalidate the liquidated damages provision in the contract.

Access not possession

3-02 There is no reference, as in many other JCT contracts, for example, JCT 98 (clause 23) and WCD 98 (clause 23), to the employer granting possession of the site to the contractor. The only mention of the word 'possession' is in the last sentence of the clause which provides expressly that the contractor shall not be entitled to exclusive possession of the site. This makes it absolutely clear that the contractor's occupation is based on a mere licence, which will be revocable upon termination of the contractor's employment whether by the employer or by the contractor himself. In either case, the contract expressly provides that the contractor is to remove all of his materials, plant or equipment from the site without delay (see clauses 32.4.1, 33.4.1 and 34.4.1).

3-03 The obligation in this contract on the employer to give access to, rather than possession of, the site is significant. If the contractor were to obtain legal possession of the site, it raises the possibility of the contractor having the right to insist upon remaining on site if his employment has been wrongly terminated or even if the point is arguable.

It is a question of some debate whether or not the contractor's licence to remain in possession is revocable or irrevocable while the contract period is still running. In other words, if the employer purports to determine the contractor's employment and this is disputed by the contractor, can he remain on site or is he compelled to leave?

An important case in this regard is *London Borough of Hounslow* v. *Twickenham Garden Developments* (1970).

FACTS

The defendants were employed by the London Borough of Hounslow to carry out the sub-structure works on a site in Middlesex. The contract was based on JCT 63 and by clause 25(1) it was provided that if the contractor should make default, *inter alia*, in failing to proceed regularly and diligently with the work, the architect could give notice by registered post or recorded delivery specifying the default, and if the contractor continued or repeated the default the council could by notice by registered post or recorded delivery forthwith determine the employment of the contractor. The council purported to do this and the contractor contested the validity of the council's action, stating that he regarded the service of such a notice as a repudiation of the contract by the council, which he, the contractor, elected not to accept.

The council unsuccessfully tried to obtain possession of the site and issued a writ claiming damages for trespass and seeking an injunction restraining the contractor from trespassing on the site. The council argued that it was entitled – irrespective of the validity or invalidity of the notices given pursuant to clause 25 – to evict the contractor and resume possession of its own property. The contractor claimed to be able to insist on performing the contract.

HELD

There was an implied term of the contract that the council would not revoke the contractor's licence to enter on the site while the contract period was still running other than in accordance with the contract. The court would not grant the council the injunction requested to expel the contractor from site because it had not been decided at that stage whether the contractor's employment had been validly determined by the council.

The effect of this decision is to produce a legal stalemate. The decision has been criticised – see *Hudson's Building and Engineering Contracts*, 11th edition at paragraph 12-090; *Keating on Building Contracts*, 7th edition, at paragraphs 11.23–11.25. Even so, the decision in the case of *Vonlynn Holdings* v. *T. Flaherty* (1988) concerning a JCT 80 contract, provides limited support for the *Hounslow* case. The position under the ICE Conditions of Contract (where the operation by the employer of the forfeiture provisions under clause 63 is dependent on the certified opinion of the engineer) is very different, the contractor being bound to vacate the site notwithstanding that the engineer's opinion is challenged by the contractor: see *Tara Civil Engineering* v. *Moorfield Developments Limited* (1989). The *Hounslow* case was criticised and not followed in the New Zealand case of *Mayfield Holdings Ltd* v. *Moana Reef Ltd* (1973). Furthermore, the *Hounslow* case has expressly not been followed in Australia: see *Chermar Productions Proprietary Ltd v. Prestest Proprietary Ltd* (1989), in which it was confirmed that the contractor's licence to occupy the site may be determined and the contractor transformed into a trespasser even if the determination of the licence involved the employer in a breach of the contract.

The criticism is well deserved. It has been suggested that the *Hounslow* case was decided at a time when a claimant could only obtain an interim injunction if he had shown a prima facie case in his favour. In the subsequent case of *American Cyanamid Company* v. *Ethicon Limited* (1975), the test applied by the House of Lords was concerned with the balance of convenience as to whether or not an injunction should be granted, and then an injunction would only generally be granted if damages were not an adequate remedy. It is unlikely that the *Hounslow* decision would be followed today.

To allow the contractor to remain on site when there has been a complete breakdown of relationships seems a nonsense. Even if the employer is in the wrong, he cannot easily be compelled to make interim payments and the architect cannot be compelled to issue necessary drawings and other information etc., so that there will be little or no progress. Also, the employer will be prevented, by the continued presence of the contractor, from employing others to finish off the work. The contractor should be required to leave the site and to pursue his claim for damages for breach of contract if he is wrongly removed from the site. While it may be contended that compelling the contractor to leave, even though he may be the innocent party, could result in damage to his reputation, nevertheless, permitting him to stay will cause a hopeless impasse in many situations.

All of this debate is resolved in favour of the employer under the MPF.

Extent of access

3-04 The contractor is not entitled to access to the whole of the site unless, exceptionally, it is necessary to have access to the whole at the same time in order for the contractor to build the project in accordance with his contractual obligations. The requirement is to give the contractor access to those parts of the site and at such times and for such periods as may be reasonably necessary for him to fulfil his contractual obligations.

The degree of access to be given by the employer will vary; for example, compare the case of a new building on a green field site with additions to an existing building which remains occupied.

3-05 If a building contract failed expressly to provide for access or possession of the site to be given to the contractor, such a provision would be implied to the extent that it was required to enable the contractor to complete by any agreed completion date. It would be necessary to imply such a term to give the contract business efficacy.

3-06 If the employer fails to give the appropriate degree of access to enable the contractor to complete on time or properly to carry out the work in accordance with any agreed programme, this will be a breach of contract by the employer. This is so even though the employer's failure was due to circumstances completely outside his control.

3-07 In practice, although the employer may have entered into a contract with the very best intentions, some difficulty in giving access may remain, e.g. a tenant who refuses to vacate without a court order. It is advisable therefore for the building contract to cater for such problems, by providing for an extension of time and for reimbursement of the contractor's loss and expense.

3-08 The failure by the employer to provide access by an agreed date may, depending on its degree and duration, amount to a serious breach of contract entitling the contractor to treat the contract as at an end and to sue the employer for damages. Apart from this, unless there is provision for an extension of time for delay in giving access, such delay, even if not amounting to a fundamental breach, will invalidate any liquidated damages clause – see *Rapid Building Group Ltd* v. *Ealing Family Housing Association Ltd* (1984). In such circumstances the employer may still claim general damages for breach of contract for delay if these can be proved. There has been some debate as to whether any such general damages claim is subject to a ceiling equal to the level of liquidated damages in the invalidated provision. The logic of such a contention is that otherwise, if the general damages were in excess of the liquidated damages figure, the employer would be profiting from his own act of delay or prevention. There is good persuasive authority that this ceiling argument is sound: see the case of *Elsley* v. *J. G. Collins Insurance Agencies Ltd* (1978) in the Supreme Court of Canada.

3-09 In the MPF, access to the site is subject to any restrictions set out in the requirements. The contractor clearly needs therefore to have these in mind when planning and programming his work. He will be entitled neither to an extension of time nor to loss and expense if he is delayed or disrupted because of restrictions which have been identified in the requirements.

Extensions of time and loss and expense

3-10 If the employer does not give access to the site or to those parts and for the periods reasonably necessary for the contractor to fulfil his obligations, this will be a breach or act of prevention on the employer's part and accordingly, if this has an effect upon the completion of the project, it can form the basis of an adjustment to the completion date (clause 12.1.8). Where it materially affects the regular progress of the project and as a consequence the contractor incurs loss and expense, the contractor can claim reimbursement (clause 21.2.1). Where the failure to provide sufficient access occurs following the initial giving of access to the site on the date stated in the appendix, this breach or act of prevention on the employer's part is clearly covered. However, where the employer fails at the outset to give access to the site on the date stated in the appendix, while such a situation falls within the extension of time provisions, there is an argument, admittedly not particularly strong, to the effect that this would not fall within the loss and expense provisions in clause 21 (see paragraph 5-89).

The site

3-11 The employer must give access to the 'Site'. This is defined (see clause 39.2) as:

> The area where the Project is to be constructed and whose boundaries are defined in the Requirements.

In practice, this is likely to be achieved by incorporating an appropriate drawing into the requirements.

Commencing execution of the project

> 9.2 Upon access to the Site being given under clause 9.1, the Contractor shall commence the execution of the Project and shall proceed regularly and diligently with the Project so as to achieve Practical Completion on or before the Completion Date.

3-12 Once access has been given under clause 9.1, the contractor must 'commence the execution of the Project'. The project is defined in clause 39.2 as:

> The works to be undertaken in accordance with the Contract, as defined in the Appendix.

Pre-start on site matters

3-13 The appendix itself begins with a very brief description of the project but then carries on 'as more fully described by the Requirements and the Proposals'. The project itself therefore will include everything which the contractor must do in accordance with his proposals to meet the requirements. This will include design work where applicable, pre-start ordering of goods or materials and other matters which, in the ordinary course, would need to be undertaken between entering into the contract and the date on which access to the site is given. Clearly, for the purposes of this clause at least, the reference to commencing the execution of the project upon access being given is intended to relate to the physical work on site relating to the project, such as setting up the site. For other purposes, commencing execution of the project might well require attending to the preliminary matters referred to, e.g. clause 1.1 (contractor to execute and complete the project in accordance with the contract including completion of the design) ; also clause 2.1 (contractor to comply with instructions issued in connection with design, execution and completion of the project).

Contractor to proceed regularly and diligently with the project

3-14 Having commenced the execution of the project, the contractor is to 'proceed regularly and diligently with the Project '.

Even without these express words it has been argued that a term to similar effect would be implied in a building contract: see *Hudson's Building and Engineering Contracts*, 11th edition, Chapter 4, paragraph 128 *et seq*. Failure by the contractor to proceed in this manner can lead to a termination of the contractor's employment under this contract (see clauses 32 and 39.2 under 'Material Breach'). This requirement to proceed regularly and diligently with the project should be of much more practical use to the employer than the obligation on the contractor to complete by a given date. In the latter case, generally no action can be taken by an employer until the completion date is past even though the contractor is clearly falling behind; whereas, in the former case, the appropriate notices etc. can be served in accordance with the contract and ultimately the contractor's employment can be terminated if necessary.

In the case of *Greater London Council* v. *Cleveland Bridge & Engineering Co Ltd and Another* (1984), at first instance, Mr Justice Staughton held that a failure by the contractor to execute the works with due diligence and expedition, entitling the employer to discharge the contractor for such failure, did not itself render the contractor in breach of contract and liable for damages. It was only a failure to use such diligence and expedition as would reasonably be required to meet the contractual deadlines, which would amount to a breach of contract by the contractor. It would seem possible to construct an argument (which it is suggested would be erroneous) that while the employer's express right to terminate the contractor's employment may be relied on, always assuming that the express contractual remedy of termination of the contractor's employment was drafted clearly enough to enable its use even where it might still be physically possible to complete on time, the employer would be taking a considerable risk in treating such a failure by the contractor before the completion date has expired as a repudiation and therefore as a 'Material Breach' and a ground for terminating the contract itself. On this basis, the employer would have to wait to see if the contractual completion date was met, or at least until it became absolutely clear that it would not be. Such an interpretation would virtually emasculate the value of such a provision. Further, the case of *West Faulkner Associates* v. *London Borough of Newham* (1994) makes it clear that the requirement to proceed regularly and diligently is not linked only to the completion date. In this case, Lord Justice Simon Brown looking at the words 'regularly' and 'diligently' said:

> 'Taken together the obligation upon the contractor is essentially to proceed continuously, industriously and efficiently with appropriate physical resources so as to progress the works steadily towards completion substantially in accordance with the contractual requirements as to time, sequence and quality of work.'

It can be readily appreciated therefore that the obligation involves not only the requirement to work steadily towards completion by the completion date but also having regard to any contractual requirements as to the sequence of working and also as to the quality of work. The obligation also extends to proceeding continuously, industriously and efficiently.

Although there is no contractual requirement for a programme, if one is produced by the contractor, his failure to keep to it may be some evidence of a failure to proceed 'regularly and diligently'.

These same words were considered by Mr Justice Megarry in the *Hounslow* case in relation to the determination provisions of JCT 63 (see paragraph 8-39).

The words 'regularly and diligently' must be read together but the contractor is required to proceed in respect of both requirements. If he fails to proceed regularly even if diligently, or alternatively if he fails to proceed diligently although regularly, he will be in breach of this provision - see the *West Faulkner* case.

Practical completion and the completion date

3-15 'Practical Completion' is a defined term in this contract and is considered below under the reference to clause 9.4. 'Completion Date' is also a defined term. Clause 39.2 states that:

> The Completion Date stated in the Appendix or fixed from time to time in accordance with clause 12 (*Extension of time*). Where the Appendix identifies that there is more than one Section then references to the Completion Date are to the Completion Date of the relevant Section.

The nature of the completion date

3-16 The date for completion is therefore to be stated in the appendix to the contract. The contract is designed for a fixed date for completion. The fact that a fixed date for completion is provided for does not, under this contract, make the time for completion of the essence of the contract so that mere failure by the contractor to complete on time is not a sufficiently serious breach to entitle the employer to treat the breach as a repudiation of the contract by the contractor. Subject to the extension of time provisions, it does however entitle the employer to liquidated damages under the contract.

If a building contract does not provide for a fixed date for completion, there will be a term implied at common law to the effect that the contractor will be obliged to complete within a reasonable time. Further, section 14 of the Supply of Goods and Services Act 1982 provides that if the time for completion is not fixed or capable of being determined by the contract, there is an implied term that the work will be carried out within a reasonable time.

If the contract contains a fixed completion date, it is a matter of interpretation, in the absence of express words, whether the time stated for completion is of the essence of the contract so that failure to achieve it puts the contractor into breach of a fundamental term entitling the employer to treat the failure as a repudiation of the contract and as releasing him from his obligations under the contract. Historically, at common law, if the contract gave a completion date, this was regarded as of the essence of the contract. However, courts of equity gave relief and did not generally regard time as of the essence and prevented the innocent party from treating the contract as at an end, though the right to claim damages for breach of contract remained.

These days, unless it is otherwise clear from the terms of the contract or the surrounding circumstances, a fixed completion date will not make time of the essence of a building contract. The standard forms of construction contract in common use in the UK, while providing for a definite completion date, do not, by their terms, make time of the essence. This is demonstrated by the fact that such contracts envisage practical completion later than the contractual or extended contractual completion date and that they contain no express ground for termination of the contractor's employment for failure to complete by the contractual completion date: see *Gibbs* v. *Tomlinson* (1992) in relation to the JCT Minor Works Form 1980. However, it is possible for the parties to agree to the completion date being of the essence despite the contract also having a liquidated damages clause: see *Peak Construction (Liverpool) Ltd* v. *McKinney Foundations Ltd* (1970).

Achieving practical completion before the completion date

3-17 The contractor is fully entitled to complete before the date for completion stated in the appendix and the employer will be obliged to issue a statement of practical completion pursuant to clause 9.4. If the employer does not want practical completion until the date for completion is reached, for example if the payment method chosen is periodic payments based on calendar dates to line up with the funding regime, clause 9.2 would require amendment.

3-18 As the contractor is entitled to complete early, does the employer owe a duty not to hinder him in achieving it? The case of *Glenlion Construction Ltd* v. *The Guinness Trust* (1987) raised this question in relation to whether or not, if a contractor provided a programme with a foreshortened completion date, the architect under a JCT 63 contract was required to supply information etc. in such a time as to enable the contractor to complete in accordance with that programme rather than by the contractual completion date.

The contract bill had provided for the contractor within a week of the date of possession to provide a programme chart of the whole of the works. The chart had to be a bar chart in approved form. The questions of law raised were as follows:

(1) Whether in so far as the programme showed a completion date earlier than the contractual completion date the contractor was entitled to carry out the works in accordance with the programme and to complete on the basis of that programme completion date.

The answer to this was that the contractor was entitled to complete early if he wished to.

(2) Whether there was an implied term of the contract that if and so far as the programme showed a completion date before the contractual completion date the employer or his architect should so perform the contract as to enable the contractor to carry out the works in accordance with the programme so as to complete them by the programmed completion date.

The answer to this was no. There was no such implied term. The contractor may be entitled to complete early but he is not obliged to do so. If such a term was to be implied it would impose an obligation on the employer but not on the contractor. In addition, in this particular contract the extension of time clause in relation to a failure to provide necessary details etc. expressly operated in relation to the contractual completion date.

The judge felt the decision was no different whether the contract actually required the contractor to provide a programme or not. The unilateral imposition of a different completion date would result in the whole balance of the contract being lost.

This does not necessarily mean that the contractor cannot recover damages for breach of contract or reimbursement of loss and expense in appropriate circumstances. Firstly, of course, if the employer is responsible for disrupting the regular progress of the works, any loss or expense flowing from this, which is attributable purely to disruption to activities not causing overall delay to completion, is recoverable. Secondly, turning to prolongation costs suffered as a result of completion being later than it otherwise would be, and if caused by the employer's delay or prevention, it is likely that the answer is in the question of what was in the contemplation of the parties. This involves questions of remoteness of damage which apply at common law to the measure of damages for breach of contract, namely the two limbs of the rule in *Hadley* v. *Baxendale* (1854). Damages are recoverable in respect of either:

- that damage which would arise in the ordinary course of things; or
- that special damage which is within the actual contemplation of the particular contracting parties at the date they entered into the contract, whether or not ordinarily in contemplation.

The employer's knowledge of the basis of the tender in terms of the contractor's proposed construction period at the date of entering into the contract may well be crucial.

A contrary view to the use of common law remoteness of damage principles is that as a claim for loss and expense is an express contractual claim which does not depend for its validity upon establishing a breach of contract on the part of the employer, e.g. delay or disruption as a result of a change instruction, coupled with a contractual requirement for the ascertainment of loss and/or expense without reference to common law principles of remoteness of damage as such, there is no room for the operation of these principles. However, the reference in some contracts (not the MPF) to 'direct' loss and/or expense does suggest that the principles of remoteness may have a role to play in implicitly excluding *indirect* loss and/or expense which would otherwise be recoverable under the second limb of the rule in *Hadley* v. *Baxendale* (1854).

In the case of *J. F. Finnegan Ltd* v. *Sheffield City Council* (1988) it was held that the appropriate date, in that particular case, from which prolongation costs could be claimed was the contractual completion date and not the contractor's earlier programmed date for completion. However, it should be appreciated that in this particular case there was evidence that the contractor had included site costs for the whole contract period in his tender rather than the shorter programmed period.

If the bonus provision in clause 14 is in operation, this adds weight to the argument that the contractor should be able to claim prolongation costs of which the loss of bonus would form part. Depending on the circumstances, the sum payable could be for the whole of the bonus which the contractor may have been able to earn, or alternatively for a proportion taking into account the possibility that he may not have achieved it (see *John Barker Construction Ltd* v. *London Portman Hotel Ltd* (1996)). If the contractor is entitled, and indeed encouraged, to complete early in return for a bonus payment, it is likely to be within the employer's contemplation that the contractor has every intention of completing early.

Use of reasonable endeavours

9.3 The Contractor shall at all times use its reasonable endeavours to prevent or reduce delay to the progress of the Project or to completion of the Project.

Delay to the progress

3-19 On a literal interpretation of the words used, it would appear that the contractor's obligation to prevent or reduce delay to progress could apply to a situation where the contractor is heading for early practical completion, i.e. practical completion before the contract completion date, and who is then faced with some delaying event. However, in context, it is difficult to accept that the contractor is required to use reasonable endeavours to reduce the delay in such a situation. He is not obliged to complete early just because that is a possibility. Additionally, if he failed to exercise reasonable endeavours in such a situation but nevertheless achieved practical completion before the contract completion date, it is difficult to see what possible damage the employer will have suffered. It should however always be borne in mind that if the threat to progress or practical completion is a matter which materially affects the regular progress of the project, and which causes the contractor to incur loss or expense, there will be a legal duty on the contractor to take reasonable steps to mitigate that loss and expense; the same applies if the claim can be framed in terms of a claim for damages for breach of contract.

Reasonable endeavours

3-20 This is often a difficult phrase to interpret. However it can be said that there is almost no situation in which the contractor would be obliged to commit himself to significant expenditure. If the contractor fails to use reasonable endeavours to prevent or reduce the delay, then he will be liable to pay or allow to the employer liquidated damages.

Not only does the use of reasonable endeavours exclude the obligation to incur financial costs, it probably also excludes taking action which would be to one's commercial disadvantage. The result is that, in practice, relatively minimal effort is required to fulfil this obligation.

The requirement to use reasonable endeavours is less burdensome than the use of best endeavours. It has been described as 'appreciably less than best endeavours': *UBH (Mechanical Services) Ltd* v. *Standard Life Assurance Co* (1986). In this case it was held that it was appropriate to balance the duty to use reasonable endeavours against all relevant commercial considerations including costs of, and the uncertainties and practicality relating to, compliance with the obligation. Further, the likelihood of success (or lack of it) following the performance of reasonable endeavours was of fundamental importance. Even the requirement to use best endeavours includes the concept of reasonableness in the way in which it operates in any given set of circumstances. It only imposes a duty to do what can reasonably be done in the circumstances, and it has been described as 'that of the reasonable and prudent Board of Directors, acting properly in the interest of their company': *Terrell* v. *Mabie Todd & Co Ltd* (1952).

In the event of delay occurring, it is clear that the contractor is not permitted to sit back and let the delay take its natural course. He must take such positive steps as are available to prevent delay. In any event this will usually be in his own best interests, especially in those situations where any costs will have to be borne by him. The better view is that the contractor would still have an obligation to reduce or prevent delay despite incurring merely trivial costs. Also relevant, as mentioned earlier, is the possible question of mitigation in this situation. If the cause of the delay is a breach of contract by the employer or a matter which could give rise to a claim for reimbursement of loss and expense under clause 21, the contractor will have a duty to take reasonable steps to mitigate the loss. He will therefore be required to take reasonable steps, even, it is suggested, the incurring of some additional cost, not to reduce or prevent the delay but to restrict the financial effects of it if this is reasonably possible. It may be going too far to suggest that the contractor may have a duty to indicate to the employer that it may be possible to reduce or prevent the delay by incurring expenditure which would have the effect of a net overall saving on the cost of the project. For example, the contractor may be in a position to advise that by resequencing or reprogramming, at a certain cost, the reduction or prevention of delay would create a saving for the employer greater than the cost incurred in adopting the suggestion. A competent committed contractor would in any event no doubt put this forward. It could be made the subject of a variation or change instruction.

Procedure leading to practical completion

> 9.4 The Contractor shall notify the Employer when in its opinion Practical Completion has occurred and, if it agrees, the Employer shall issue a statement recording the date of Practical Completion. Where the Employer does not agree that Practical Completion has occurred, it shall notify the Contractor of the work that it requires to be completed before Practical Completion will occur. When the Contractor considers such work has been completed, it shall notify the Employer and, when satisfied that it has been completed, the Employer shall issue the statement.

Summary

3-21 This clause deals with the mechanism for the issue of a statement of practical completion. It is for the contractor to notify the employer when he is of the opinion that practical completion has occurred. If the employer agrees then he issues a statement to that effect. If the employer does not agree he notifies the contractor of the work still required to be completed. When the contractor considers that work has been completed he notifies the employer who, if satisfied, issues a statement. The notifications under this clause, being communications between the parties, are required to be either in writing, or in accordance with any electronic communication procedure specified in the appendix, or by any other means which has been agreed by the parties. (See clause 38.1 and comments on it in paragraph 9-03 *et seq.*).

This clause adopts the sensible approach of requiring the contractor to set the procedure in motion, c.f. clause 16.1 of WCD 98 where the obviously necessary role of the contractor in this process is not formally recognised. The employer's response to the contractor's notification is not governed by a timetable, the use of such timetables being generally eschewed in this contract. Bearing in mind the importance of the date of practical completion (see below), some contractors may well have preferred such a timetable, no doubt with a deemed practical completion in the event of the employer failing to meet it. In practice, if common sense does not prevail, the contractor has the option of seeking a rapid decision from an adjudicator. The same option is open if the contractor does not agree with any notification by the employer that practical completion has not occurred.

Practical completion

This is a defined term. Clause 39.2 provides that:

Practical Completion takes place when the Project is complete for all practical purposes and, in particular:

- the relevant Statutory Requirements have been complied with and any necessary consents or approvals obtained,
- neither the existence nor the execution of any minor outstanding works would affect its use,
- any stipulations identified by the Requirements as being essential for Practical Completion to take place have been satisfied, and
- the health and safety file and all 'as built' information and operating and maintenance information required by the Contract to be delivered at Practical Completion has been so delivered to the Employer.

Where the Appendix identifies that there is more than one Section then, unless stated otherwise, references to Practical Completion are to be read as references to the Practical Completion of the relevant Section.

MINOR OUTSTANDING WORKS

3-22 The above definition, providing that minor outstanding works which would not affect the intended beneficial use of the project do not prevent practical completion being achieved, reflects what is probably the generally accepted meaning of practical completion.

Most standard forms of construction contract by their express terms treat completion as something less than entire completion. They do not require the contract works to be completed in every detail before contractual completion is achieved. They permit small details to be completed during a maintenance or rectification period. Completion is qualified by reference to its being practical.

Under the general law, practical completion probably means that the contract works have been completed to the extent that nothing important or which significantly affects their use for their intended purpose remains outstanding. In the case of *Emson Eastern Ltd (in receivership)* v. *E. M. E. Developments Ltd* (1991), Judge John Newey QC sitting on Official Referee's Business considered the meaning of practical completion in a JCT 80 contract. He reviewed a number of cases including *The Lord Mayor Aldermen & Citizens of the City of Westminster* v. *J. Jarvis & Sons Ltd and Another* (1970) and *H. W. Nevill (Sunblest) Ltd* v. *William Press & Son Ltd* (1981). He preferred his own view expressed in the *Nevill* case in which he held that the (similar) JCT 63 contract gave the architect a discretion to certify practical completion:

> '... where very minor de minimis work had not been carried out, but that if there were any patent defects ... the Architect should not have given a statement of Practical Completion ...'.

The judge stressed that construction work was not like manufacturing goods in a factory. It was virtually impossible to achieve the same degree of perfection as could a manufacturer.

OTHER PRE-CONDITIONS TO THE ISSUE OF A STATEMENT

3-23 The clause 39.2 definition also provides that if the requirements state that something is essential before practical completion can take place, then this must be achieved as a pre-condition. The definition goes on to expressly provide that if the contract requires the health and safety file and all 'as built' information and operating and maintenance information to be delivered by practical completion this is again a pre-condition.

Contractors need to carefully examine the requirements for any such matters which could delay the issue of a statement. Employers need to be sure that if there are to be any such pre-conditions a clear statement that practical completion will not take place without the condition being fulfilled is in the requirements.

NO RESCINDING STATEMENT FOR DEFECTS DISCOVERED LATER

3-24 Once practical completion has been achieved and an appropriate statement under the contract issued to that effect, it cannot later be cancelled by reason of the discovery of a defect which renders the works unusable for their intended purpose: see *The Lord Mayor Aldermen & Citizens of the City of Westminster* v. *J. Jarvis & Sons Ltd and Another* (1970).

QUALIFIED STATEMENT

3-25 It is not unusual for employers to issue a statement of practical completion which is qualified by a reference, often by schedule, to defects which are required to be remedied. One or more of these, or all when taken together, may mean that the works have not contractually achieved practical completion in accordance with the generally accepted meaning of that phrase in building contracts or, where the contract defines it, in accordance with that definition. In such a case, is it open to argument that no statement of practical completion has been issued at all? If that argument were to succeed it would have significant implications in connection with contract insurances in respect of the works, liquidated damages, release of retention (not relevant for the MPF) and the commencement of any rectification or defects liability period. Of all of these, the requirement for insurance of the works is probably the most important as the others, once the dispute is resolved by adjudication or otherwise, are likely retrospectively to be sorted out tolerably well. However, in connection with the insurance, e.g. if the contractor ceases cover in respect of the works on the basis that the purported statement of practical completion is valid, nothing which is done later by an adjudicator, arbitrator or judge following a finding that there was no statement of practical completion issued, is likely to be able to adequately retrieve the situation if the works have been damaged by an insurable event.

It is probable that such a qualified statement of practical completion, even though not strictly envisaged by the wording of the contract, will nevertheless still be a statement of practical completion under it. In the case of *George Fischer Holdings Limited* v. *Multi Design Consultants Limited and Davis Langdon & Everest and Others* (1998), Judge Hicks QC considered this point among a number of points raised in the case. It was argued before him that the purported statement of practical completion was not in fact such a statement at all because it included the words 'subject to the enclosed Schedule of Defects and Reserved Matters'. The judge said:

> 'It may first be observed that any submission that this was not a contractual certificate would lie ill in the mouth of a professional advisor who had chosen to use the words "certify" and "practical completion" and to describe it as issued "under the terms of the ... contract", even if the basis for such a submission were in other respects stronger than it is in this case.'

He rejected the substance of the argument and held that what had been issued was indeed a certificate of practical completion, though he emphasised that that was independent of the question whether practical completion had in fact been achieved.

DATE WHEN PRACTICAL COMPLETION IS ACHIEVED

3-26 In the normal course of events, the date on which practical completion is achieved will almost certainly be earlier than the date on which the statement itself is issued. Some contracts, e.g. WCD 98 in clause 16.1, provide that practical completion of the works is deemed for all the purposes of the contract to have taken place on the day stated. In other words, the key for the operation of practical completion under the contract is not the date of issue of the statement but the date on which the project was practically completed. Such words are not needed in clause 9.4 as the statement will confirm the opinion of the contractor as to the date of practical completion.

SIGNIFICANCE OF THE STATEMENT

3-27 The date of the issue of the statement of practical completion is relevant for a number of purposes. For example, it is important that in respect of any insurance of the works against loss or damage, that insurance should be maintained at least until the date of the issue of the statement of practical completion and should not expire on the date of practical completion itself. This provides certainty and avoids the situation in which loss or damage is occasioned to the works before the date of the issue of the statement of practical completion but after the day named in it as the date of practical completion. What might otherwise be an uninsured period is avoided by ensuring that the insurance is kept in place at least until the date of the issue of the statement.

SECTIONS

3-28 Where the work is divided into sections (see clause 39.2 for definition) by means of entries in the appendix to the contract, references to practical completion throughout the contract are to be read as references to practical completion of any relevant section.

Practical completion of the project

Clause 39.2 provides:

> Practical Completion of the Project occurs upon Practical Completion or, when there is more than one Section, when all the Sections have achieved Practical Completion.

This makes it clear that if the project is divided into sections, practical completion of the project as a whole takes place once all sections have achieved it.

Damages for delay – clause 10

Clause 10 deals with the employer's entitlement to liquidated damages.

Liquidated damages

> 10.1 If the Contractor fails to achieve Practical Completion by the Completion Date, it shall be liable to pay the Employer liquidated damages calculated at the rate stated in the Appendix for the period from the Completion Date to the date of Practical Completion.

3-29 Most standard forms of building contract provide for the payment of a sum by the contractor to the employer by way of liquidated damages for delay in completion of the contract works.

By means of such a provision an agreed sum is payable as damages for delay in completion. In building contracts the sum may be expressed simply as so much per week or per day of delay, or alternatively, in a suitable case, e.g. industrial units, and where the contractual framework permits it, as so much per week or per day for each incomplete unit.

The main advantage to the employer of such a provision is that he is not required to prove his actual loss as a result of the delay.

The existence of a valid liquidated damages clause can be of particular benefit to an employer where the cost of delay cannot easily be measured, e.g. a school or a hospital. In such a case the employer would have difficulty in actually proving a loss beyond perhaps the loss of the use of capital money invested in the project with a delayed return together with something in respect of administration costs.

Provided that the stipulated sum represents, at the time of entering into the contract, a genuine attempt at a pre-estimate of the likely loss, it is not a penalty and will, other things being equal, be upheld by the courts: see *Dunlop Pneumatic Tyre Company Ltd* v. *New Garage & Motor Co Ltd* (1915) and, more recently *Philips (Hong Kong) Ltd* v. *Attorney General of Hong Kong* (1993). This is so even if, in the result, no actual loss is suffered at all: see *BFI Group of Companies Ltd* v. *DCB Integration Systems Ltd* (1987). The courts have traditionally construed liquidated damages and extension of time clauses in printed forms of contract strictly *contra proferentem*, which means in effect against the employer as they are generally regarded as inserted primarily for the benefit of the employer: see *Peak Construction (Liverpool) Ltd* v. *McKinney Foundations Ltd* (1970). But more recently there has been evidence of a renewed robustness in the courts in support of upholding liquidated damages provisions wherever reasonably possible: see *Philips (Hong Kong) Ltd* v. *Attorney General of Hong Kong* (1993). Further, when dealing with liquidated damages and extension of time clauses in respect of JCT 80, Mr Justice Coleman in *Balfour Beatty Building Ltd* v. *Chestermount Properties Ltd* (1993) said:

> 'In this respect the contract is not so ambiguous or so unclear as to call for application of the *contra proferentum* rule...'

There has been a noticeable move by the courts in favour of allowing 'penalty provisions' to be enforced where commercial (rather than consumer) parties are of equal or near equal bargaining positions, so that such clauses are hardly being imposed *in terrorem;* instead the parties are left to make their own bargain. See for instance, the Australian case of *Amey-UDC Finance Ltd* v. *Austen* (1986), High Court of Australia; *Indian Airlines Ltd* v. *GIA International Ltd* (2002). However, more recently the Court of Appeal, in a non-building case, appears to have reverted to the more traditional approach of requiring the sum stipulated for to be a genuine pre-estimate of likely loss – see *Jeancharm* v. *Barnet Football Club* (2003).

Loss of entitlement to liquidated damages

3-30 An employer may lose the right to liquidated damages in a number of ways:

(1) Where the sum stipulated for is in truth a penalty. A penalty clause is unenforceable in English law. For instance, if the amount stipulated for goes beyond what could conceivably be a genuine pre-estimate of the likely loss due to delay; or where the same sum is stipulated for in differing circumstances

where it is clear that the actual loss must vary so that the pre-estimate is not genuine, then it will amount to a penalty: see *Ford Motor Co.* v. *Armstrong* (1915).

(2) By the employer so delaying the contractor as to prevent the contractor completing by the agreed completion date. If, in such circumstances, the contract contains no provision for an extension of time covering such delay, then the employer cannot insist on completion by the agreed date and will lose his right to liquidated damages: see *Dodd* v. *Churton* (1897); *Peak Construction (Liverpool) Ltd* v. *McKinney Foundations Ltd* (1970); *Rapid Building Group Ltd v. Ealing Family Housing Association Ltd* (1984). In such a case, the fixed completion date will be lost and the contractor will have a reasonable time within which to complete the contract works. The employer may still have a right to claim unliquidated damages, subject to such a claim being capped at the level of the invalidated liquidated damages clause (see earlier paragraph 3-08). The MPF does provide for an extension of time in such a situation (see clauses 12.1.8).

(3) By the breakdown of the contractual machinery for calculating liquidated damages. A liquidated damages clause needs to be operated from one specific date until another, namely from the expiry of the contractual completion date until actual completion is achieved. If therefore the starting date has not been specified or the finishing date is lost, it can lead to difficulties in calculation and thereby also in the operation of the liquidated damages provision.

(4) Failure of the employer to comply with a condition precedent. For example, in certain contracts, e.g. JCT 98, the employer may first need a certificate from the architect that the contractual completion date has passed and the works remain uncompleted: see *Token Construction Co Ltd* v. *Charlton Estates Ltd* (1973).

(5) Wilful failure by the employer to properly administer the machinery of an extension of time clause: see *Peak Construction (Liverpool) Ltd* v. *McKinney Foundations Ltd* (1970). A number of cases decided in the middle to late 1980s, dealing with third party certifiers, e.g the architect, under the contract, suggest that the mere failure, even if due to incompetence and even possibly negligence, on the part of the architect in administering the extension of time machinery under the contract may not be enough to invalidate the liquidated damages provision. The contractor can generally seek immediate adjudication/arbitration/litigation, to put matters right: see for example *Temloc Ltd* v. *Errill Properties Ltd* (1987); *Lubenham Fidelities & Investment Co Ltd* v. *South Pembrokeshire District Council and Another* (1986); *Pacific Associates Inc and Another* v. *Baxter and Others* (1988). More recently there have been cases suggesting that perhaps there may be an implied term to the effect that the architect or other third party certifier will administer the extension of time clause in a reasonable, fair and impartial manner. If this is correct, then any failure by the architect or other third party certifier to do so will amount to a breach of contract by the employer enabling the contractor to claim damages rather than simply seeking to have the architect's decision reviewed and altered: see *John Barker Construction Ltd* v. *London Portman Hotel Ltd* (1996); *Balfour Beatty Civil Engineering Ltd* v. *Docklands Light Railway Ltd* (1996); *Beaufort Developments Ltd* v. *Gilbert-Ash NI Ltd and Others* (1997). If it is the employer or his representative who negligently fails to operate the extension of time provisions properly, it is certainly arguable that the liquidated damages provision will be invalidated.

Claim for general damages

3-31 Should the liquidated damages clause become invalidated the employer will still be able, subject to proof, to claim damages at common law if any delay in completion by the contractor amounts to a breach of contract: *Rapid Building Group Ltd* v. *Ealing Family Housing Association Ltd* (1984). But note the possible limitation on this referred to earlier (paragraph 3-08).

Liquidated damages as a limitation clause

3-32 On occasions, where the figure inserted in the appendix in respect of liquidated damages is less than the actual loss suffered by the employer as a result of the contractor's breach of contract in failing to meet the completion date or any extended time for completion, the liquidated damages clause in effect operates as a limitation of liability in favour of the contractor. This is not always appreciated.

Procedures

3-33 Refreshingly, there is no requirement for the employer either to notify the contractor that the contractor has not achieved practical completion by the completion date (which must be self-evident), or to notify the contractor in writing that he may seek to operate the liquidated damages provision (c.f. clause 24 of WCD 98).

3-34 A further benefit to the employer is that if he intends to deduct liquidated damages, this can be reflected in the calculation of the sum to become due under the employer's payment advice (see clauses 22.5.4 and 22.6.2). As it is part of the calculation of the sum due, it avoids any need for the service of a withholding notice on the contractor under the provisions of section 111 of the Housing Grants, Construction and Regeneration Act 1996. This is, however, just an option for the employer so that he can instead withhold the sum from the amount due under a payment advice if he wishes to. This would be the only option if the right to recover liquidated damages arose after a payment advice had been issued. In such a case an appropriate notice of withholding would be required. This is dealt with under clause 23 (see paragraph 5-163 *et seq*.).

The appendix

The appendix entry contains provision for the completion date for the project, the rate of liquidated damages and the rate of bonus. The entries can relate to the project as a whole or to sections of the project. The appendix entry provides for a daily rate for liquidated damages. It does not, as it does for instance in relation to the bonus provision, provide that 'Where no rate is specified the rate shall be NIL'. This is surprising and may have been a drafting oversight.

If no figure is inserted, does this leave the employer free nevertheless to claim general (unliquidated) damages in respect of the contractor's breach of contract in failing to complete by the contract completion date? In the case of *Temloc Ltd* v. *Errill Properties Ltd* (1987), in a JCT 80 contract under the appendix item for 'liquidated and ascertained damages', the employer inserted '£nil' and the period over which the damages was to be run, e.g. week/month etc., was left blank. Completion was delayed. The employer argued that the insertion in the appendix properly interpreted meant that clause 24 dealing with liquidated and ascertained damages was excluded from the contract altogether. Therefore, said Errill, they could claim unliquidated damages for breach of contract for delay. This argument was emphatically rejected by the Court of Appeal who made the point that it was open to the parties to agree what they liked regarding damages for non-completion provided that it did not amount to a penalty. If they agreed £1 per week they could. Therefore if they agreed £nil per week then this was effective.

Lord Justice Croom-Johnson pointed out that clause 24 was headed 'Damages for non-completion', presumably to cover all types of damage for delay in completion. In other words damages for late completion was meant to be governed by clause 24 which provided for liquidated and ascertained damages and not for unliquidated damages. Clause 24 provided an exhaustive remedy for damages for delay. The fact that £nil was inserted did not mean that the liquidated damages clause became ineffective. It remained valid.

This decision has been criticised, (see 39 BLR 31) on the basis that it is inherently unlikely that an employer such as a property developer would intend by inserting £nil to waive altogether any rights for damages for delay.

The meaning to be given to such an entry in the appendix will depend upon a proper construction of the contract as a whole. In the MPF the appendix does not indicate what the position is to be if no entry is made. It does point to it being more likely that there was an intention in foregoing liquidated damages, not to also forego the right to claim unliquidated damages. As in the *Temloc* case, the heading of clause 10 which provides for liquidated damages is headed 'damages for delay' rather than 'liquidated damages for delay'. It is suggested that the uncertainty is such that if an employer wishes to claim unliquidated damages rather than liquidated damages, clauses 10.1 and 10.2 should be amended and the appendix item against clause 10 deleted in its entirety. From a contractor's point of view, it might be sensible to include an additional sub-clause as 10.3, stating that the provisions of clause 10.1 and the appendix provide an exhaustive remedy for damages for delay in completion. Perhaps the MPF should have put the matter beyond doubt. For example, clause 47(4)(b) of the ICE 7th Edition provides that if there is no entry in the appendix or alternatively if 'nil' is inserted, then 'to that extent damages will not be payable'. This would rule out a claim for general damages.

Repayment

10.2 Where liquidated damages have been paid to the Employer and the Completion Date is subsequently adjusted in accordance with clause 12 (*Extension of time*), the Employer shall be liable to repay to the Contractor any liquidated damages to which the Employer is no longer entitled.

3-35 This clause provides for repayment of liquidated damages by the employer following further adjustments to the completion date. Presumably, this should be reflected in the next payment advice.

3-36 The requirement is for the employer to repay the liquidated damages which have been obtained and to which the employer is no longer entitled. The employer's entitlement will cease following the adjustment to the completion date. As there is no stipulated time for repayment, it can probably be included in the next payment advice, though some might argue that the requirement is to repay the sum immediately. There seems no reason why, for the purposes of the requirements of section 110(1) of the 1996 Act which requires all construction contracts to provide an adequate mechanism for determining both when a payment becomes due and a final date for payment in relation to sums which become due, the requirement for immediate repayment, however stringent this may appear, should not be effective. By virtue of section 110(1) the parties are free to agree the period between when a payment becomes due and the final date for its discharge. If clause 10.2 does require immediate repayment, the parties are agreeing that the due date and the final date are one and the same. This will have the unfortunate consequence of putting the employer into breach of contract unless repayment is made immediately, presumably on the same day. Literally 'immediately' means now. No doubt it will be argued that 'immediately' means as fast as the employer's reasonable payment machinery may permit. This can, however, only produce uncertainty and would open up the provision generally to the argument that it was uncertain and did not amount to an adequate payment mechanism as required by section 110. If the requirement is for immediate repayment, the result is that the contractor will be entitled, by virtue of clause 24, to simple interest at the rate of 5% in excess of the 'Base Rate' (as defined in clause 39.2) from and including the day after the adjustment of the completion date until payment is made.

Taking over parts of the project – clause 11

3-37 Clause 11 provides for the circumstances in which the employer may take over a part or parts of the project before the project itself or of any section of it, in which the part is to be found, reaches practical completion.

Summary

3-38 The employer may take over a part or parts of the project before they reach practical completion if the contractor consents. That consent is not to be unreasonably delayed or withheld. This provision is similar to what, in other JCT contracts, e.g. WCD 98 and JCT 98, is called partial possession. In the MPF the drafting is simple and more direct.

The part taken over is treated as having achieved practical completion and there is a proportionate reduction in the level of liquidated damages for which purpose the part taken over must be valued. It is worth noting that unlike the other JCT contracts mentioned above, the MPF does not expressly address the question of the rectification period in respect of parts taken over, nor the insurance implications in respect of loss or damage to the works after takeover.

Comparison with sectional completion

3-39 Taking over part of the project is of course a different thing to sectional completion. Where it is known before the contract is entered into that part of the works are required to be completed before other parts, it is necessary to use the sectional completion option in the contract. The taking over of part under clause 11 is consensual and is for use where it is a requirement which arises only after the contract has been entered into.

Taking over part with consent

> 11.1 The Employer may take over any part or parts of the Project prior to Practical Completion of the same with the consent of the Contractor, which consent shall not be unreasonably delayed or withheld.

3-40 The taking over of part of the project is a consensual process, unlike sectional completion which is a pre-agreed contractual requirement. However, the consent of the contractor is conditioned to the extent that it is not to be unreasonably delayed or withheld.

Contractor's consent

3-41 In most instances the contractor will be prepared to consent as it will have the effect of achieving practical completion of the part taken over. However, some of the advantages which flow from the similar partial possession provisions to be found in many other JCT contracts such as WCD 98 and JCT 98, are not applicable, e.g. the rectification of defects period does not start to run. There is of course no retention to be the subject of early release.

Occasionally there may be reasonable grounds on which the contractor can withhold consent, e.g. where consent would result in difficulties of access or working which would inevitably cause the contractor to overrun on the remainder of the contract into a period of time when his resources were to be fully committed elsewhere.

Practical considerations

3-42 In deciding whether to seek partial possession the employer must carefully consider the practicability of it. While it may pose few problems in some situations, e.g. individual housing or industrial units, it may raise very considerable difficulties in others. Of particular importance are common services such as light, heat and power where to take over part of a building may require taking over the common services for the whole. Examples where there could be potential difficulties in relation to common parts are central staircases or roof structures in relation to office blocks. Also, there could be common roads, accesses, etc. to be considered. If only part of a common service is to be taken over there may be difficulties both in defining and valuing it.

Statement of part taken over

> 11.2 Where the Employer takes over any part of the Project prior to Practical Completion, it shall issue a statement identifying the part of the Project taken over, the date when it was taken over and the value of that part.

3-43 If the employer does take over any part of the project he must issue a statement which must identify:

- the part taken over;
- the date when it was taken over;
- its value.

The precise identity of the part taken over can be important both in terms of it having achieved practical completion (see clause 11.3) and in connection with insurances in respect of loss or damage to the project or parts of it. Similar comments can be made in respect of the fixing of the date of the taking over.

The fixing of the date and the requirement to provide the value for the part will also be important in connection with the operation of the liquidated damages provision (see below).

Reduction of liquidated damages

> 11.3 From the date identified in the statement issued under clause 11.2, the part of the Project that is taken over shall be treated as having achieved Practical Completion and the rate of liquidated damages stated in the Appendix in respect of the Project or the Section or Sections of which it forms part (where applicable) shall reduce by the same proportion that the value of the part bears to the Contract Sum or the value of the Section or Sections, as stated in or calculated in accordance with the Pricing Document.

3-44 As from the date identified in the statement issued under clause 11.2, the taken over part is to be treated as having achieved practical completion. Clause 11.3 then goes on to deal with the effect of this upon the liquidated damages provision. It may be thought surprising that something as apparently important as part of the project achieving practical completion should be tucked away in a sub-clause dealing with the apportionment of liquidated damages. It might have been thought that the reference to the part taken over achieving practical completion should stand alone somewhere. However, under this contract, unlike some other JCT contracts (mentioned above), practical completion of part is not relevant in respect of matters such as retention and the commencement of the period for rectification of defects (though it is still relevant in connection with insurance issues). The main effect therefore of part of the project achieving practical completion is the provision for the reduction in liquidated damages.

3-45 The rate of liquidated damages stated in the appendix for the project, or any relevant section or sections which include the part taken over, is to be reduced by the same proportion that the value of the part taken over bears to the contract sum or the value of the section or sections. This valuation is based upon the contract sum or value of the relevant section or sections stated in or calculated in accordance with the contract pricing document. Clause 39.2 defines the pricing document as 'The document identified in the Appendix containing the contract sum analysis and particulars of the manner in which the Contract Sum is to be paid to the Contractor'. The contract sum is defined as 'The amount stated in the Appendix'.

A requirement that the reduction in liquidated damages is to be the same proportion as the value of the part bears to the contract sum or any relevant section, might not always produce a satisfactory result. Firstly, as it is the contract sum stated in the appendix that is the base figure, any increase in the total value of the project as a result of changes will not be taken into account. If there are significant variations to the project, coupled with large parts of the project being taken over, the effect could well be that the liquidated damages figure is reduced, possibly to zero, even though there is a significant part of the project which has still not reached practical completion.

The operation of this reduction or apportionment formula makes no allowance for the different effects in terms of financial loss which any delay in completion of outstanding parts may have. For instance, certain retail units may be taken over with only, say, £20 000 worth of landscaping outstanding. Here any delay in the outstanding part of the works may cause little or no loss in terms of the unit opening for trade and maintaining it. On the other hand, such units may be taken into possession for the employer to arrange to have them fitted out ready for trade, leaving the public access roads still to be finished off by the original contractor at a cost of, say, £15 000. Here any delay in finishing off the balance of work beyond the employer's fitting out period, could prevent trading and the estimated loss could be just as much as if no part had been taken over at all and yet the liquidated damages rate will have been significantly reduced. The operation of this formula is therefore inherently unlikely to produce a liquidated damages figure which is based upon a genuine pre-estimate. As a result it has been suggested by some that a contract with this type of clause in it may cause the liquidated damages provision to be a penalty. However, if in general terms it remains an attempt, in the circumstances, to achieve a genuine pre-estimate, then it is likely to be upheld even if it is lacking in precision: see for example *Philips (Hong Kong) Ltd* v. *Attorney General of Hong Kong* (1993).

Matters not covered

3-46 A number of matters which ought to have been dealt with consequentially on the taking over of part have not been addressed.

No effect on rectification period

3-47 So far as remedying defects of a part taken over is concerned, clause 17, dealing with the rectification of defects, does not specifically address the question of a taken over part. The definition of the 'Rectification Period' in clause 39.2 states that it is 'The twelve-month period commencing on the date Practical Completion of the Project occurs'. The definition of practical completion, also in clause 39.2, confirms that a section of the works will be covered. However, nothing is said about a taken over part. On a literal interpretation, the twelve-month period for the rectification of defects does not start running until the project, or a section of it, has achieved practical completion. The taken over part, it would appear, does not have a rectification period applying to it until the section of which it forms a part or, if there are no sections, the whole of the project has reached practical completion. This hardly seems fair. However, the contractor may find it difficult to use this consequence as a ground for 'reasonably' withholding his consent, making it dependent on operating the rectification period in relation to the part taken over as from its practical completion. The reason for this is that the contractor should be aware when entering into the MPF that the rectification period is geared to practical completion of sections or the project and not to parts taken over.

Insurance position

3-48 As for the insurance aspects, other JCT forms, such as those mentioned above, expressly provide that the policy covering the insurance of the works against loss or damage will cease to apply to the relevant part as from the date it is taken over. It also provides that the part taken over will, in the event that it forms part of an existing structure, be covered by any policy of the employer which covers loss or damage to the existing structure. In the MPF, clause 27, which deals with insurances, does not state what is to be covered and on what terms. It provides for the appendix to the contract to indicate the nature of such insurances as are required. If there is a possibility of part being taken over, it would be wise for both parties to check in advance of entering into the contract that this situation is clearly and adequately catered for in their insurance arrangements.

Bonus

3-49 Nothing is said in the contract about the relationship between any bonus provision under clause 14 (which operates between the date of practical completion and the contract completion date at the rate stated in the appendix) in circumstances where part of the project or the relevant section is taken over. It might have been anticipated that some sort of formula would have been adopted to ensure that the contractor gets an appropriate part of the bonus at take over of the part rather than having to wait until the project or relevant section achieves practical completion. Should the contractor wish to withhold his consent to the taking over of part on this ground he may have some difficulty for the reasons given above in relation to the rectification period.

Extensions of time – clause 12

Introduction

3-50 Clause 12 deals with extensions to the completion date where completion of the project is affected by any of the stated matters. There is a requirement for the contractor to notify the employer of causes of delay to progress of the project and there are review provisions.

When compared with other JCT contracts such as WCD 98 and JCT 98, the provisions in this contract dealing with extension of time are very different. Though there is a list of causes which can give rise to an adjustment to the completion date, the list is far shorter. Essentially, this is achieved by combining a number of employer delays into a general clause (see clause 12.1.8) and also eliminating some neutral events (such as exceptionally adverse weather conditions, work carried out by a local authority or statutory undertaker in pursuance of his statutory obligations, civil commotion, strikes or lockouts, etc.). Although not expressly included, these so-called neutral events may nevertheless, exceptionally, be brought in under the *force majeure* provision (clause 12.1.1, discussed later).

3-51 Where the delay is what might be called a neutral delay, i.e. not a direct fault of either the employer or the contractor, such as *force majeure*, many instances of loss or damage caused by a specified peril such as fire or flood, changes in statutory powers affecting the execution of the project, terrorist activity (being delays which might otherwise be at the contractual risk of the contractor – see for example *Percy Bilton Ltd* v. *Greater London Council* (1982) per Lord Fraser at pages 13 and 14), the extension of time clause benefits the contractor by relieving him of any liability to pay liquidated damages. On the other hand, if the event which causes the delay is the fault of the employer or someone for whom he is responsible, e.g. other direct contractors working on the site, then the effect of extending the contractual completion date to a new later date will keep intact the liquidated damages clause for the benefit of the employer, which would not be the case if the employer's delay was not covered by the extension of time provision.

3-52 The procedural aspects of clause 12, such as notice provisions, supporting documentation, notification by the employer of adjustments to the completion date and their review, take up a considerable amount of the clause and could with advantage have been significantly reduced and simplified. This no doubt reflects not only the natural concern of the parties in establishing clearly when liquidated damages will or will not be applied, but also, and more unfortunately, a lack of confidence of the parties in one another to operate extensions of time provisions in a fair and sensible manner. The problem with procedural detail is that it can become itself the subject of uncertainty and dispute.

Introductory words

> 12.1 The Contractor shall be entitled to an adjustment to the Completion Date to the extent that, having regard to the principles set out in clause 12.7, completion of the Project or any Section (if applicable) is or is likely to be delayed by . . .

3-53 The introductory words of clause 12.1 provide that the contractor is entitled to an adjustment to the completion date where completion of the project or of any section is affected or likely to be affected by one of the stated matters. The use of the word 'entitled' is inappropriate in relation to those matters which are the employer's responsibility, e.g. clauses 12.1.6, 12.1.7 and 12.1.8. If the cause of the delay is the employer's responsibility, as stated above, the purpose of the extension of time provision is to keep the employer's liquidated damages entitlement intact, and it is the employer who wants an extension of time to the completion date rather than the contractor.

In determining whether to adjust the completion date and by how much, regard is to be had to clause 12.7 (dealt with later at paragraph 3-100 *et seq.*).

The events giving rise to an adjustment to the completion date

3-54 Clauses 12.1.1 to 12.1.8 inclusive set out the various events which, should they delay the project, can result in an adjustment to the completion date for the project. These will be considered in turn.

Force majeure

> 12.1.1 . . . force majeure;

3-55 *Force majeure* is an event which, if it causes delay to the completion date, can give rise to an extension of time.

Meaning

3-56 *Force majeure* has been described as having reference to all circumstances independent of the will of man, and which it is not in his power to control. It has a meaning wider than *vis major* (act of God). It may well not cover an event, however catastrophic, which has been brought about by the negligence of the contractor: see *J. Lauritzen A/S* v. *Wijsmuller BV (the Super Servant Two)* (1989). The event in question must be beyond the control of the party relying on it.

There appear to be two interrelated principles in connection with the application of *force majeure* clauses. Firstly, no party will be entitled to rely upon an event as amounting to *force majeure* unless he can demonstrate that the occurrence of the event was beyond his control, and secondly, that there were no reasonable steps that he could have taken to avoid or mitigate the consequences of the event. Applied together they create a formidable presumption against a *force majeure* clause being available to any party who has in any way contributed to the event or failed to deal reasonably with its consequences. This has led, for example, to employers having to accept responsibility for industrial action among their workforce – *B & S Contracts & Design Ltd* v. *Victor Green Publications Ltd* (1984).

A leading case on *force majeure* is *Lebeaupin* v. *Richard Crispin & Co* (1920). As long ago as 1920 Mr Justice McCardie in this case said of *force majeure* clauses:

> 'This phrase "force majeure" has been introduced into many English commercial contracts within recent years. It is employed not only with increasing frequency, but without any attempt to define its meaning or any effort to co-ordinate the phrase to the other provisions of documents.'

The same judge looked at the meaning of *force majeure* on the Continent and its meaning in English contracts. He confirmed that war, inundations, and epidemics could be cases of *force majeure*; even a strike of workmen could in given circumstances. So could any direct legislative or administrative interference such as an embargo.

However, in *Matsoukis* v. *Priestman & Co* (1915) Mr Justice Bailhache said:

> 'The defendants claimed an allowance of certain other days in respect of bad weather, a shipwrights' strike, breakdown of machinery, football matches and a funeral. So far as the shipwrights' strike is concerned it comes within the very words of the exceptions clause. As to delay due to breakdown of machinery it comes within the words "force majeure" which certainly cover accidents to machinery. The term "force majeure" cannot, however, in any view, be extended to cover bad weather, football matches or a funeral. These are the usual incidents interrupting work, and the defendants, in making their contract, no doubt took them into account.'

It is important to appreciate that the meaning to be given to a *force majeure* clause will depend upon the remainder of any clause in which it is to be found and the contract as a whole. In the *Lebeaupin Case* Mr Justice McCardie said:

> 'I take it that a "force majeure" clause should be construed in each case with a close attention to the words which precede or follow it, and with a due regard to

> the nature and general terms of the contract. The effect of the clause may vary with each instrument.'

The meaning to be given to *force majeure* will be judged in the light of the general background and terms of the contract which contains this phrase. It will therefore be construed against and take its meaning from other stated events. In some other JCT contracts which use a more detailed listing approach in their extension of time provisions, the meaning of *force majeure* would be more restrictive, being considered in the context of these other listed events, e.g. strikes and weather, which might arguably otherwise be within the meaning of that phrase in the MPF.

It is suggested that in the context of the MPF form, the event must be something which in relation to large and complex construction projects, can be regarded as truly abnormal in nature or extent and does cover every situation in which, even though through no fault of the contractor, completion is delayed.

Some contracts seek to expand on the meaning of *force majeure* in terms of its definition. Those typically associated with private finance initiative schemes can be extremely lengthy; others seek to give to the words what the draftsmen believe is their general meaning. For instance:

> 'Force majeure means the non-performance of a relevant act due to circumstances beyond the control of the person relying upon it, which are abnormal and unforeseeable whose occurrence and consequences could not have been avoided through the exercise of all due care.'

Exceptionally adverse weather

3-57 In practice, in the MPF, the main application for the *force majeure* event is bound to be in connection with exceptionally adverse weather conditions. With the absence of an exceptionally adverse weather conditions event in clause 12.1, the *force majeure* provision will frequently be relied on. However, it is worth remembering that the definition of 'Specified Peril' (see below paragraph 3-60) includes lightning, storm, tempest and flood. It may therefore be that extreme instances of exceptional weather, or even not so exceptional, e.g. a routine storm, may well fall under clause 12.1.2.

To amount to *force majeure*, the weather conditions must not just be adverse but exceptionally adverse to qualify. The contractor can expect adverse weather conditions and should therefore be able, to some degree, to allow for it in his programme. However, if the weather conditions are exceptionally adverse, and therefore truly abnormal, either because of their intensity or their duration, such that their occurrence and effects could not be foreseen or anticipated, there is a chance that they will be covered by the *force majeure* provision.

Where exceptionally adverse weather forms the basis of a claim to an extension of time based on *force majeure*, it will no doubt still give rise to the usual debating points such as:

(1) Should the contractor have to provide meteorological evidence for the previous 5, 10 or 15 years in order to demonstrate that the weather conditions were exceptionally adverse?
(2) To what extent, if any, should regard be had not only to the period of exceptionally adverse weather conditions but also to any exceptionally beneficial

weather conditions occurring shortly before or after which may mean that *overall* progress has not been affected by such weather? E.g. an exceptionally mild February which is particularly beneficial to the progress of the contractor's external works, during which month however there were two days of unprecedented snowfall which prevented progress being made on those two particular days.

It is for the employer to determine (reviewable in adjudication and/or litigation) a fair and reasonable adjustment to the completion date (see clauses 12.4.1 and 12.6) and it is difficult, and often unhelpful, to try and lay down fixed rules or principles.

Local authorities and statutory undertakers

3-58 There is no reference, in the list of matters, to the carrying out by a local authority or statutory undertaker of work pursuant to its statutory obligations or any failure to do so. When a local authority or statutory undertaker carries out work pursuant to statutory obligations, they are not engaged, and do not require to be authorised by, the employer, and clause 12.1.6 will not therefore apply. In such situations, any delay caused by them, to the extent that it is unforeseeable in nature or extent, and truly abnormal, will arguably fall under clause 12.1.1.

However, with the increase in privatisation of former statutory bodies who operate on an increasingly commercial basis, there will often be a direct contract with the employer, at any rate so far as the carrying out of commercial activities rather than statutory duties is concerned. In such cases any delays caused may result in clause 12.1.6 (interference by 'Others' on the site) applying.

Conclusion

3-59 In conclusion, with the provision of a *force majeure* clause where there are relatively few neutral delays included in an accompanying list, the likelihood is that this will not necessarily be too disadvantageous to the contractor. Most neutral events, i.e. those for which the contractor cannot be held to blame or for which he would not ordinarily be regarded as responsible, if they are truly unforeseeable in nature or extent and have a significant impact on the project, are likely to be covered.

Specified perils

> 12.1.2 ... the occurrence of one or more of the Specified Perils;

3-60 This covers loss or damage to the project due to the occurrence of one or more of what are known as 'Specified Perils'. 'Specified Peril' is defined in clause 39.2 as:

> Fire, lightning, explosion, storm, tempest, flood, escape of water from any water tank, apparatus or pipe, earthquake, aircraft or other aerial devices or articles dropped therefrom, riots and civil commotion.

Where therefore the project is delayed by reason of the occurrence of any of these events, an adjustment to the completion date is available.

Specified peril caused by contractor's negligence

3-61 The list of specified perils includes many of the insurable risks, e.g. fire, flood. Is it possible for the contractor to obtain an adjustment to the completion date under this clause where the occurrence was brought about by his own negligence or that of someone for whom he is responsible? It might have been thought that as the contractor must not benefit from his own negligence or breach of contract he might lose his entitlement to any adjustment. A contracting party cannot be seen to profit from his own breach of contract. However, in the case of *Surrey Heath Borough Council* v. *Lovell Construction Ltd* (1988) in which it was held at first instance, under a WCD 81 contract, that the negligent destruction by fire by a sub-contractor of part of the works was a breach by the contractor of an implied term similar to the express obligation in clause 5.5 of the MPF, alternatively an implied obligation to carry out work in a careful manner, it was assumed by Judge Fox-Andrews QC, though without argument on the point, that an extension of time granted under a similarly worded provision to that now under consideration was appropriate. The contract under consideration did, however, expressly provide for insurance in the joint names of the employer and contractor, thus preventing the operation of subrogation rights by the insurers and thereby preventing the insurers from claiming from its own insured. In the House of Lords case of *Co-operative Retail Services Limited and Others* v. *Taylor Young Partnership and Others* (2002), it was held that the effect of such an insurance scheme, including the requirement for joint names insurance, meant that a contractor who, it was assumed for the sake of argument, had negligently caused fire damage, had no liability to the employer. It was also accepted, without argument on the point, that the contractor was entitled to an extension of time.

In *Scottish & Newcastle Plc* v. *G.D. Construction (St Albans) Limited* (2002) (see the masterly judgment of Mr Justice Aikens), the Court of Appeal overturned a surprising first instance decision of Judge Seymour that while fire damage could be a specified peril, as defined, it was not a specified peril if it was caused by the negligence of the contractor. That first instance decision could not easily be reconciled with the *Co-operative Retail* case referred to above. If the first instance decision had been right, a negligently caused fire, flood, etc. would not be a specified peril and so could not fall within the specified perils event in many JCT extension of time clauses in such contracts as WCD 98 and JCT 98. However, it is now clear that if any of the listed perils is caused by the contractor's negligence, the contractor will still be able to seek the benefit of an extension of time to the completion date under those contracts.

In the JCT contracts mentioned, the definition of specified perils is related to the insurance clauses in those contracts. In insurance policies of the kind envisaged, a specified peril includes one which is caused by the contractor's negligence. As the same defined phrase is used in the extension of time provisions it is hardly surprising that it is given the same meaning and therefore includes contractor negligence.

However, in the MPF, the type and extent of insurances are to be agreed between the parties on a project-by-project basis (see clause 27) and this could well have a bearing on whether clause 12.1.2 does encompass the contractor's negligence and that of his sub-contractors. If insurance is taken out in respect of specified perils, it is likely that the precise meaning given to that phrase will be found in the insurance policy and it will be defined differently. For example, it will certainly be related to the causing of loss or damage, whereas the MPF definition is not so linked (see under the next heading). If the starting point is that the contractor ought not to benefit from his own negligence and breach of contract, it would take clear words, which are not to be found in the MPF itself, to remove the employer's right to compensation from the contractor by giving the contractor an adjustment to the completion date for delay to the project caused by the contractor's own breach of contract and negligence. There must be real doubt therefore whether, as currently drafted, the MPF entitles the contractor to an adjustment to the completion date on the occurrence of a specified peril brought about by the negligence of himself or his subcontractors.

If contrary to the view expressed above, specified perils include those which result from the contractor's negligence, an argument seems untenable that, in such circumstances, no adjustment to the completion date need be made because the contractor has failed to use reasonable endeavours to prevent or reduce delay (see clause 9.3). The contractor's reasonable endeavours relate to preventing *the delay* rather than preventing *the cause* of the delay.

No reference to loss or damage

3-62 In other JCT contracts such as WCD 98 (clause 25.4.3) and JCT 98, the delay to progress is related to 'loss or damage occasioned by any ... of the Specified Perils'. In the MPF all that is required is that the occurrence of the specified peril causes completion of the project to be delayed. So a flood (even if off-site - see next paragraph) which causes no loss or damage but which causes delay is enough; similarly a leaking water tank which causes no loss or damage but which has to be repaired or replaced, thereby causing delay, would suffice; riots not causing loss or damage would also be included. The effect is to extend the application of the provision to events not covered by the other JCT contracts mentioned. The significance of this is no doubt reduced by the fact that in the other JCT contacts mentioned, such situations could arguably fall within the *force majeure* relevant event in those contracts.

The absence of any reference to loss or damage being occasioned does not sit well with the definition of 'Specified Peril'. While the occurrence of fire or flood etc. is understandable, the occurrence of e.g '... aircraft or other aerial devices ...' without reference to loss or damage makes no sense.

Not restricted to the site or the project

3-63 It should be noted that the occurrence is not expressly required to be physically connected with the site or even to the project so long as the delaying effect is connected. Any occurrence of a specified peril which causes a delay to completion can fall within this clause. So, a fire at a supplier's factory might suffice; so might a flood or explosion at a distant location which affects the transport of materials to the site.

Exercise of Government powers

> 12.1.3 ... the exercise after the Base Date by the United Kingdom Government of any statutory power that directly affects the execution of the Project, other than alterations to Statutory Requirements as referred to by clause 4.5;

3-64 This provides that if, after the base date, the UK Government exercises any statutory power which directly affects the execution of the project, an adjustment to the completion date is available. A similar event listed in the extension of time provisions in other JCT contracts, e.g. WCD 98 and JCT 98, derives from the '3-day week' in the early 1970s when industries, including the construction industry, were put on intermittent 3-day week working as a result of the miners' strikes. It might just as easily have been omitted and treated as falling within the *force majeure* event.

The key to its operation will be how the word 'directly' is construed. It will almost certainly include situations where, as a result of the exercise of any such statutory power, the contractor's ability to obtain necessary labour or materials for the project is affected. It will also, of course, include difficulties in obtaining fuel or energy which are needed for the carrying out of the project. However, if the exercise of such a power makes the execution of the project more expensive but does not directly affect progress, a decision by the contractor to slow down progress to alleviate its effects on cost would not be a direct effect.

If the exercise of the statutory power has the effect of altering the 'Statutory Requirements' (defined in clause 39.2), then any amendment required to the requirements or proposals may give rise to an entitlement to an adjustment of the completion date under clause 12.1.5 in accordance with clause 4.5 and the definition of 'Change' referred to earlier (paragraph 2-63).

Terrorism

> 12.1.4 ... the use or threat of terrorism, as defined by the Terrorism Act 2000, and/or the activities of the relevant authorities in dealing with such a threat;

3-65 This clause is aimed at delay caused by terrorism or the threat of it. It is widely worded to extend not only to the direct consequences of a terrorist act but also delays attributable to threats, e.g. a hoax telephone call. In addition it extends to delay caused by the actions of relevant authorities in dealing either with terrorism itself or the threat of it.

The clause defines terrorism by reference to the Terrorism Act 2000. Section 1 of that Act provides:

> '(1) In this Act 'terrorism' means the use of threat of action where –
> (a) the action falls within subsection (2),
> (b) the use or threat is designed to influence the government or to intimidate the public or a section of the public, and
> (c) the use or threat is made for the purpose of advancing a political, religious or ideological cause.
>
> (2) Action falls within this subsection if it –
> (a) involves serious violence against a person,
> (b) involves serious damage to property,
> (c) endangers a person's life, other than that of the person committing the action,
> (d) creates a serious risk to the health or safety of the public or a section of the public, or
> (e) is designed seriously to interfere with or seriously to disrupt an electronic system.

(3) The use or threat of action falling within subsection (2) which involves the use of firearms or explosives is terrorism whether or not subsection (1)(b) is satisfied.

(4) In this section –
  (a) 'action' includes action outside the United Kingdom,
  (b) a reference to any person or to property is a reference to any person, or to property, wherever situated,
  (c) a reference to the public includes a reference to the public of a country other than the United Kingdom, and
  (d) 'the government' means the government of the United Kingdom, of a Part of the United Kingdom or of a country other than the United Kingdom.

(5) In this Act a reference to action taken for the purposes of terrorism includes a reference to action taken for the benefit of a proscribed organisation.'

Changes

12.1.5 . . . any Change;

3-66 Any change can give rise to an adjustment to the completion date. 'Change' is defined in clause 39.2 and this definition is discussed later in commenting on clause 20 (see paragraph 5-10). It includes instructions given by the employer which alter the requirements or the proposals and which give rise to alterations in design, quality or quantity. It also covers employer's instructions altering any restriction or obligation in the requirements or proposals as to the manner in which the contractor is to execute the project, as well as imposing additional restrictions or obligations. Finally, and importantly, the contract requires certain matters to be treated as giving rise to a change.

Matters which the contract requires to be treated as a change

3-67 These include:

- Clause 4.2 (employer's instructions when making his choice between discrepant provisions in the requirements in a manner different to that proposed by the contractor).
- Clause 4.5 (changes to statutory requirements which had not been announced at the base date and which necessitates amendments to the requirements or proposals).
- Clause 5.4 (employer accepting alternative materials or goods to those specified of lesser quality or standard). This appears to mean that if stated kinds or standards of materials or goods are not procurable and the contractor proposes a lesser kind or standard, which the employer accepts, this will, if it involves a delay to completion, attract an entitlement to an extension of time; whereas if the kind or standard is higher it will not. This must be unintended. Presumably the reference in clause 5.4 to the acceptance by the employer of the alternative materials or goods being treated as a change within clause 20 was in order to obtain a reduction in price. It is suggested that the proviso to the definition of change in clause 39.2 to the effect that any matter which the contract requires to

be treated as a change will not be if it is as a result of negligence or default on the part of the contractor, does not apply in this situation. It may be that the employer can consider the particular merits of the contractor's application in determining what is a 'fair and reasonable adjustment' under clauses 12.4 and 12.6.

- Clause 8.2 (proposed amendments to the requirements or proposals to overcome problems with ground conditions or man-made obstructions which could not reasonably have been foreseen by an experienced and competent contractor).
- Clause 13.2 (where the employer issues an instruction in relation to acceleration).
- Clause 16.1 (instructions for the contractor to open up for inspection or to test work etc. unless included for in the contract or the inspection or test discloses non-compliance by the contractor).
- Clause 27.8 (additional work required as a consequence of loss or damage due to terrorism where insurance for terrorism cover has ceased to be available).

No provisional sums

3-68 It should be noted that as there is no provision for the inclusion of provisional sums in this contract, there can be no instructions pursuant to provisional sums and accordingly no consequential adjustment to the completion date.

Agreeing the adjustment

3-69 By virtue of clause 12.7.1, any agreed adjustments in respect of the completion date reached by the operation of clause 20 which deals with change instructions, are to be implemented. Clearly, the thrust here is for adjustments to the completion date to be sorted out by agreement prior to an instruction being issued.

Other persons on site authorised by the employer

> 12.1.6 ... interference with the Contractor's regular progress of the Project by Others on the Site;

3-70 This covers interference with the contractor's regular progress of the project by 'Others' on the site. 'Others' is defined in clause 39.2 as:

> Persons whose presence on the Site has been authorised by the Employer, other than the Contractor, its sub-contractors and suppliers and any other persons under the control and direction of the Contractor.

These other persons presumably include the employer's advisors appointed pursuant to clause 15.2 when on site, even though also mentioned expressly in clause 12.1.8. Clause 9.1 makes it clear that the contractor does not have exclusive possession of the site so he can expect that others authorised by the employer may also be working on or visiting the site. Clearly, to the extent that the requirements expressly provide information about the activities of others on the site which is sufficient to enable the contractor to take them into account in planning, programming and pricing his own work, such activities will not be regarded as interfering with the contractor's regular progress. Where there is no such information, or insufficient information or indeed where the activities of others in some way go beyond what could reasonably have been anticipated, then an adjustment to the completion date will be possible.

A common enough situation on projects is where at some point it is intended that a separate contractor, whether engaged by the employer or by someone else such as an incoming purchaser or tenant, comes in to fit out the completed shell of a building. This clause may well be relevant in such a situation. However, if the level of disruption is likely to be unpredictable and significant, the more usual approach adopted would be for the contractor to achieve a level of completion which would enable the employer to take over the relevant part of the project under clause 11.

It should be noted that the interference is related to some site presence; so, for example, if someone authorised by the employer were to visit off-site fabrication work before it reached a critical stage of manufacture, any unforeseeable delay caused to the project would not be covered under this head, though it might well fall within clause 12.1.8.

Contractor validly suspending

12.1.7 . . . the valid exercise by the Contractor of its rights under section 112 of the HGCRA 1996;

3-71 This deals with delay caused by the contractor's exercise of his statutory right to suspend performance of his contractual obligations for non-payment pursuant to section 112 of the Housing Grants, Construction and Regeneration Act 1996. The JCT decided that despite there being a statutory provision for effectively extending the contractual time as a result of any such suspension, it would be sensible to provide a specific event in its contracts in order to maintain a complete code in respect of delaying events.

Comparison with the statutory extension of time

3-72 However, there appear to be differences between the statutory extension in section 112(4) and the entitlement under this clause. Section 112(4) provides:

> 'Any period during which performance is suspended in pursuance of the right conferred by this section shall be disregarded in computing for the purposes of any contractual time limit the time taken, by the party exercising the right or by a third party, to complete any work directly or indirectly affected by the exercise of the right.'

It is arguable that section 112(4) relates the period of the extension of the contractual period directly to the period of the actual suspension. The period of suspension may cause a lesser or greater delay to completion than the period itself. On this basis the period of the statutory extension does not appear to depend on the delay caused to the completion date by the suspension, whereas under this clause it is the delaying effect of the suspension which is taken into account. By way of example, therefore, a suspension of, say, 10 days, after which time the employer pays in full, might lead to a situation where an innocent sub-contractor who was thereby prevented from working on the site, elected to remove his tower crane to relocate it to another contract in order to comply with his contractual obligations in relation to that other contract and which he would have been able to do by finishing his work had there been no such suspension. It may be that the sub-contractor cannot bring the tower crane back onto site for 4 weeks and it may not be practicable (even if contractually valid, which is doubtful) for the contractor to secure the services of another sub-contractor and tower crane in order to reduce the delay. The actual delay caused by the suspension could therefore be very much greater than the 10-day suspension itself.

The situation could be reversed, with the 10-day period of suspension resulting in no delay to completion. Under this clause the contractor would not be entitled to an adjustment to the completion date whereas, under the statutory provision, which would still apply, the contractor would be.

3-73 Where the suspension is of such length that the contractor acts reasonably in winding down the site, any delay attributable to having to gear up again will be a consequence of the suspension and therefore covered by this clause.

Other breaches or acts of prevention of the employer

> 12.1.8 ... any other breach or act of prevention by the Employer or its representative or advisors appointed pursuant to clause 15.2;

Delay to the completion date as a result of any breach or act of prevention by the employer or his representative or advisors is a general catch-all provision intended to cater for any acts of prevention or employer breaches of contract not already listed. As a catch-all provision it avoids the risk that by attempting, but failing, to exhaustively list out individually all the possible employer delays, an employer delay may occur which has escaped the list, thereby invalidating the liquidated damages provision in the contract.

Breach or act of prevention

3-74 The 'act of prevention' is likely to take its meaning from the preceding reference to 'any other breach' so that it envisages some positive default on the part of the employer, his representative or advisor. The fact of prevention is not sufficient. It must be unwarranted in nature. It is likely to include an act of omission as well as commission, e.g. a failure to provide the contractor with necessary information to enable him to fulfil his obligation to complete within the contract or adjusted contract period.

It is not just the employer who is referred to. It extends to the employer's representative appointed pursuant to clause 15.1 (see later paragraph 4-09). The employer's representative expressly acts on behalf of the employer and he has full authority to act as the employer under the contract. In addition, this provision covers advisors appointed pursuant to clause 15.2. That clause enables the employer to appoint others to advise in connection with the project and requires the contractor to co-operate with them. Importantly, clause 15.2 expressly states that those advisors have no authority to act on behalf of the employer and the contractor is clearly made contractually aware of this fact. This could on occasions cause issues to arise as to whether any delay 'caused' by such a representative was in truth the result of the contractor treating the advisor as an agent or representative of the employer. For example, if the employer's structural engineer carries out site inspections of certain elements of the work and at some point asks the contractor not to cover work up before it is inspected, the effect of which is likely to cause delay, the contractor should be wary of simply complying. While he has, by clause 15.2, an express duty to co-operate with the structural engineer, this would not extend to allowing the progress of the project to be delayed. The appropriate action for the contractor would be to immediately contact the employer's representative for confirmation, or otherwise, of the structural engineer's request.

Repudiatory breaches

3-75 Although a breach of contract, for which the employer is responsible and which causes delay to the project, may form the basis of an adjustment to the completion date for the project, this does not mean that in all circumstances the contractor must accept that this is the only consequence of it. If the breach is sufficiently serious in nature to amount to repudiatory conduct on the part of the employer, the contractor will have the option of bringing the contract to an end at common law and claiming damages for breach of contract; or alternatively treating the breach as a 'Material Breach' which includes any repudiatory breach of contract (see clause 39.2) and giving the contractor the right to terminate his own employment under the contract (clause 33). This can be a reason for sometimes expressly including in the contract empowering provisions entitling the employer to do what might otherwise amount to a repudiatory breach. A good example of this is a failure to give initial access to the site. Such a failure is clearly a breach of contract. Coming as it does at the very beginning of the contract, it is regarded as fundamental and as a very serious breach (see earlier paragraph 3-08). Having an express power to defer the giving of access, coupled with an express provision making it a ground for an extension of time, would prevent this from happening. If the contract expressly empowers one party to do something, that cannot be a breach of the contract. A similar point may be made with respect to other conduct which would amount to a breach if the contract did not expressly permit it, e.g. the employer ordering a postponement of work. If any such breach evidences a disregard of his contractual obligations on the part of the employer, it is capable of amounting to repudiatory conduct.

The MPF does not adopt this means of including what would be repudiatory breaches within the list of events. Clause 2 requiring the contractor to comply with all written instructions issued by the employer in connection with the design, execution and completion of the project, does not enable the employer to breach either express or implied terms of the contract under cover of such an instruction. In the circumstances it might have been better to employ a more detailed list of extension of time matters expressly covering such serious matters together with the general saving clause found in clause 12.1.8 as in, for example, JCT 98 and WCD 98.

Clause 12.1 proviso

> Provided always that there shall be no adjustment to the Completion Date in respect of any matter where it is specifically stated by the Contract that such matter will not give rise to a Change.

3-76 Clause 12.1 contains a proviso that there is to be no adjustment to the completion date in respect of any matter that is specifically stated by the contract as not giving rise to a change. This should be read in conjunction with clause 2.2 which states that where the contract provides that an instruction is not to be treated as giving rise to a change, the contractor shall not be entitled to an adjustment to the completion date as a consequence of complying with the instruction. However, it is not just instructions which this proviso covers. It applies in respect of any matter that is specifically stated by the contract as not giving rise to a change. So, for example, if, in connection with the design submission procedure in clause 6, the employer comments upon any design document and the contractor does not under clause 6.8 disagree with it, the comment will not be treated as giving rise to a change (see clause 6.9).

The following is a complete list of such matters:

- Clause 4.3 – instructions on discrepancies within the proposals;
- Clause 4.5 – alterations to statutory requirements announced before the base date;
- Clause 5.4 – the contractor proposing alterations to kinds or standards of materials or goods where those described in the contract are not procurable;
- Clause 6.9 – contractor not notifying employer of his disagreement with employer's comment on design documents under clause 6.8;
- Clause 8.1 – contractor's proposals to overcome problems on encountering ground conditions or man-made obstructions;
- Clause 16.1 – opening up for inspection or test where materials or goods are found not to be in accordance with the contract;
- Clause 16.3 – instructions issued under clause 16.2 as a consequence of work, materials or goods not being in accordance with the contract;
- Clause 27.1 – contractor implementing remedial measures required by insurers due to non-compliance with the Joint Fire Code.

Procedures and timetable for adjustments to completion date

3-77 Clauses 12.2 to 12.8 inclusive provide for the procedures and timetable for adjusting the completion date should any of the matters listed in clauses 12.1.1 to 12.1.8 delay completion of the project.

3-78 If the contractor becomes aware that progress is delayed or is likely to be delayed from any cause at all he must notify the employer forthwith stating the cause and its anticipated effect upon progress and completion.

If the contractor also considers that the cause is one of the events covered by clauses 12.1.1 to 12.1.8, then he must also provide documentation to demonstrate the effect on progress and completion. That documentation must be revised as appropriate so that the employer is always aware of the anticipated or actual effects of the cause.

Where the contractor has given the appropriate notification and the cause of the delay is one of those listed, the employer has 42 days from receipt of the notification to notify the contractor either of a fair and reasonable adjustment to the completion date, or of the reason why he considers that the completion date should not be adjusted. Any adjustment made is based on the documentation provided by the contractor, though the employer may also take account of other information which may be available to him. Any such notification of an adjustment given by the employer can be reviewed by him at any time in the light of further documentation from the contractor or if the effects of any identified cause of delay become clearer.

Within 42 days of practical completion the contractor is to provide documentation in support of any further adjustment to the completion date which he considers fair and reasonable, and the employer must then undertake the review within a further 42 days of receipt of that documentation. The review has regard to any documentation provided and will result in either notification to the contractor that a further fair and reasonable adjustment to the completion date has been made or, alternatively, that the current completion date is confirmed.

In making any adjustment to the completion date the employer must:

- Implement any agreements regarding adjustments to the completion date resulting from the contractual provisions for acceleration, cost savings and value improvements and any changes;
- Have regard to any breaches by the contractor of his duty to use reasonable endeavours to prevent or reduce delay;
- Make a fair and reasonable adjustment even though the delay may also be due to the concurrent effect of a cause which is not within 12.1.1 to 12.1.8, more particularly where the contractor is also culpably in delay.

No subsequent adjustment to the completion date can result in an earlier completion date than one that has already been notified, except by agreement with the contractor.

Figure 3.1 shows the procedural aspects of adjustments to the completion date.

Notification

> 12.2 Whenever the Contractor becomes aware that the progress of the Project is being or is likely to be delayed due to any cause, it shall forthwith notify the Employer of the cause of the delay and its anticipated effect upon the progress and completion of the Project or any Section (if applicable).

3-79 Whenever the contractor becomes aware that progress of the project or any section is being or is likely to be delayed due to any cause, there is a requirement to notify the employer forthwith of that cause and its anticipated effect upon progress and completion of the project or section.

Any cause

3-80 This is not just referring to a cause which is included in the list of causes for which an adjustment to the completion date can be made. It includes any cause which does or is likely to create a delay to progress. This clearly therefore includes the contractor's own defaults and failures. So, for example, if the contractor has by an oversight failed to order materials to ensure their delivery to site at the appropriate time, this must be notified to the employer if the progress of the project or any section is being or is likely to be delayed because of it, and the anticipated effect upon progress and completion of the project must be provided. This 'owning up' requirement is also contained in other JCT contracts, e.g. WCD 98 and JCT 98. In some 25 years or more of practising law in this area, this author has yet to see or even hear of any such notification in respect of the contractor's own default.

Delays or likely delays

3-81 Whether the delay has already occurred or is only likely, the contractor is to notify the employer and indicate the cause as well as the anticipated effects upon progress

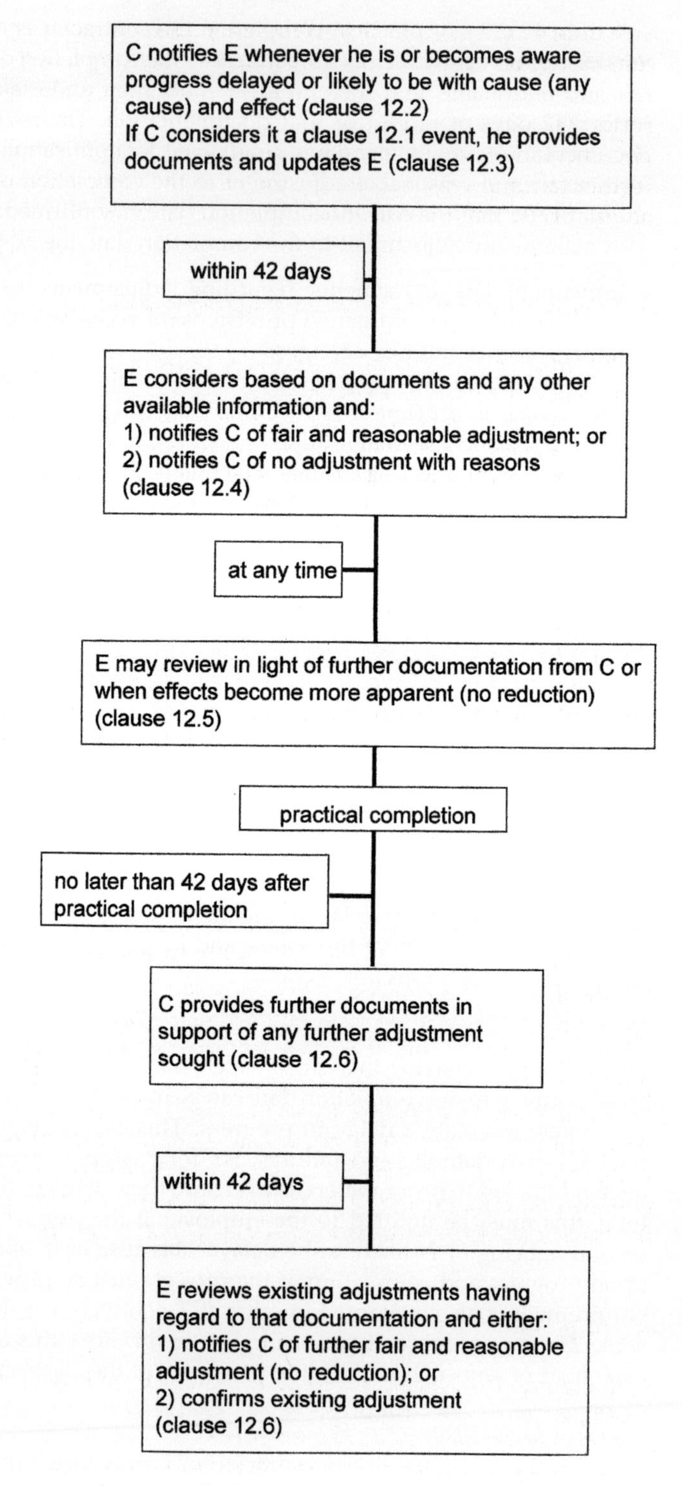

Fig. 3.1 Adjustments to the completion date.

and completion. This is important as it gives the employer the opportunity of considering action which may reduce or negate any delay.

Forthwith

3-82 The notification must be given 'forthwith'. This must mean that notification is required to be given immediately the contractor becomes aware that progress of the project or section is being or is likely to be delayed.

Contractor to provide and update supporting documentation

12.3 Where the Contractor considers the cause of the delay is one of those identified in clauses 12.1.1 to 12.1.8 it shall also:

.1 provide supporting documentation to demonstrate to the Employer the effect upon the progress and completion of the Project or any Section (if applicable) and

.2 revise any documentation provided so that the Employer is at all times aware of the anticipated or actual effect of the cause upon progress and completion of the Project or any Section (if applicable).

Supporting documentation

3-83 If the contractor considers that a notified cause of delay is covered by the list in clause 12.1.1 to 12.1.8, the contractor must also provide supporting documentation to demonstrate the effect upon progress and completion of the project or any section, and must then revise the documentation as necessary so that the employer is kept aware of the anticipated or actual effects of the cause upon progress and completion.

Time for provision of supporting documentation

3-84 While it is clear from clause 12.2 that the notification of delay or likely delay is required immediately the contractor becomes aware of it, there is no express reference in clause 12.3 dealing with when the initial supporting documentation must be provided. Presumably, this is on the basis that as it relates to proof of the effects of the cause of delay rather than the fact that it has or is likely to happen (which latter is identified by means of the notification), if the contractor wants the adjustment to the completion date he will provide the supporting documentation, particularly as the last paragraph of clause 12.4 makes it clear that any adjustment is to be calculated by reference to the documentation provided. Even so, it might have been prudent to include a requirement for the contractor to provide whatever supporting documentation he has within a stated time. Failure to provide this early on may prejudice the employer's ability to assess the anticipated or actual effect of the cause contemporaneously or as near to the point of it occurring as possible.

Employer's response

12.4 Where the Contractor has notified the Employer under clause 12.2 and the cause of delay is identified as being one of those in clauses 12.1.1 to 12.1.8 the Employer shall, within 42 days of receipt of the notification, either:

.1 notify the Contractor of such adjustment to the Completion Date as it then calculates to be fair and reasonable; or

.2 notify the Contractor why it considers that the Completion Date should not be adjusted.

Any adjustment made to the Completion Date shall be calculated by reference to the documentation provided by the Contractor in accordance with clause 12.3. The Employer may take account of other information available to it.

Adjustment to completion date

3-85 This clause provides that the employer is to calculate a fair and reasonable adjustment to the completion date where the contractor has notified the employer of the delay or likely delay and where the cause is identified as being one of those listed in clauses 12.1.1 to 12.1.8.

Is the notification requirement a condition precedent to the making of an adjustment to the completion date?

3-86 Is the requirement for notification by the contractor a condition precedent to the operation of the clause? It is suggested, though tentatively, that it is not, certainly in relation to those causes of delay listed in clauses 12.1.5, 12.1.6, 12.1.7 and 12.1.8 which are inserted primarily for the benefit of the employer, to preserve his right to claim liquidated damages.

3-87 To construe the requirement for notification as a condition precedent would, in the absence of such notification (quite possibly as a result of a deliberate decision on the part of the contractor), leave the contract without a mechanism for adjusting the completion date (the clause 12.6 post practical completion review provision not appearing to cater for unnotified events as it considers only previous adjustments), arguably invalidating the liquidated damages provision. This cannot be the intended effect of an absence of due notification even though, unlike in other JCT contracts (see, for example, clause 25.3.3.1 of WCD 98), there is no express facility in the post practical completion review clause 12.6, for the employer to adjust the completion date in the absence of notification. In the case of *London Borough of Merton* v. *Stanley Hugh Leach Ltd* (1985) it was held, under a JCT 63 contract with admittedly less prescriptive wording, that the notice of delay required from the contractor by that contract's extension of time clause (clause 23) was not a condition precedent to the contractor's entitlement to an extension of time.

It is nevertheless certainly arguable on the wording of clause 12 that the notification requirement is to be construed as a condition precedent to any adjustment to the completion date and that its absence or lateness is intended to result in liquidated damages applying during a period in which it was the employer's breach of contract which was preventing completion. Such a construction would be likely to render the liquidated damages provision a penalty. Support for this proposition is to be found in the Australian case of *Gaymark Investments Pty Ltd* v. *Walter Construction Group Ltd* (1999), though the decision has been criticised (see Ian Duncan Wallace in *Construction and Engineering Law*, Vol. 7, issue 2, p. 23 *et seq.*), not least because it can be argued that it is not the application of the prevention principle (i.e namely, the absence of an extension of time machinery for an employer delay invalidating the liquidated damages provision) which is in operation here, but rather the contractor seeking to benefit from his own breach of contract in failing to duly notify the cause of delay.

On the assumption that the notification requirement is not a condition precedent, the contractor's failure to give notice in breach of contract could be waived by the employer who would then under the final sentence of the last paragraph of clause 12.4 calculate an adjustment based on any information available to him. If as a result of the contractor's failure to notify, either the employer is prejudiced in his ability to make an accurate assessment of the delay, e.g. if his opportunity to investigate the matter is now limited, or if he has lost the chance to issue appropriate instructions, or to take other measures which could have reduced the overall delay, this could no doubt be taken into account when assessing what amounts to a fair and reasonable adjustment to the completion date.

3-88 Presumably the contractor will identify the cause as falling within those listed in the notification. While this may be done by means of describing the cause, it would obviously be sensible for the contractor to expressly identify in the notification the listed cause or causes by reference to the clause numbering.

Time-scale

3-89 Interestingly, the employer has 42 days from receipt of the *notification* in which to calculate a fair and reasonable adjustment to the completion date, or alternatively to notify the contractor why he considers that there should be no adjustment. The 42 days is related to the receipt of the *notification* rather than the receipt of *supporting documentation*. However, as referred to above, at the end of the 42-day period the adjustment to be made is dependent upon the documentation provided and if no documentation or insufficient or inadequate documentation is provided, this will be reflected in any adjustment (or lack of it).

A possible difficulty is that, based upon the wording, the 42 days from receipt of the notification are stated to be a maximum. In other words, there is nothing to prevent the employer making the adjustment at any time from receipt until the expiry of the 42-day period, presumably based upon whatever documentation has been provided at that time. Precipitous action by the employer prior to the expiry of the 42-day period would not necessarily be fully remedied under clause 12.5 which provides that the notification of an adjustment under clause 12.4 *may* be reviewed by the employer at any time in the light of further documentation or the effects of any identified cause of delay becoming clearer. Any reconsideration sought by the contractor might therefore have to wait until the post practical completion review under clause 12.6, during which time the employer would of course be entitled to deduct and hold liquidated damages.

3-90 Whether the employer adjusts the completion date or decides not to, the contractor must be notified.

The adjustment

3-91 The employer must calculate a fair and reasonable adjustment to the completion date and any adjustment is to be calculated by reference to the documentation provided by the contractor. In addition, the employer may take account of other information available to him. It would have been fairer to the contractor if the employer was required to take account of any relevant available information of which he was aware, rather than letting it apparently be discretionary. As it is discretionary rather than compulsory, it will at least avoid what might otherwise have been an argument as to what information could properly be regarded as available to the employer. Being discretionary, an employer's decision not to take account of such information could not be challenged by the contractor in adjudication or in the courts (see next section on interim review of adjustment).

Interim review of adjustment

> 12.5 Any notification given under clause 12.4 may be reviewed by the Employer at any time in the light of further documentation from the Contractor or when the effects of any identified cause of delay become more apparent.

3-92 Any notification given under clause 12.4 may be reviewed by the employer at any time in the light of further documentation from the contractor or upon the effects of any identified cause of delay becoming more apparent.

3-93 As noted above, this is not compulsory but is discretionary and extends not just to the contractor's documentation but to the effects of any identified cause of delay becoming clearer. It may therefore apply even though the documentation remains unrevised. The passage of time may make clearer the calculation of a fair and reasonable adjustment.

3-94 It is important therefore to consider whether, being discretionary, it is possible for any failure of the employer to review, or any disagreement between the parties as to the extent of any adjustment made in the event of such a review being carried out, to be submitted to an adjudicator or the courts. A decision on the part of the employer not to exercise his discretion cannot, it is suggested, be adjudicated upon or reviewed by the courts. To have an adjudicator or a judge stand in the shoes of the employer and decide whether or not to operate a discretionary provision in the contract would in effect be either to take away that discretion or alternatively to condition it upon it being, for example, fair and reasonable to exercise it or not to exercise it as the case may be. Accordingly, once the initial obligatory calculation of an adjustment has been made in respect of a particular notification, the contractor is at the mercy of the employer until the next obligatory review, which takes place after practical completion. This is of course very different from disputing the initial calculation of the employer made within 42 days of the notification. That dispute would have to centre around the documentation provided by the contractor at that time and on such other information as the employer had available to him. An adjudicator could not review that calculation based upon either documentation provided by the contractor later or because the effects of the cause of the delay had become clearer.

3-95 If the employer does however carry out such a review, then, as any further adjustment must be calculated on a fair and reasonable basis having regard to any further documentation supplied or upon the delaying effects of the cause becoming clearer, such review will no doubt itself be reviewable by an adjudicator or the courts.

It is important just to note that by clause 12.8, it is not possible to reduce an adjustment already notified. It can only therefore apply in order to calculate an additional adjustment. It is likely therefore to be the contractor who is pushing for this.

Post practical completion review

> 12.6 No later than 42 days after Practical Completion of the Project the Contractor shall provide documentation to support any further adjustment to the Completion Date that it considers fair and reasonable. Within 42 days of receipt of that documentation the Employer shall undertake a review of its previous adjustments to the Completion Date. The review

> shall have regard to that documentation and the Employer shall either notify such further adjustment to the Completion Date as is fair and reasonable or confirm the Completion Date previously notified.

Final review

3-96 The employer has an obligation to undertake a review of previous adjustments to the completion date, provided however that the contractor has provided documentation in support of any further adjustment that the contractor considers would be fair and reasonable. There is no stand-alone obligation on the employer to carry out this review. This is surprising. Even if the contractor does not provide any documentation which might justify any further adjustment to the completion date, the employer may have information available to him or alternatively the effects of any previously identified cause of delay may become clearer. In both such instances there ought to be an obligation on the employer to take this into account in the post practical completion review. The express reference in clause 12.6 to the review having regard to 'that' documentation only reinforces what appears to be a significant gap in the review machinery. Either the interim review under clause 12.5 should have been obligatory (in the light of further documentation from the contractor or the effects of identified causes of delay becoming clearer), or the final review in clause 12.6 should have expressly countenanced the employer taking account of further information available to him or the effects of any identified cause of delay becoming clearer than was the case when an earlier calculation was made. Indeed, any final review should have required the employer to take into account all relevant circumstances.

The position is made worse in that although clause 12.6 refers to the employer carrying out a review, it is hardly a review at all. The only basis upon which a further adjustment to the completion date can be made is upon receipt of documentation (presumably not previously submitted) in support from the contractor. In the absence of this there is no way in which what may well be a fairly preliminary calculation made on the initial notification by the contractor, can be reviewed at all. Why is it that on the initial notification the employer *may* (it should probably have been 'shall') calculate the adjustment by reference to other relevant available information, and on the discretionary interim review in clause 12.5 the employer can make use of the benefit of hindsight which may reveal as fact the actual effects of the identified cause of delay, and yet these cannot form part of the final review where they would be particularly pertinent?

This review therefore is dependent upon the contractor providing documentation in support of any further adjustment.

3-97 It is just about arguable that on the basis that the contractor does provide supporting documentation, the employer can then have regard to other information available to him, and take into account the actual effects of any identified cause of delay. This is because the clause provides that 'The review shall have regard to that documentation'. This may not be exclusive in the sense that while regard shall be had to 'that documentation', it does not rule out taking into account other matters. However, if this was the intention, it should have been made very much clearer.

Time-scales

The documentation from the contractor must be provided within 42 days of practical completion. Within 42 days of receipt of that documentation the employer must undertake the review having regard to the documentation provided. The employer must either:

- Notify such further adjustment as is fair and reasonable; or
- Confirm the previously notified completion date.

Effect of missing time limits

3-98 The contractor is required to provide any supporting documentation within 42 days of practical completion. What if he misses this time limit? It is arguably a condition precedent to the review taking place. The result could be that no review can take place. This would be a very serious problem for the contractor as he may not be in a position to adjudicate, or litigate, in relation to the earlier adjustments if his case for a further adjustment is dependent on additional documentation not available to the employer when the original assessment was made. There would be no dispute as to the adjustment made on the original documentation provided (e.g. *Edmund Nuttall Ltd* v. *R. G. Carter Ltd* (2002).

3-99 What if the employer fails to notify the adjustment within 42 days of receipt of the contractor's documentation? Clearly any failure to do this will be a breach of clause 12.6. Does it however invalidate the liquidated damages provision, on the basis that it is a penalty, as, literally, the employer could now deduct liquidated damages for a period when the contract provides for a fair and reasonable adjustment? On a literal construction, the employer is not permitted to make the adjustment after the 42-day period. However, having regard to the more liberal approach to the construction of liquidated damages clauses (and therefore linked extension of time clauses) (see earlier paragraph 3-29), while the late adjustment may be a breach of clause 12.6 by the employer, it is unlikely to invalidate the liquidated damages clause. However, an argument that, after the expiry of the 42-day period, the employer has no contractual basis upon which to make an adjustment and is *functus officio* in this respect, is nevertheless likely to be attempted.

Some specific points to which the employer must have regard in adjusting the completion date

12.7 In considering any adjustment to the Completion Date the Employer shall:
.1 implement any agreements about the Completion Date reached in accordance with clauses 13 (*Acceleration*), 19 (*Cost Savings and Value Improvements*) and 20 (*Changes*);
.2 have regard to any breach by the Contractor of clause 9.3;
.3 make a fair and reasonable adjustment to the Completion Date notwithstanding that completion of the Project may also have been delayed due to the concurrent effect of a cause that is not listed in clauses 12.1.1 to 12.1.8.

3-100 Clause 12.7 provides that in calculating any adjustment to the completion date the employer is to consider certain matters.

Previously agreed adjustments – clause 12.7.1

3-101 Clause 12.7.1 confirms that where as a result of an acceleration (see clause 13), or any cost savings and value improvements (see clause 19) or any changes (see clause 20) there have been agreed adjustments to the completion date, the employer must of course implement the agreement. Such adjustments will not therefore be dependent upon any notification process. In all three cases, the agreed adjustment to the completion date must be expressly stated in the relevant instruction (see clauses 13.2, 19.3 and 20.5). For clarity, the instruction should state the then current completion date together with revised date as a result of the agreement reached. It should not just say, for example, that 3 weeks will be deducted from or added to, as the case may be, the 'Completion Date'.

Clarification is also required as to the status of any pending contractor notification under clause 12.2, which is awaiting an employer response. This issue should be addressed and agreement reached either that in the light of the agreed adjustment any pending notification is withdrawn or modified together with confirmation by the contractor that he is not aware of any other cause of delay etc.; or alternatively, and perhaps more problematically, that the notification is to remain in force. In the latter case, the employer will have to consider what amounts to a fair and reasonable adjustment of the completion date after it has either been reduced or extended by agreement. Suppose for instance, that the site has been flooded so that critical foundation works cannot proceed for 6 weeks. This may be covered by clause 12.1.1 (*force majeure*) or 12.1.2 (the occurrence of a specified peril). The contractor has therefore given a notification to the effect that progress has been halted and that the expected delay to the completion date at the time of notification is, shall we say, also 6 weeks. Then a proposed change instruction produces an agreement which provides for a 2-week increase to the contract period. The contractor has agreed to this as, although as a result of the instruction the lead time for different materials is 5 weeks longer than for those originally specified, he regards only two of these weeks as critical, the other 3 weeks overlapping the critical delay caused by the flooding. The agreement to the 2-weeks' increase to the completion date is reflected in the confirming instruction without regard to the outstanding notification. Will a fair and reasonable adjustment for the flooding therefore be 6 weeks, which the contractor would contend for, or 4 weeks? The employer is unaware of the contractor's reasons for agreeing the 2 weeks' extension, so to him the 2-weeks' increase will now look as though it forms part of the same 6-week period covered by the flooding. It is better to avoid any argument and specifically address the matter of outstanding notifications.

Contractor's failure to use reasonable endeavours to prevent delay – clause 12.7.2

3-102 In considering any adjustment to the completion date, the employer 'shall ... have regard to any breach by the Contractor' of his obligation, under clause 9.3, to at all times use reasonable endeavours to prevent or reduce delay to progress or completion of the project. It might have been more accurate to provide that the employer be entitled to have regard to such matters rather than be obliged to. However the point is clear that in calculating a fair and reasonable adjustment, the employer can take account of any failure by the contractor to use its reasonable endeavours. The obligation to use reasonable endeavours to prevent or reduce delay to progress or completion is discussed earlier (see paragraph 3-19 *et seq.*).

Concurrent effect of causes of delay – clause 12.7.3

3-103 Any consideration of a fair and reasonable adjustment to the completion date is to take place even though the completion of the project might 'also have been delayed due to the concurrent effect of a cause that is not listed in clauses 12.1.1 to 12.1.8'. The employer is therefore required to make a fair and reasonable adjustment to the completion date even though, at the time when a cause of delay which is included in the list in clauses 12.1.1 to 12.1.8 is having effect, there is at the same time a 'cause' of delay which is not within the list which is having concurrent effect. Note that in fact it is not a concurrent delay which is being referred to but a delay which is having concurrent effect. This arguably means that the delays need not have commenced at precisely the same time but can be overlapping. They can commence at different times, e.g. a delay attributable to a non-listed cause, for example, the contractor carrying out remedial works following the discovery of defects which it is necessary to carry out before further site work can continue, during which delay an instruction for a change (which involves waiting for the delivery of materials to the site) is issued which would equally prevent further site work being carried out. At any rate it is probably the intention of this clause to allow overlapping rather than truly concurrent (i.e. starting at precisely the same time) delays to qualify for an adjustment to the completion date. Having said this, it can by no means be guaranteed that this is how a court would interpret the reference to delays 'due to the concurrent effect of a cause that is not listed' in clause 12.7.3. This is not an easy concept and this is not an easy clause to analyse. It may not prevent a 'first in time' causation argument being raised to restrict the application of clause 12.7.3 to situations of true concurrency (i.e. where the two events have effect at the very same moment in time (see below).

Probably, the general intention is to ensure that where, at any given period of time, either or any of two or more causes of delay would, standing alone, be an effective cause of delay, then if at least one of them is a listed cause, the contractor is to get an extension of time.

The non-listed causes of delay are almost inevitably going to be delays for which, in contractual terms, the contractor is responsible, e.g. lack of productivity, failure to order materials or let sub-contracts in good time, and so on. The probable rationale for this approach is based on the well-known argument that if the employer seeks liquidated damages for the contractor's breach in not completing by the contract completion date, it is for the employer to demonstrate such breach. If therefore during a period of delay for which the employer is seeking liquidated damages, e.g. contractor's overrun as a result of failing to order materials in time, there exists another 'cause' of delay, such as the activities of others on site authorised by the employer and for whom the contractor is not responsible, which would, had there been no contractor delay, equally have caused delay to completion, the employer fails to establish that it is the contractor's breach which has caused the overrun. That being so there should be no liquidated damages payable and accordingly an extension of time should be given. This method of dealing with the concurrent effect of more than one 'cause' of delay, is often then reversed when it comes to a contractor's claim for loss and expense where, because it is for the contractor to establish that the disruption to progress has been brought about by a cause for which the employer is responsible, if equally there is a 'cause' for which the contractor is responsible, then the contractor should not succeed in a claim for loss and expense to the extent that the same loss and expense would have been equally attributable to the cause for which he is responsible if it stood alone.

This approach is undoubtedly a kind of compromise designed to reduce arguments based upon actual causation. Whether it will turn out to actually be expedient remains to be seen. There is an argument that unless the two or more causes are truly concurrent in the sense that they start at exactly the same time, there can really only be one effective cause of the delay and that is the one which first occurred (the 'first in time' causation argument). On this analysis, any later 'cause' is not actually an effective cause as the delay is already running and will continue to run as a result of the earlier true cause. Without the inclusion of a provision such as that in clause 12.7.3 therefore (assuming it is effective), it could on this basis be argued that the reference in clause 12.2 to the project being 'delayed due to any cause' would simply not apply if a pre-existing cause is already running and having effect. The fact that without the pre-existing cause the later 'cause' would have had this effect is theoretical rather than actual. This argument against the concurrent effect of causes of delay is evident in the judgment of Judge Richard Seymour QC who in the case of *Royal Brompton Hospital* v. *Hammond and Others* (2000) said:

> 'However, it is, I think, necessary to be clear what one means by events operating concurrently. It does not mean, in my judgment, a situation in which, work already being delayed, let it be supposed, because the contractor has had difficulty in obtaining sufficient labour, an event occurs which is a Relevant Event and which, had the contractor not been delayed, would have caused him to be delayed, but which in fact, by reason of the existing delay, made no difference. In such a situation although there is a Relevant Event, the completion of the Works is (not) likely to be delayed *thereby* beyond the Completion Date...
> ...The Relevant Event simply has no effect upon the Completion Date.'

It has to be said that while this has a certain logic to it and is significantly easier to operate in practice, it does not appear to be the generally accepted approach in the industry.

The approach adopted in clause 12.7.3 appears to be an attempt to echo that contained in the Society of Construction Law Protocol for Determining Extensions of Time and Compensation for Delay and Disruption, published October 2002 (with corrections in April 2003) which provides in clause 1.4.1 that:

> 'Where Contractor Delay to Completion occurs concurrently with Employer Delay to Completion, the Contractor's concurrent delay should not reduce any EOT (extension of time) due.'

The approach of the Protocol to the concurrent effect of different causes of delay has been heavily criticised, e.g. see Ian Duncan Wallace QC (*Construction and Engineering Law*, vol. 7, issue 2, p. 16 *et seq*.), being described as a 'dogmatic and one-sided ruling (which) seems logically quite unsupportable'. The criticism is aimed at the move away from what would be the normal contractual position of dealing with actual causation; for example, where the contractor is in culpable delay and because of this, what would otherwise be a necessarily urgent instruction (if delay is to be avoided) required from the employer in relation to a change ceases to be urgent, perhaps giving the employer further time to investigate and reflect upon how to deal with the situation. The contractor would argue that had he not been in delay, he would have been held up because of the absence of that instruction. An example, given in the Ian Duncan Wallace article referred to, is that of a contractor in culpable delay, and therefore carrying out work at a time when had it not been for the culpable delay the contractor would not have been on site at all as he would have completed the project, when there occurs a national strike or exceptionally adverse weather which, had it not been for the contractor's delay, could have had no impact at all upon the completion date. This is distinguished from other events; for example, discovery of adverse soil conditions which would have been encountered at some time in any event and would have been likely to have caused delay. In terms of the MPF, a similar example might be a contractor's culpable delay during which period the works are damaged by the occurrence of a specified peril, e.g. storm damage. Other examples are easy to imagine. In such a case, as a matter of causation, it may be argued that the contractor should not gain an extension of time where the listed event would never have occurred but for the contractor's own breach.

The vital question must be whether clause 12.7.3 is effective in achieving what appears to be the probable intention of its drafters (assuming it is intended to reflect the Society of Construction Law Protocol). Firstly, it may not prevent a 'first in time' causation argument as has already been mentioned; secondly, is the requirement to 'make a fair and reasonable adjustment'. Although the clause says that the obligation to make a fair and reasonable adjustment exists even though the project may have also been delayed due to the concurrent effect of a non-listed cause, namely a contractor's culpable delay, it does not entirely rule out the possibility that in the type of examples given above, particularly those relating to the listed events which would never have occurred but for the contractor's culpable delay, the employer can justifiably form the view that a fair and reasonable adjustment to the completion date would exclude any consideration of the listed events. If a court considered that the Protocol's pro-contractor contentions referred to above were a departure from what would otherwise be the accepted contractual principles of causation, it might form the view that much clearer words are required for clause 12.7.3 to have this effect.

No earlier completion date

> 12.8 Except by agreement with the Contractor (including as set out in clauses 13 (*Acceleration*) and 19 (*Cost saving and value improvements*), no adjustment to the Completion Date shall give rise to an earlier Completion Date than one that has already been notified.

3-104 Clause 12.8 provides that unless the employer and contractor agree, no adjustment to the completion date can produce an earlier date than one that has already been notified. Examples of the employer and contractor agreeing a completion date earlier than the date notified will of course include agreements as to acceleration (clause 13) and in relation to cost savings and value improvements (clause 19).

3-105 There can accordingly be no possibility of an adjusted completion date being replaced with an earlier one as a result of an instruction for a change which omits work or otherwise reduces the construction period. In other JCT contracts, e.g. WCD 98 and JCT 98 (see clauses 25.3.2 and 25.3.3.2 respectively) such a possibility is expressly envisaged. Even so, the wording of clause 12.8 does not appear to rule out the possibility that where since the last notification of an adjustment to the completion date, there has been, for example, a change instructed, the effect of which reduces the time for construction, alongside another unconnected instructed change (and possibly even the occurrence of other listed causes) which would delay what would otherwise be the completion date, the employer could, in determining a fair and reasonable adjustment, look to the overall effect. A less arguable example may arise where, for example, the contractor is seeking an adjustment to the completion date under the *force majeure* provision in clause 12.1.1 on the basis of the occurrence during the last week of February of exceptionally adverse weather conditions. Such weather conditions may have created, say, a 1-week delay but, because of exceptionally benign weather conditions during January and the earlier part of February, the contractor's progress has been considerably greater than it otherwise would have been had the ordinary adverse weather conditions to be expected at this time of year actually been experienced.

Acceleration – clause 13

Introduction

3-106 Clause 13 provides a mechanism for accelerating practical completion to a date earlier than whatever at the relevant time is the completion date. As the result of any acceleration is to reduce the period for completion, any attempt to make it a unilateral requirement of the employer is unlikely to be satisfactory. Accordingly, it is essentially a consensual provision. If the parties cannot agree, there is little in effect that the employer can do about it.

3-107 Some other standard forms of JCT contract also provide for acceleration, for example, the JCT Management Contract 1998 and the JCT Construction Management Trade Contract 2002. Both of these as well as the MPF obviously start the process with the employer/client taking the initiative. However, in both the Management Contract and Construction Management Trade Contract acceleration provisions, the employer/client, through the architect/contract administrator or construction manager respectively, has to provide detailed information which is not the case with the MPF.

Under the Management Contract clause 3.6, the architect/contract administrator issues a preliminary instruction in which there must be set out the exact nature of the employer's wishes in regard to the completion date. Under the Construction Management Trade Contract clause 2.6, the construction manager issues an instruction which must state the required period of the reduction and also, sensibly, must also state what extensions to the completion period have already been given at the date of the instruction, together with an accurate estimate of any extensions of time applied for by the trade contractor which are likely to be given but have not yet been confirmed.

In both of those contracts, there is an opportunity for the management contractor (and effectively works contractors) and/or the trade contractor respectively, to make a reasonable objection. In practical terms it would be difficult for the employer or the construction manager respectively to argue with this as the alternative for the management contractor or trade contractor would be to load the lump sum or cost reimbursement required to achieve an acceleration. In addition, in the case of the management contract, the management contractor could also offer a minimal reduction in the time for completion. In the case of the construction management trade contract, the construction manager's instruction indicates the period of reduction required and there appears to be no halfway house for the trade contractor in his quotation to offer anything less; the option is either a reasonable objection or a quotation which shows how the reduction asked for is to be achieved. The MPF has no provision for a reasonable objection.

Both the management contract acceleration provisions and the construction management trade contract acceleration provisions are considerably more detailed and procedural than in the MPF. The MPF adopts a very different approach. It is a very brief facilitation clause and is deliberately kept simple.

The MPF merely invites proposals with no requirement as to what is being sought. Clearly to have an effective consensual process, the employer will, in his invitation, have to make known at least in general terms what he is seeking.

The invitation and the proposal

13.1 Where the Employer wishes to investigate the possibility of achieving Practical Completion at a date earlier than the Completion Date it may invite proposals from the Contractor. The Contractor shall either:

.1 make such proposals accordingly, identifying the time that will be saved and any additional costs that would be incurred; or

.2 explain why it is impracticable to achieve an earlier date.

3-108 If the employer wishes to investigate the possibility of achieving earlier practical completion he may invite proposals from the contractor. In response the contractor must either:

- Make proposals identifying the time that will be saved and any additional costs that would be incurred; or, alternatively
- Explain why it is impracticable to achieve an earlier date.

If the contractor chooses the latter route and the employer disagrees with it, it is difficult to see how the employer can contractually do much about it. The contractor could not be made to make such a proposal and any loss to the employer as a result of the contractor's breach of contract in not being able to demonstrate impracticality would be difficult to establish, particularly in terms of its remoteness. In any event, a contractor could of course respond by actually making a proposal but offering a minimal reduction of time and possibly also seeking large additional costs. Effectively, as said earlier, it is more consensual in nature.

If the contractor makes a proposal, any additional costs would need to take account of the fact that if a bonus provision is in operation (clause 14), the contractor would receive an additional benefit from that source.

3-109 Unlike the other JCT acceleration provisions referred to previously, here there is no time-scale applied so that the contractor's proposal or explanation as to why it is impracticable to achieve an earlier date will be required within a reasonable time in all the circumstances. As the invitation and proposal are communications between the parties, they are required to be either in writing, or in accordance with any electronic communications procedure specified in the appendix, or by any other means which has been agreed by the parties. (See clause 38.1 and comments thereon in paragraph 9-03 *et seq.*).

Employer's reaction

13.2 The Employer may accept any proposals made by the Contractor or seek revised proposals. If the Employer accepts any proposals it shall issue an instruction identifying the agreed adjustment to the Completion Date and additional costs (if any) and that instruction shall be treated as giving rise to a Change.

3-110 The employer can accept the contractor's proposals or seek revised proposals. Although not needing to be mentioned in the clause, the employer could also of course either do nothing or expressly reject the proposal without seeking any revision. If the employer decides to accept the proposal, then an instruction must be issued which, by clause 2, must be in writing. An oral instruction would not be binding on the contractor. The instruction must identify the agreed adjustment to the completion date and any additional costs. It is to be treated as giving rise to a change. As the agreement to accelerate will include a revision to the completion date, which the employer will adjust as a result of the instruction accepting the proposal (see clause 12.7.1), the position in relation to any outstanding contractor notification of delay under clause 12.2 should be addressed and clarified.

3-111 If the employer rejects the proposals or revised proposals, there is no provision enabling the contractor to claim his costs of preparation, however reasonable these may be. In this respect the position is similar to that in relation to the employer's request for a quotation in respect of a proposed change (see clause 20).

Limits on employer's ability to reduce completion date

13.3 Save as set out in clauses 13 (*Acceleration*) and 19 (*Cost saving and value improvements*) the Employer may not instruct the Contractor to achieve Practical Completion on a date earlier than the Completion Date.

3-112 Apart from operating the acceleration provisions in clause 13 and any instruction given under clause 19.3 where the employer agrees the contractor's suggestions for cost savings or value improvements (see paragraph 4-118) there is no other express facility for the employer to instruct the contractor to achieve practical completion on an earlier date than the completion date. Clause 13.3 confirms this. It would be the case in any event without an express clause as one party to a contract cannot unilaterally alter its terms, including the completion date, unless there is express provision within the contract enabling a party to do this. In the MPF it is a sensible provision bearing in mind that clause 2.1 dealing with instructions, requires that the contractor 'shall comply with all written instructions issued by the Employer in connection with design, execution and completion of the Project, except to the extent that the terms of the Contract restrict the Employer's right to issue any particular instruction'. Clause 13.3 puts the matter beyond doubt.

Bonus – clause 14

3-113 Clause 14 introduces the possibility of the contractor receiving a bonus for early completion. It is intended for use where prior to entering into the contract a provision for a bonus payment is agreed. It is not for use in respect of a post-contract agreement, though its use then can of course always be agreed between the parties.

The bonus provision

14.1 If the date of Practical Completion is earlier than the Completion Date the Employer shall be liable to pay to the Contractor a bonus calculated at the rate set out in the Appendix for the period from the date of Practical Completion to the Completion Date.

3-114 If the parties agree that the contract should provide for a bonus provision in the event of practical completion being earlier than the completion date, this is achieved by including a sum calculated at the rate set out in the appendix in respect of the period from the date of practical completion to the completion date. In other words, it is a rate based upon a period of time. The appendix requires it to be a daily rate. If no rate is specified then the rate is to be nil so that in effect clause 14 will not then apply. The bonus can relate to sections as well as to the project as a whole.

3-115 The possible effect on the bonus provision where part of the project or a section is taken over under clause 11 has been discussed earlier (see paragraph 3-49). Also discussed earlier (see paragraph 3.18) is the matter of loss of bonus due to the employer's breach or act of prevention forming part of a contractor's loss and expense claim recoverable under clause 21.

Chapter 4
Control

Content

4-01 Clauses 15 to 19 inclusive fall under the general heading of 'Control'. They cover the following topics:

- The employer's representative
- Testing and compliance
- Rectification of defects
- Pre-appointed consultants and named specialists
- Cost savings and value improvements.

The employer's representative – clause 15

4-02 Clause 15 deals with the employer's obligation to keep in place at all times a representative. Clause 15.2 on the other hand has nothing to do with an employer's representative except, perhaps, by way of being distinguished from it. Clause 15.2 states that the employer may appoint others to advise in connection with the project. If the employer does so, the contractor is to co-operate with them. Unlike the employer's representative, such advisors have no authority to act on behalf of the employer.

Employer's representative

15.1 The Employer shall ensure that at all times a person is appointed to act as its representative who shall exercise all of the powers and functions of the Employer under the Contract. The appointment of a representative shall take effect upon, and may be revoked at any time by, notification to the Contractor.

Obligation to appoint

4-03 Under the contract, the employer does not have the option of whether or not to appoint a representative. It is obligatory, and a failure to do so will be a breach of the contract. No doubt in practice the contractor will generally have been made aware of the identity of the proposed employer's representative before the contract has been entered into. If this was not the case the employer must make the appointment immediately the contract has been formed.

The identity of the employer's representative could be of some importance to the contractor and could even have an effect upon the contractor's tender price.

However, the contractor does not of course have any control over either the appointment or any replacement which the employer may wish to make.

The choice of employer's representative

4-04 The employer's representative should be, and remain, familiar with the detail of the project. The person appointed could be from within the employer's own organisation, such as a director or manager. Alternatively it could be someone from outside the organisation, connected with the project, such as one of the professional team. Having regard to the extent of the authority which clause 15.1 gives to the representative, the employer needs to feel confident in the choice. Unlike some other contracts, such as WCD 98 (see Article 3), the MPF does not place any restrictions on the representative's authority nor give the employer the option to notify the contractor of any restrictions on authority. Accordingly, the representative is fully authorised to make important decisions in relation to payments, including loss and expense, valuations, adjustments to the completion date, instructing changes, and even giving the appropriate notices in connection with the termination of the contractor's employment. So far as the contract is concerned, the representative has all of these powers without having to refer anything back to the employer organisation.

4-05 If someone outside the employer's organisation is appointed as the employer's representative, it is highly likely that there will be terms of engagement and these may well include restrictions on the authority to be exercised under the contract. However, it is important to note that, unless perhaps the contractor is specifically made aware of these restrictions (and such restrictions would be a breach of clause 15.1), the contractor is fully entitled to accept the lawful actions of the representative made pursuant to the contract as being fully authorised. It may be that some employers will wish to have the facility for notifying the contractor of restrictions on his representative's authority. If so they will need to amend clause 15.1.

4-06 Clause 15.1 provides that the appointment shall be of 'a person'. The contract does not define what is meant by a person but ordinarily it would include not only an individual but also an incorporated body such as a company. It would not include an unincorporated association such as an ordinary partnership. One of the main purposes in having an employer's representative is that, so far as the contract is concerned, a single point of contact will be available which the contractor can rely on as being the employer speaking. The appointment of someone other than a single individual will thwart this objective. For instance, the appointment of a company as employer's representative could lead to the situation where two or more individuals appear to be acting on behalf of the company and therefore of the employer, and this can cause unnecessary confusion and misunderstanding.

Agency

4-07 The person appointed, whether a director or employee of the employer or someone from outside, will be acting in law as an agent of the employer in their dealings with the contractor under the contract. Their appointment will therefore be subject to the general law of agency.

The relationship of principal and agent arises where the principal consents to the agent acting on his behalf. This confers authority upon the agent and gives him power to act.

Under the general law, an agency relationship is one where the agent has power to change the legal relations of another, called in law the principal. The agent's power to effect in law the legal relationship between the principal and third parties is primarily in the area of contract and acts connected with the performance of a contract. However, it can be relevant in connection with the law of property and also in other areas such as the law of tort.

A common feature of the law of agency is the general rule that an agent generally has power to bind his principal while himself dropping out of the transaction and incurring neither rights nor liabilities in relation to it.

The agent's power to affect the legal position of his principal rests upon his authority. This may be actual authority or apparent authority. Actual authority is that which is expressly given to the agent by the principal as a result of the agreement reached between them and to which they alone are the parties. Apparent (sometimes called ostensible) authority is the authority of an agent as it appears to others, including the third party, e.g. contractor under the MPF form. Under the doctrine of apparent authority, the principal may be bound to third parties because the agent appeared to have authority to act even though as between principal and agent there was in fact no such authority granted. It is for this reason that under clause 15.1 of MPF, even if the employer's representative has not expressly been given the extent of authority reflected in this clause, and even if the agreement between principal and agent expressly limits or excludes any such authority, it will be of no concern to the contractor, who can rely upon clause 15.1, at any rate unless the limits on such authority have been notified to him.

Notification of the appointment to the contractor

4-08 The appointment of the employer's representative is not effective until it has been notified to the contractor. As notification is clearly a communication, clause 38.1 dealing with communications will apply. This provides that communications are to be in writing, or alternatively in accordance with the procedure (if any) specified in the appendix to the MPF for electronic communications, or by any other means agreed in writing between the parties.

The appointment can similarly be revoked by notification at any time. If this does happen, clearly a replacement representative must be appointed at once to avoid the employer breaching clause 15.1.

Express authority of the employer's representative

4-09 Clause 15.1 states that the client's representative 'shall exercise all of the powers and functions of the Employer under the Contract'. Taken literally it produces odd results. For example:

- Clause 15.1 itself gives the employer the power to appoint a representative. Literally therefore, the employer's representative could appoint a person to act as the employer's representative and so on *ad infinitum*. Clearly it is not appropriate to have more than one representative; nor generally would it be expected that the employer's representative should have the authority to appoint his own replacement. It would be surprising if the clause is intended to mean this.

- There are clauses of the MPF under which the employer incurs a liability, e.g. to pay money or to indemnify the contractor. Clearly in one sense the employer has a function under the contract, namely to pay or to indemnify the contractor, but it cannot be intended that the employer's representative can be sued personally. In these cases the general principle of the law of agency would apply, that, in effecting legal relations between principal and third party, the agent drops out of the picture incurring neither rights nor obligations.

Breach of warranty of authority

4-10 It should be noted, however, that in the rare event of the employer's representative actually holding himself out to the contractor as having some form of authority which is in excess of his actual or apparent authority, he may find himself directly liable to the contractor for breach of warranty of authority. This is in effect a specific type of collateral contract where the agent offers to warrant his authority in return for the third party's dealing with his principal. The fact that the agent acted in good faith is no defence to such a claim. So far as the MPF form is concerned, the scope of the authority given to the employer's representative is so wide that it would only be in the most unusual circumstances that this could occur, e.g. if the employer's representative issued an instruction not authorised by the contract, claiming wrongly that he had the specific authority of the employer to take this course of action. Another possibility is that of the employer's representative without express authority seeking some form of extra contractual compromise, e.g. waiving the employer's entitlement to liquidated damages in exchange for the contractor dropping a loss and expense claim. In cases such as this, the contractor would be well advised to ask the employer's representative to provide evidence of the specific actual authority from the employer to do something which the contract does not expressly provide that the employer may do.

No provision for contractor's representative

4-11 It may have been sensible for the MPF form to have included a short clause requiring the contractor to appoint a representative to represent the contractor under the contract to avoid any ambiguity or misunderstanding as to who it is who speaks for the contractor.

Employer may appoint others

15.2 The Employer may appoint others to advise in connection with the Project and may notify the Contractor of their appointment and their role on the Project. The Contractor shall cooperate with such other advisors but they shall have no authority to act on behalf of the Employer.

4-12 Clause 15.2 provides that the employer may appoint others to advise in connection with the project. This is self-evident. It goes on to say that the employer may notify the contractor of their appointment and their role on the project. The employer is not therefore obliged to notify the contractor that advisors have been appointed. Even so, the contractor, whether he has received notice or not, is to co-operate with these advisors. In practice, it is extremely important that if any such advisors could, in carrying out their functions, potentially cause disruption to the contractor or prevent him from completing by the completion date, the employer should notify the contractor giving sufficient details and information regarding their role, preferably as part of the tender documentation, so that the contractor can programme his work to allow for any foreseeable impact the activities of such advisors may have on the project. An example might be in connection with planned testing, e.g. piles or special equipment. Clearly, on occasions the need to obtain an advisor may not arise until during the course of the project. Even so, if their involvement could affect the contractor's progress, the contractor should be notified and details given as to their role so that the contractor at least has the opportunity, if appropriate, of reprogramming to minimise any disruption or delay.

It is interesting to note that the MPF contains a definition of 'Others' in clause 39.2 which provides:

> Persons whose presence on the Site has been authorised by the Employer, other than the Contractor, its sub-contractors and suppliers and any other persons under the control and direction of the Contractor.

As in clause 15.2 the reference is to 'others' with no capital 'o', this definition does not apply. The MPF refers to 'Others' in relation to the employer's responsibility for them on site (see clauses 12.1.6 and 21.2.2 - extension of time and loss and expense where others on site interfere with regular progress; and 26.1 and 26.2 - employer responsible for others on site causing injury or damage). Clearly the advisors referred to in clause 15.2 could also fall within the definition of 'Others' when on site with the employer's authority. However, they may well of course have off-site activities and the contractor is nevertheless required to co-operate with them whether they are on or off site.

Advisors have no authority to act on behalf of employer

4-13 It is made clear in clause 15.2 that these advisors have no authority to act on behalf of the employer. Accordingly the contractor would be most unwise to treat any such advisors as having such authority. If they purport to act on behalf of the employer in a way which requires the contractor to take any action, the contractor should seek confirmation from the employer or the employer's representative before acting.

Even so, note that if such advisors, by an act of prevention, in some way delay completion of the project, the contractor will be entitled to an adjustment to the completion date (clause 12.1.8) and to recover loss and expense (clause 21.2.1).

Testing and compliance - clause 16

Summary

4-14 Clause 16 deals with instructions for the uncovering of executed work for inspection or testing, the testing of materials or goods and the giving of further instructions where work, materials or goods are found not to be in accordance with the contract. It provides a very powerful tool in the hands of the employer. There are broadly equivalent clauses in WCD 98 - clauses 8.3 and 8.4.

Briefly, what clause 16 does is to give the employer power to instruct the contractor to open up for inspection and testing executed work, materials or goods supplied for the project where this has not already been provided for in the contract, more particularly in the requirements and/or the proposals and/or the pricing document. If the result of the inspection or test is that the work, materials or goods are found to comply with the contract, the instruction is treated as giving rise to a change which will of course entitle the contractor to payment and an appropriate adjustment to the completion date.

If the work, materials or goods are found not to be in accordance with the contract the instruction is not treated as giving rise to a change. In addition, following this discovery, the employer is given further significant powers and may:

- Instruct its or their removal from the site in whole or in part;
- After consultation with the contractor instruct that it or they may still be used subject to compensation for the employer;
- After consultation with the contractor instruct any further necessary work as a consequence of the removal or use of non-conforming work, materials or goods;
- Instruct further opening up inspection or testing as is reasonable to establish whether similar work, materials or goods are in accordance with the contract.

Instructions in relation to any of the above are not treated as giving rise to a change.

Instructions following failure to carry out work in a good and workmanlike manner not covered

4-15 Work or materials provided by the contractor may be in accordance with the contract and yet the work may not have been carried out in a proper and workmanlike manner, e.g. use of a tower crane which constitutes a trespass over adjoining air space; or damage to the soil (see for example *Greater Nottingham Co-operative Society Ltd* v. *Cementation Piling & Foundations Ltd* (1989). Mention has already been made of the inappropriate way in which this obligation is dealt with in the MPF (see commentary to clause 5.5, paragraph 2-119). It is also unfortunate that the MPF does not provide the employer with express powers to issue instructions, at no cost to the employer, which may be necessary as a consequence of such a failure. This is covered in most of the other JCT forms, e.g. WCD 98 and JCT 98 (clause 8.5 in both).

However, it must be said that while an employer may well regard this as a sensible and appropriate power to have, it could nevertheless be potentially extreme in nature. There would inevitably be room for argument. For example, supposing the contractor in excavating ready for foundations, adversely affects the surrounding sensitive soil conditions in a manner which breaches the obligation to carry out work in a proper and workmanlike manner. The result is that the employer's existing foundation design is rendered inadequate, and additional works, say piling, will be necessary if the building is to proceed. However, it would also be possible at considerably less cost for the building to be slightly relocated on the site so that the foundations will not be affected by the damage to the soil. Would the employer have been able to instruct the contractor to install piles at very great cost in such circumstances? Put another way, firstly, would cost be relevant in determining what was reasonably necessary, and secondly, would the employer be entitled to insist upon the contractor complying with his contractual obligations when to modify them slightly would avoid considerable cost? Both issues would be relevant in determining whether the employer's instruction was reasonably necessary as a consequence of the contractor's failure. Despite these issues, some employers are likely to amend the MPF to cater for this situation.

Clause 16 will now be considered in more detail.

Instructions to open up or test

> 16.1 Where the Employer instructs the Contractor to open up for inspection or to test any work executed or materials or goods supplied for the Project and that opening up, inspection or test is not provided for by the Contract, the instruction shall be treated as giving rise to a Change unless the opening up, inspection or test discloses that the work, materials or goods are not in accordance with the Contract. Where the work, materials or goods are found not to be in accordance with the Contract the instruction shall not be treated as giving rise to a Change.

Inspection or test

4-16 The wording of clause 16.1 appears to have gone a little astray. The clause begins logically by referring to the employer's instruction to the contractor to 'open up for inspection or to test any work executed or materials or goods supplied'. It then refers to the matter of whether it is already provided for in the contract as, if it is not, it will be treated as giving rise to a change. Unfortunately, in so doing the clause refers to 'that opening up, inspection or test is not provided for by the Contract'. It should have referred to 'that opening up for inspection or test'. The clause then goes on to say that, if not already provided for in the contract, the instruction shall be treated as giving rise to a change unless the 'opening up, inspection or test discloses' non-compliance with the contract. Clearly, the opening up itself has no function in determining whether or not work, materials or goods are in accordance with the contract. The opening up is simply a facility in order for the inspection or test to make that finding. It probably makes little difference in practice but it is unfortunate that the drafting is not consistent.

Inspection or test already provided for by the contract

4-17 The instruction is only treated as giving rise to a change where the opening up for inspection or test is not already provided for by 'the Contract'. Literally, the contract includes the contract conditions, as well as the requirements, proposals and pricing document. It would have been more appropriate to refer specifically to the requirements or proposals (c.f. by way of example, clause 8.3 of WCD 98). In one sense as the contract includes the conditions, clause 16.1 itself provides for inspections and tests with the literal result that it could never give rise to a change even if the result was that the work, materials or goods are found to be in accordance with the contract. Clearly this cannot be the intention and the words will no doubt be construed in an appropriate purposive fashion to avoid this construction.

4-18 The reference to the opening up for inspection or test already being provided for by the contract, resulting in it not being treated as giving rise to a change, is on the basis that it will have already been both priced and programmed by the contractor. This reveals a significantly different approach to that in the equivalent clause in WCD 98 – clause 8.3. While clause 8.3 provides that in such circumstances nothing will be added to the contract sum in respect of such opening up or making good in consequence, it still permits the contractor to seek loss and expense and an adjustment to the completion date, though no doubt the extent of loss and expense recoverable and the adjustment to the completion date to be made will take into account any allowance which the contractor could properly have made in planning and programming. In clause 16.1, as the opening up for inspection or testing can never be treated as giving rise to a change, even where it discloses that the work, materials or goods are in accordance with the contract, there is no possibility of the contractor seeking any loss or expense or adjustment to the completion date. Contractors need to be aware therefore that where the contract documents provide for inspections or test they cannot seek loss and expense or adjustments to the completion date or of course the cost of the opening up, testing or making good in consequence, all of which will need to have been allowed for in the pricing and planning of the project. This could pose real problems for the contractor, who may not know from the contract documents whether the timing of the instruction will affect a critical path for the project's progress.

Scope of the opening up for inspection or test

4-19 Clause 16.1 allows for the inspection and testing of both executed work and of materials or goods supplied for the project. In this respect it is wider than clause 8.3 of WCD 98 which does not extend to the opening up of materials or goods as opposed to executed work.

4-20 It is clear that it is for the contractor to carry out the test. The employer is not permitted to undertake or arrange for the undertaking of the test. So far as the inspection is concerned, clause 16.1 does not expressly state who can inspect so presumably the employer can and of course will want to inspect or arrange for the inspection.

4-21 Bearing in mind the important financial consequences which flow from the results of any inspection or test, it may be very important to determine whether or not an instruction refers to a single inspection or test or in fact to a series of different inspections or tests. For example, if the instruction calls for an identical test on six critical welded joints, two of which are discovered to be unsatisfactory and four of which are satisfactory, are there six tests, two of which are at the contractor's expense and four at the employer's expense? Or is there just one single test which has disclosed non-complying work and which will therefore be wholly at the contractor's expense? While it must clearly be a question of degree in all the circumstances of any particular case, generally the determining factor is likely to be the nature of the test, i.e. inspection of critical welds in the above example being one test rather than the same test being repeated a number of times.

The inspection or test covers 'materials or goods supplied for the Project'. It does not refer to goods which are yet to be supplied. It is probably the case that in order to be supplied, the materials or goods must have at least been delivered to the site. It may also mean that the property in the materials or goods must legally have passed to the employer. Until this has happened, there may exist only an agreement to supply. This has important consequences where the employer would like to have materials or goods tested prior to paying for them. It may be that there is some specialist plant or equipment being manufactured off-site, the ownership in which has not passed but which the employer wants tested before it is transported to site for installation. On this interpretation, no instruction for testing under this clause can validly be given. This interpretation is supported by the wording in clause 16.2.1 which refers to the employer instructing the 'removal from site' of non-complying work, materials or goods.

If it is known in advance that off-site testing or inspection is required, the obligation can no doubt be clearly set out in the requirements so that the contractor can make the necessary arrangements, e.g. when entering into the sub-contract for the supply of such materials or goods.

No doubt the extent to which the word '*supplied*' is to be considered in its physical and/or legal sense will be a matter of argument. Perhaps it should have been made clearer. In clause 8.3 of WCD 98, the testing applies to 'any test of any materials or goods (whether or not already incorporated in the Works)' and this can certainly arguably extend to materials or goods whether ownership has passed or not and also whether on site or off site.

Instructions giving rise to a change

4-22 Where the instruction is given in relation to opening up for inspection or testing which is not already provided for by the contract, the question of whether it is to be treated as giving rise to a change will depend on the outcome of the inspection or test. The fact that this will be unknown at the time of the instruction does not sit easily with the various provisions relating to a change in clause 20. For instance, clauses 20.2.1 and 20.3 refer to the possibility of the value of the change and any adjustment to the completion date being agreed prior to the instruction being issued. This is not appropriate here although as a matter of sensible practice the parties can of course agree to do this without prejudice to whether the instruction is to be treated as a change or not.

Duration of power to issue instructions

4-23 Clause 16.1 does not specifically state whether the instruction can be given after as well as before practical completion. However, other clauses of the contract suggest that such an instruction cannot be given after practical completion. For example, as the instruction could be treated as giving rise to a change, clause 20 dealing with changes, provides, for example in clause 20.9, that the contractor is to provide particulars of any further valuation it considers should be made in respect of a change within 42 days of practical completion. This suggests that practical completion is the cut-off date for instructions which can be treated as giving rise to a change. If this is correct, it is unfortunate that the employer cannot issue such an instruction during the rectification period, which lasts for twelve months from the date of practical completion. Such a facility could well be very useful. It may be that if during the rectification period work, materials or goods are discovered to be not in accordance with the contract, clause 16.2.4 allows the employer to instruct further opening up for inspection or testing to see if similar work, materials or goods are in accordance with the contract, and this cannot be treated as giving rise to a change (see clause 16.3). Even this power, however, is not entirely clear when regard is had to the other sub-clauses of clause 16.2, which are relevant in construing the scope of clause 16.2.4, and which hardly seem appropriate once practical completion has been achieved. If clause 16.2.4 is not adequate to give this power, the employer would have to arrange for any necessary inspection or test himself and at his own cost, though where any such inspection or test was reasonable in nature and extent and was consequent upon the contractor's failure to provide work, materials or goods in accordance with the contract, the employer could claim (whatever the result of the inspection or test) the cost from the contractor as special damages for breach of contract.

Non-complying work, materials or goods – employer's instructions

> 16.2 Where work, materials or goods are not in accordance with the Contract the Employer may: . . .

4-24 Where work, materials or goods are found to be not in accordance with the contract, clause 16.2 provides the employer with the power to issue instructions covering a number of matters and these are discussed below. It is worth pointing out at the outset, however, that clause 16.2 stands independently of clause 16.1. In other words clause 16.2 applies not only where non-compliance with the contract is discovered as a result of an inspection or test, but also where it is discovered by any other means, e.g. as a result of a routine site inspection on behalf of the employer or a malfunction of some kind occurring during construction. The reference in clause 16.2.4 to 'further opening up, testing or inspection as is reasonable' seems somewhat inconsistent with this but nevertheless on balance does not seem to detract from the point if the clause is considered in its overall commercial context.

4-25 No instruction under clause 16.2 is to be treated as giving rise to a change (clause 16.3).

Instructions to remove non-complying work, materials or goods from the site

> 16.2.1 . . . instruct their removal from the Site, either wholly or partially;

4-26 Under clause 16.2.1 the employer may issue instructions requiring the removal from site of any work, materials or goods not in accordance with the contract. The clause refers to removal 'either wholly or partially'. This means that part only of the non-complying work may be required to be removed with the remainder being retained, probably under clause 16.2.2.

4-27 It is not sufficient for the instruction simply to condemn the work, materials or goods even if that instruction also refers expressly to clause 16.2.1. The employer must expressly require its removal. In *Holland Hannen & Cubitts (Northern) Ltd* v. *Welsh Health Technical Services Organisation and Others* (1981), in a case under JCT 63 containing a similar provision in clause 6(4) of that contract, Judge John Newey said:

> 'In my opinion, an architect's power is simply to instruct the removal of work or materials from the site on the ground that they are not in accordance with the contract. A notice which does not require the removal of anything at all is not a valid notice under clause 6(4).'

Instructions for non-complying work, materials or goods to be used

> 16.2.2 ... after consultation with the Contractor, instruct that they may be used on the Project, but subject to the Contractor becoming liable to pay the Employer an appropriate amount calculated in accordance with the prices and principles set out in the Pricing Document and without the Contractor having any entitlement to an adjustment to the Completion Date and/or to the payment of loss and/or expense;

Consultation with the contractor

4-28 While the employer must consult with the contractor before issuing such an instruction, the contractor's consent is not required. Bearing in mind that the contractor becomes liable to pay the employer 'an appropriate amount', the exercise of this power could place the contractor in a difficult position. Such an instruction would enable the employer to in effect change the requirements or proposals and seek an appropriate reduction in price while depriving the contractor of any option to replace the work, materials or goods with those which will comply with the contract. As between the contractor and a sub-contractor or a supplier, the contractor may prefer to reject the work or goods and seek replacement work or goods and possibly compensation in addition. This may be the better option for the contractor. The requirement for the 'appropriate amount' to be based upon the prices and principles set out in the pricing document would make it difficult for the contractor to argue that the calculation should take into account the value of having lost the option of rejecting the work or goods under any sub-contract or supply contract.

4-29 Must the contractor comply with the instruction? The meaning of the word 'may' in clause 16.2.2 is critical. Does it mean that the contractor may use the work etc. but is not bound to ? If so he has the option to replace with complying work, etc. This would seem the most sensible interpretation particularly having regard to the potentially extreme powers open to the employer under clause 16.2.3. The biggest problem with this interpretation, however, is that if the contractor has the option of whether or not to comply with the instruction, the contractual requirement for prior consultation seems meaningless. If therefore the word 'may' has been used simply to make the point that the contractor is being allowed to use non-complying work etc., the employer, having duly consulted with the contractor, can instruct the use of such non-complying work and there is no way in which the contractor can challenge the decision.

Contractor becoming liable to pay the employer an appropriate amount

4-30 If the option provided in this clause is followed, the contractor will be liable to pay the employer an appropriate amount calculated in accordance with the prices and principles set out in the pricing document.

The actual method of calculation will depend on the prices and principles set out in the pricing document. This could range from stage payments, through a contract sum analysis, which is a basic elemental breakdown of the price, to a fully fledged detailed system of valuation supported by bills of quantities. It will be particularly difficult to calculate an appropriate amount where stage payments are used. It may be that in many instances a fair valuation of the difference between the complying

instruction that non-conforming work etc. shall remain, the question of whether such further works instructed under clause 16.2.3 are necessary is linked only to the consequences of the use of non-conforming work and not whether it was reasonable to issue an instruction to retain that work in the first place.

The result could be potentially devastating for the contractor. Take an extreme example: A contractor is obliged to provide and install a specified window system. The wrong window system is installed although of its kind it is of satisfactory quality. The employer decides to retain it. However, it does not interface properly with the cladding system so the employer issues an instruction to remove the existing cladding system and provide a slightly different one which produces the proper weather-tight interface. Strictly speaking it was a necessary consequence of the decision to retain the non-conforming work etc. However the result is absurd if the cost significantly outweighs that of replacing the window installation itself. The decision to retain non-complying work etc. should have required the consent of the contractor, not to be unreasonably delayed or withheld, wherever the cost to the contractor of that decision is greater than replacing the non-complying work etc.

WHERE NON-COMPLYING WORK ETC. IS TO BE REMOVED

4-37 Clause 16.2.3 deals not only with instructions following a decision to use non-complying work but also following a decision to instruct its removal under clause 16.2.1. This could on occasions lead to complying work becoming abortive, having to be taken down and replaced as part of the removal and replacement of the non-complying work etc. This would not need an instruction under clause 16.2.3 as it would in any event form part of the contractor's obligation to carry out and complete the works in accordance with the contract.

It is just possible that the employer might instruct the removal of non-complying work etc. but not its replacement and, instead, choose some alternative way of proceeding. This is most unlikely to be a necessary consequence of the removal as it is the exercise of an option and would properly be the subject of a change instruction. It is not generally easy to envisage situations following removal of non-complying work which require *necessary,* rather than sensible or desirable, instructions for further works over and above that which the contractor will be contractually obliged to undertake as part of the removal and replacement in any event.

Opening up, testing or inspection

> 16.2.4 . . . instruct such further opening up, testing or inspection as is reasonable in all the circumstances to establish to the reasonable satisfaction of the Employer that other similar work, materials or goods are in accordance with the Contract.

4-38 Clause 16.2.4 gives the employer power to give instructions for further opening up for inspection or testing as are reasonable in all the circumstances to establish to his reasonable satisfaction that other similar work etc. is in accordance with the contract.

4-39 The same minor drafting point, namely that the reference should be to 'opening up for inspection or testing', as was mentioned in connection with clause 16.1 (see paragraph 4-16) applies here. In addition, whereas in clause 16.1 the reference is always to 'opening up, inspecting and testing,' here, no doubt inadvertently, the reference is to 'opening up, testing or inspection'. Little probably flows from this transposition.

THE RATIONALE

4-40 The clause is similar in effect to clause 8.4.3 of WCD 98. The policy behind this clause appears to be reasonable. A failure of work, materials or goods to be in accordance with the contract will generally amount to a breach of contract by the contractor (some may say only a temporary disconformity - see Lord Diplock's speech in *Hosier & Dickinson Ltd* v. *P & M Kaye Ltd* (1972)). The employer may reasonably enough wish to find out if there are any further similar failures, especially where there are health and safety or structural considerations involved. Even if the opening up etc. shows that there has been no further failure, it may reasonably be argued that the employer should not have to meet the cost associated with an instruction which was reasonable in nature and extent, including the contractor's loss and expense, as he would if the inspection or test was instructed under clause 16.1 (even if it was the contractor's breach of contract which necessitated the opening up etc). Under clause 16.2.4 in such circumstances, the instruction is not treated as a change, so the contractor meets the cost. On the other hand, it could be seen to involve a disproportionate degree of risk for the contractor, and consequently for sub-contractors, particularly in specialist areas, where the employer may not be equipped to exercise such potentially drastic powers in the most sensible way. However, a requirement that the instruction should be reasonable in all the circumstances will normally ensure that any such instruction is issued responsibly.

THE EXTENT OF THE OPENING UP FOR INSPECTION OR TESTING

4-41 An area of potential difficulty will be in relation to just how much further opening up for inspection or test is reasonable and also its nature. Where health and safety issues or structural integrity is relevant, then no doubt the extent and nature of testing which is reasonable will be different to that where this is not the case. For example, if during the construction of a hotel with balconies, a balcony is found not to comply with the contract in a way which is fundamental to its structural integrity, it may well be reasonable to instruct that all balconies are to be tested. The position may be similar if there are a number of critical welding joints and one has been found to be faulty. In other situations some form of sample testing may well be the reasonable option. There is no doubt that contractors and sub-contractors have expressed concern about the scope of such powers as exist under this clause. For example, in both WCD 98 and JCT 98, a code of practice is included in the contract conditions to help in the fair and reasonable operation of the similar provisions in those contracts. That code sets out criteria which include such matters as:

- The extent to which the problem can be established as unique and therefore not likely to occur in similar elements;
- Whether it is safe and appropriate to carry out sample testing rather than full-scale testing;
- The significance of the non-compliance;
- The consequences of any similar non-compliance in terms of health and safety and the integrity of the building;
- The level and standard of supervision and control of the works by the contractor;

- Consideration of the contractor's and sub-contractor's records;
- Any relevant codes of practice or similar advice in connection with such matters;
- Any failure of the contractor to carry out tests forming part of the requirements or proposals;
- The reason for non-compliance when this has been established;
- Any technical advice obtained by the contractor in respect of the non-compliance;
- Current recognised testing procedures;
- The practicality of progressive testing;
- If alternative testing methods are available, the time required for and the consequential costs of the alternative methods;
- Any proposals the contractor may have.

No doubt as the need to use such provisions is fortunately rare, it was sensibly thought to be unnecessary to include them in the MPF.

16.3 No instruction issued under clause 16.2 shall be treated as giving rise to a change.

4-42 This clause provides that no instructions issued under clause 16.2 shall be treated as giving rise to a change. The consequences of this have already been dealt with in commenting on the various sub-clauses of clause 16.2 (see paragraph 4-24 *et seq.*).

Rectification of defects – clause 17

Background

4-43 Clause 17 is the rectification of defects clause which becomes effective from project practical completion. Similar provisions are to be found in most, if not all, standard forms of construction contract.

Many standard forms, including other JCT forms, place the defects rectification clause immediately following the clause dealing with practical completion. That has of course some logic to it. In the case of the MPF, practical completion is dealt with in clause 9.4 under the general heading of 'Time', whereas the rectification of defects, clause 17, is placed under the general heading of 'Control'. This is an equally logical way to allocate the two provisions.

4-44 Typically, such a clause will refer to a stated period for which it is to run, commencing with the date of practical completion. Often it is six or twelve months.

Minor outstanding items

4-45 Often under building contracts such provisions require the contractor to remedy defects appearing within the rectification or defects liability period but do not expressly refer to the completion of any known outstanding items which exist at the time when practical completion is achieved. This can sometimes lead to a strict interpretation of what amounts to practical completion, namely, that nothing at all must remain to be done however minor in nature. The MPF however (see clause 39.2) defines practical completion so as to permit the issue of the appropriate statement where minor outstanding works remain, provided that neither their existence not their execution will affect the use of the project. This is a sensible approach which ties in with how the courts have tended to construe the phrase in practice – see *H. W. Nevill (Sunblest) Ltd* v. *William Press & Son Ltd* (1981), in which Judge John Newey QC said, of the JCT 63 contract under consideration, that it gave the architect a discretion to certify practical completion

> '... where very minor de minimis work had not been carried out, but that if there were any patent defects ... the Architect should not have given a certificate of Practical Completion'.

Common law rights unaffected

4-46 Unless a contract by its terms expressly and unequivocally states to the contrary, the contractor's obligation to attend to defects etc. discovered within the defects liability period will not exclude or limit the employer's legal right to recover damages for losses suffered, if any, as a result of defective work. It simply means that the employer can call for the physical presence on site of the contractor to carry out the remedial work. Many such provisions, e.g. WCD 98 clause 16.2, are expressed in such a way as to give the contractor the right, subject to exceptions, to return to remedy defects. This can be of considerable importance to a contractor. Firstly, in the absence of such a right the employer could get the defects remedied by another contractor and, subject to the employer's duty to mitigate his loss (see below), the reasonable cost of doing so would be payable as damages by the contractor for breach of contract. The opportunity therefore for the contractor himself to attend to the defects can save him money. Secondly, it helps minimise the risk to the contractor of his reputation being tarnished if he can himself attend to the defects rather than having a third party examine his work.

Employer's duty to mitigate and contractor's right to return

4-47 If in such circumstances the employer unreasonably refuses to allow the contractor back to remedy defects, and has the remedial works carried out by others, this could well be a failure to mitigate loss on the employer's part, which may prevent the employer recovering in full the costs of such remedial works from the contractor – see for example *City Axis Ltd* v. *Daniel P. Jackson* (1998) and also *Pearce & High Ltd* v. *John P. Baxter* (1999). While clause 17.1 of MPF is not expressed so as to give the contractor a right to return to remedy defects, simply giving the employer the right to call the contractor back (c.f clause 16.2 of WCD 98 which provides that such defects 'shall be specified by the Employer in a Schedule of Defects which he shall deliver to the Contractor as an instruction of the Employer'), nevertheless as the clause goes on to state that the employer may engage others to do the work only if the contractor has failed to comply with an instruction to rectify the defect, it implicitly gives the contractor such a right.

Summary of clause 17

4-48 The rectification period lasts for twelve months following the date of practical completion. The employer may issue an instruction to the contractor to remedy defects at no cost to the employer. 'Defect' is defined by clause 39.2, by reference to any fault in the project due to the contractor's failure to comply with his contractual obligations, together with the consequences of that fault. If the contractor does not comply within a reasonable time with the instruction to rectify the defect, the employer can engage others to do so at the contractor's expense.

After the expiry of the period and when all defects subject to a making good instruction have been remedied, the employer issues a statement to that effect.

Where the defects have not been remedied within a reasonable time of the expiry of the rectification period, the employer must issue a statement identifying:

- Those defects that he intends to engage others to rectify, with a proper estimate of the cost;
- Those defects that he does not intend to rectify with particulars of an appropriate deduction in the calculation of the final payment.

Clause 17 is without prejudice to any other remedies of either party.

4-49 The rectification of defects provision in MPF is shorter than that to be found in most other JCT contracts. In particular no mention is made of a schedule of defects to be served upon the contractor within the rectification period or within 14 days of its expiry - see for example WCD 98 clause 16.2. Under the MPF clause 17, the employer issues instructions as and when he chooses to do so except that it must be within the rectification period and, unlike clause 16.2, cannot be issued up to 14 days after its expiry.

Instructions to remedy defects

> 17.1 During the Rectification Period the Employer may instruct the Contractor to remedy any Defect. The Contractor shall comply with any instructions within a reasonable time and at no cost to the Employer and, should it not do so, the Employer may engage others in accordance with clause 2.3.

4-50 This provides for the employer to issue instructions during the rectification period for the contractor to remedy defects within a reasonable time and provides for the employer to engage others if the contractor fails to do so.

Rectification period

4-51 The Rectification Period is defined by clause 39.2 as:

> The twelve-month period commencing on the date Practical Completion of the Project occurs.

Defect

4-52 'Defect' is defined by clause 39.2 as:

> Any fault in the Project that arises as a consequence of a failure by the Contractor to comply with its obligations under the Contract, together with the consequences of that fault.

This is a wide definition of what amounts to a defect. It enables the instruction to encompass not only the remedying of defective design or defective work, but also to remedy the consequences of that fault. While this reference to the consequences of the fault is not altogether clear in meaning or scope, it is probably intended to make it clear that the rectification encompasses such matters as the need to remove and replace good work as part of remedying the defective work. There must be some limit however. For example, if because of the faulty work there was a need for significant redesign of other work this is probably outside the scope of the remedying of the defect, even though the contractor may still have a liability to the employer if this is the foreseeable result of the contractor's failure to comply with his contractual obligations. In such a case the contractor may well, on occasions, wish to have the opportunity of carrying out such work and may therefore argue that it is encompassed within the definition of defect.

Clause 17.1 does not give any indication as to when the defect must become apparent. Some other JCT contracts refer to defects which appear within the rectification period - see for example clause 16.2 of WCD 98. It is probably sensible not to refer to their appearance being within the rectification period but even where clauses do this, the accepted view appears to be that defects known to exist prior to practical completion will still be included - see for example *Keating on Building Contracts*, 7th Edition, paragraphs 18–176. Further, under the JCT Minor Works Form 1980 it has been held that defects appearing before practical completion are covered where the words used were 'which appear within three months of the date of practical completion' - see *William Tomkinson & Sons Ltd* v. *The Parochial Church Council of St Michael and Others* (1990).

Minor outstanding works

4-53 The statement of practical completion can be issued even where there are minor outstanding works (see earlier paragraph 3-22). Such outstanding works do not easily fall within the definition of a 'Defect'. They are hardly a fault due to the contractor's failure to comply with his contractual obligations. It might have been anticipated that there would be an express provision requiring the contractor to finish off such outstanding work during the rectification period. Even in its absence it must remain the contractor's obligation to complete the works in accordance with his obligations so that the outstanding work will have to be carried out, presumably within a reasonable time (and at a convenient time for the employer) following practical completion of the project.

Contractor to comply with instructions within a reasonable time

4-54 What amounts to a reasonable time within which a defect is to be rectified will depend of course upon all the circumstances including:

- Its importance to the proper functioning of the project;
- The time required to gear up;
- The consequences of any delay, e.g. further deterioration;
- The convenience to the employer or other occupiers of the project.

At no cost to the employer

4-55 Clause 17.1 provides that the contractor will comply with the instruction to remedy the defect at no cost to the employer. This must mean the cost based on the actual work required to remedy the defect. It would not, for example, extend to indemnifying the employer in respect of consequential losses such as loss of use, loss of or reduction in rents or compensation to purchasers or tenant as a result of the defect. However, these losses would be recoverable under the general law as damages for breach of contract on the part of the contractor to the extent that such consequential losses were reasonably foreseeable or within the actual contemplation of the contractor at the time of entering into the contract. The preservation of the employer's general contractual rights is made doubly clear by the provisions of clause 17.4 stating that the provisions of clause 17 are without prejudice to any other rights or remedies which the parties may possess.

Failure by the contractor to remedy the defect

4-56 If the contractor fails to remedy the defect within a reasonable time the employer can engage others to do the work under clause 2.3, which gives the employer the general power to engage others to give effect to an instruction which the contractor has failed to comply with (see paragraph 2-36 *et seq*.). Note should also be taken of clause 17.3.1 where, if the defects have not been remedied within a reasonable period of the expiry of the rectification period, the employer can state his intention, by means of a statement, to engage others and to make a proper estimate of the cost (see paragraph 4-59).

Statement of remedying of defects

17.2 After the expiry of the Rectification Period and when all Defects that the Contractor has been instructed to remedy under clause 17.1 have been remedied the Employer shall issue a statement to that effect.

4-57 This provides that after the expiration of the rectification period and when all defects that the contractor has been instructed to remedy have been remedied, the employer must issue a statement to that effect. If no statement can be issued because there remain outstanding defects, clause 17.3 operates with the employer issuing what is in effect a modified statement.

Defects not remedied by contractor – employer's options

17.3 Where there are Defects that the Contractor has been instructed to remedy under clause 17.1, but which have not been remedied within a reasonable period of the expiry of the Rectification Period the Employer shall issue a statement identifying:

.1 those Defects that it intends to engage others to rectify, together with a proper estimate of the cost of undertaking those rectification works, and

.2 those Defects that it does not intend to rectify, together with particulars of the appropriate deduction it intends to make in the calculation of the amount due to the Contractor.

4-58 The purpose of clause 17.3 is to avoid undue delay in the employer calculating the final payment. If the end of the rectification period is reached and there are outstanding defects, it enables the employer to take the initiative in dealing with these in the ways described so that the accounting procedures for the project can be closed. Clause 17.3 provides that if the defects which the contractor has been instructed to remedy have not been remedied within a reasonable time of the expiry of the rectification period, the employer is to issue a statement indicating his decision as to which defects:

- He will engage others to rectify (clause 17.3.1); and
- He does not intend to rectify (clause 17.3.2).

This clause will only relate to those situations where the contractor has been instructed to rectify the defect. In other words, the employer will not be able to exercise the options contained in this clause where he has simply not required the contractor to remedy the defect at all. The employer's options under this clause are linked to defects not having been remedied 'within a reasonable period of the expiry of the Rectification Period' and the reference to a reasonable period must be a reference to the reasonable time within which the contractor must remedy the defect following an instruction from the employer to do so.

The options for the employer contained in clause 17.3 do not therefore apply to the situation where, during the rectification period, the employer decides not to have the contractor remedy the defect, for whatever reason. He cannot as an alternative decide to retain the faulty work and claim an appropriate deduction. Clause 16.2.2, which gives the employer this option in respect of non-complying work, materials or goods, does not appear to be able to operate after practical completion of the project.

Employer engaging others to rectify

4-59 If the employer chooses the option of engaging others to remedy the defect under clause 17.3.1, he must issue a statement within a reasonable period of the expiry of the rectification period stating that he intends to do so and also providing a proper estimate of the cost of undertaking those rectification works. It is most surprising that clause 22.6 (dealing with the calculation of the final payment advice), does not expressly provide for this estimate to be taken into account in the calculation of a final payment advice to determine the total amount to which the contractor is entitled (in the way it is for the appropriate deduction under clause 17.3.2 – see clause 22.6 and particularly clause 22.6.3). It could well be a drafting oversight. Assuming it is nevertheless implicitly covered by clause 22.6, and this is by no means certain, the estimate will presumably be final so far as the calculation of the total amount due to the contractor is concerned, and will not be subject to correction once the actual figure is determined. However, it is not possible to be certain as the MPF does not say what the status of this estimate is meant to be. It can be envisaged therefore that there could well be arguments as to what amounts to a proper estimate. The estimate can always, of course, subject to the time limits in clause 22.7, be challenged by the contractor in adjudication and/or litigation.

4-60 Depending on how the estimate under clause 17.3.1 is to be treated under the MPF (see previous paragraph), it could of course work against the employer's interests. He may make an estimate of the likely costs of undertaking the outstanding rectification works, and subsequently discover that the nature or extent of the fault is different or greater than that anticipated, e.g. where more good work has to be removed and reinstated than anticipated. If this happens, can the employer claim the additional costs from the contractor outside of the accounting mechanisms in the contract? Clause 17.4 expressly preserves the parties' other legal rights or remedies. There will have been a breach of contract by the contractor. There is nothing to indicate in clause 17.3, or elsewhere in the contract, that any such estimate is intended to be in full and final settlement of all the employer's losses as a consequence of that breach. It may be therefore that the employer could make such a claim for damages, giving credit for the estimate. This conclusion could produce the unfortunate result that while an overestimate, provided it was reasonable, would appear to bind the contractor, if the same reasonable estimate turns out to be inadequate to compensate the employer, he can take action for breach of contract to recover any additional losses, provided they are not too remote. None of this is certain. This clause is in need of some clarification.

4-61 If the employer, having made proper estimate, changes his mind and decides not to engage others to rectify the defect, it will in practice probably be treated as a clause 17.3.2 situation with the estimate being equal to the appropriate deduction under that clause. It should be borne in mind, however, that the estimate and appropriate deduction are by no means the same thing. The estimate is likely, for instance, to include the cost of removing and replacing abortive work which is necessary in order to rectify the defect, whereas under clause 17.3.2, the appropriate deduction where the defective work is to be left in place would not include the cost of removing and replacing good work, which will never actually happen. Contractors may need, perhaps, to keep an eye on this possibility.

Defects which the employer does not intend to rectify

4-62 The employer can choose not to rectify outstanding defects. If this is his decision the statement issued within a reasonable period of the expiry of the rectification period must identify this intention together with particulars of the appropriate deduction which the employer intends to make in the calculation of the amount due to the contractor in the final payment advice.

The calculation of the deduction will be made in accordance with clause 22.6 (see particularly clause 22.6.3).

The appropriate deduction

4-63 Under clause 17.3.2, if the defects are not to be rectified, the employer has to provide particulars of the appropriate deduction he intends to make. This deduction is part of the calculation of the amount due to the contractor in the final payment advice. It could mean, of course, that if the deduction together with any other deductions in calculating the amount due results in the employer having overpaid the contractor, there will be no actual physical deduction capable of being made and the employer will have to make a claim for this sum or any balance in respect of it.

4-64 The fact that the employer does not intend to rectify the defect does not necessarily mean that the appropriate deduction is based purely on the prices and principles set out in the pricing document (whatever that may mean) which is used in the calculation of the appropriate amount under clause 16.2.2 (see paragraph 4-31). Under clause 17.3.2, it may be that the employer does not intend to rectify because he has instead compensated a purchaser or tenant who will themselves rectify. That compensation may therefore reflect the fact that the cost of rectification could include costs in addition to those purely based upon the difference between the contract value of what should have been provided and that which has been provided under the contract. For example, it could well include the cost of removing good work and replacing it, which is necessary in order to rectify the defect. The particulars of the appropriate deduction required by clause 17.3.2 will of course have to identify this fact.

Preservation of other rights and remedies

17.4 The provisions of clause 17 are without prejudice to any other rights or remedies the parties may possess.

4-65 Clause 17.4 states that the provisions of clause 17 are without prejudice to any other rights or remedies which the parties may possess. This would probably have been the situation without such a clause but it is probably sensible, particularly from the employer's point of view (being the one more likely to be making claims as a result of defective work), to make the position absolutely clear.

Pre-appointed consultants and named specialists - clause 18

Introduction and background

4-66 Clause 18 deals with two of the most important contractual issues in relation to the procurement of building projects. These are, firstly, where, as with the MPF, there is contractor design, the issue of the employer's professional team and its relationship with the contractor at the time the contractor takes on design obligations; and, secondly, the right which the employer reserves to appoint specialists to provide design or work and materials and who are to become sub-contractors to the main contractor.

Before considering the provisions of clause 18, it is appropriate to summarise the legal and practical background to these matters.

The professional team

4-67 Employers on major projects, whether the contractor is to take on a minor or major role in the design of the project, will appoint a professional team. In some circumstances they may only be concerned with such preliminary matters as scheme viability, planning considerations, site investigations, etc. Employers generally go beyond this and engage professionals to carry out further services, for example, the preparation of tender documentation, advising on the contractors to be invited to tender and evaluating tenders received. In addition, the majority of employers will engage professionals (or use in-house professionals) to undertake design, including by way of drawings and specifications. The extent of the design work will vary, possibly from outline scheme design through to significant detailed design.

4-68 So far as the contractor is concerned, he may either be required to complete the design prepared on behalf of the employer, without having responsibility for the employer's design, or alternatively be asked to adopt the employer's design as his own. Even in the case of the former there is a question whether the contractor, in order to complete the design using reasonable skill and care, will have a duty, as part of the exercise of that reasonable skill and care, to check the employer's design in order to ensure that, nothing in it, which could have been detected using reasonable skill and care, will prevent the completed design from working – see *Co-operative Insurance Society Limited* v. *Henry Boot Scotland Limited* (2002) (see paragraph 2-82). In the latter situation a further relevant matter is that if a contractor is to accept responsibility to the employer for breaches by the professionals under their respective appointments with the employer, the contractor will wish to ensure that his own professional indemnity insurance covers him in this situation. While some policies will provide this cover, it may only be on the basis that the contractor checks the design and possibly other services provided by such professionals.

4-69 Some contracts, such as the MPF (clause 18.4), make the contractor solely responsible for the services provided by named pre-appointed consultants (i.e. appointed by the employer to provide services before the MPF is entered into), with however in the case of the MPF, a saving provision to the effect that the contractor is not responsible for the contents of the requirements or for the adequacy of any design in them. This is discussed later (see paragraph 4-85 *et seq*.).

Even where the contractor is to have no responsibility for the services provided by the professional team to the employer, it often makes sense for the contractor to engage relevant members of that professional team to carry out further services in connection with the project for which the contractor accepts responsibility and which the professionals are well placed to provide. This might be in the form of terms of appointment between the contractor and the relevant professionals. However, if the contractor is to accept responsibility to the employer for any of the services provided to the employer by the professional team under their terms of appointment with the employer, it will be imperative that the contractor has a suitable contractual arrangement with the professionals concerned for the additional reason that if the contractor finds himself in breach of an inherited obligation, e.g. the negligent design of someone in the professional team, he will in turn wish to ensure that he has a remedy against the relevant professional. The way in which this is achieved in practice is through the use of a novation agreement entered into by the employer, the relevant professional and the contractor.

NOVATION GENERALLY

4-70 A modern and succinct explanation of a novation can be found in the judgment of Lord Justice Staughton in *Linden Gardens Trust Ltd* v. *Lenesta Sludge Disposals Ltd* (1992) when this case was passing through the Court of Appeal. He said (57 BLR 76):

> 'This is the process by which a contract between A and B is transformed into a contract between A and C. It can only be achieved by agreement between all three of them, A, B and C. Unless there is such an agreement, and therefore a novation, neither A nor B can rid himself of any obligation which he owes to the other under the contract. This is commonly expressed in the proposition that the burden of the contract cannot be assigned, unilaterally. If A is entitled to look to B for payment under the contract, he cannot be compelled to look to C instead, unless there is a novation. Otherwise B remains liable, even if he has assigned his rights under the contract to C.
>
> Similarly, the nature and content of the contractual obligations cannot be altered unilaterally. If a tailor A has contracted to make a suit for B, he cannot by assignment be placed under an obligation to make a suit for C, whose dimensions

may be quite different. It may be that C by an assignment would become entitled to enforce the contract – although specific performance seems somewhat implausible – or to claim damages for its breach. But it would still be a contract to make a suit that fitted B, and B would still be liable to A for the price.

A contract that A will build a house for B, and follow his instructions on such variations as the contract may allow, cannot be converted by an assignment into an obligation to follow the instructions of C. In *Kemp* v. *Baerselman* (1906) 2 KB 604 it was held that a contract to supply as many eggs as K might require for his manufacturing business could not require further performance when K had transferred his business to a company.'

4-71 A pure novation is one where the employer steps out of the terms of appointment with the professional and the contractor replaces the employer under that appointment. The professional would therefore owe exactly the same duties, but to someone else. All three parties need to enter into the novation agreement for it to work, as it is not possible without such an agreement for someone in the position of the employer, who has obligations to the professional, to substitute someone else, namely the contractor, to perform those obligations, e.g. the payment of fees. This marks the difference between a novation and an assignment. Generally speaking, a contracting party is permitted to assign benefits under a contract to a third party, at any rate to the extent that it really does not matter to the person undertaking the obligation whether it is carried out in favour of the other party or a third party. The obligation remains the same. Lord Justice Staughton in the *Linden Gardens* case referred to above explained (at 57 BLR 77) the nature of assignment as follows:

'This consists in the transfer from B to C of the benefit of one or more obligations that A owes to B. These may be obligations to pay money, or to perform other contractual promises, or to pay damages for breach of contract, subject of course to the common law prohibition on the assignment of a bare cause of action.

But the nature of and content of the obligation, as I have said, may not be changed by an assignment. It is this concept which lies, in my view, behind the doctrine that personal contracts are not assignable: see *Chitty on Contracts* (26th Edition Vol paragraph 1416, citing *Tolhurst* v. *The Associated Portland Cement Manufacturers* (1900) Ltd [1902] 2 KB 660 at page 668:

"The benefit of a contract is only assignable in 'cases where it can make no difference to the person on whom the obligation lies to which of two persons he is to discharge it'." '

More is said about assignment when commenting on clause 29 (see paragraph 7-01 *et seq.*).

A novation is regarded as 'pure' if, as between A, B and C, C steps into A's shoes and A steps out of the picture altogether. However, what often happens in practice is that B, while now owing obligations to C, nevertheless still owes obligations of one kind or another to A so that A does not completely step out of the picture. This can cause difficulties in interpreting the scope of the novation agreement and is dealt with below.

NOVATION IN THE DESIGN AND BUILD CONTEXT

4-72 For many years now, employers have sought to engage their own professionals prior to the appointment of the design and build contractor and then to novate them across to the contractor on the basis that the contractor is then to accept entire responsibility to the employer for the services already provided by the professional, e.g. design.

In essence, following a true novation, the professional would be treated as having undertaken the obligations under his terms of appointment in favour of the employer, in favour of the contractor instead, and as being bound by the terms of it as though the contractor was the original contracting party in place of the employer. Accordingly, the professional would then release the employer from further performance of the employer's obligations under, for example, the obligation to pay fees. Additionally, the professional would release the employer from all claims and demands in respect of the terms of engagement which instead would have to be made against the contractor.

However, in practice, in many instances the employer does not discharge the professional from liability for claims and demands in connection with defaults in performance prior to the novation, and a pure novation is not achieved.

Most novation agreements will expressly state whether the professional's obligations to the employer have been discharged or not, but if this is not stated it is then a matter of interpretation whether the professional's obligations to the contractor replace those to the employer. Where the contractor's obligations under the design and build contract are different in nature or scope to those of the professional, this would tend to indicate that some at least of the professional's liabilities to the employer under the original appointment will be retained. For example, the design and build contractor will no doubt have obligations in connection with design, and indeed will possibly expressly agree to inherit any of the design obligations of the professional under the appointment; but there may be other obligations under the appointment in respect of which it would be inappropriate for the design and build contractor to accept responsibility. Obvious examples will be in relation to matters such as initial scheme viability, or evaluation and advice to the employer on proposed tenderers; similarly in relation to tenders received. Clearly it is not appropriate for the contractor to have a liability to the employer in relation to advising the employer on which contractor's tender to accept.

A pure novation will, and other types of novation often do, seek to place the contractor in the employer's shoes with the apparent effect that the professional owes these other duties, which are not in truth reflected in the design and build contract, to the contractor on the assumption that the contractor was always the other party to the appointment. This can cause significant difficulties in the interpretation of such agreements. An example of some of the difficulties which can arise can be seen in the Scottish case of *Blyth & Blyth Ltd* v. *Carillion Construction Ltd* (2001) in the Outer House of the Court of Session. It is sufficiently interesting to be worth detailed consideration.

4-73 *BLYTH & BLYTH LTD* v. *CARRILLION CONSTRUCTION LTD* (2001)

The employer (THI Leisure (Fountainpark) Ltd) entered into a deed of appointment with consulting engineers Blyth & Blyth Ltd for pre-construction and construction services in connection with a leisure development building. THI also entered into an amended WCD 81 contract with Carillion Construction Ltd. In particular it should be noted that this contract was amended in typical fashion to impose upon the contractor the responsibility for design and related services carried out by the employer's professionals prior to the design and build contract being entered into.

The case came before the court when, following a deed of novation between Blyth, THI and Carillion, Blyth sought payment of its fees from Carillion. Carillion counterclaimed in respect of a number of alleged breaches by Blyth of its appointment, including in relation to alleged pre-novation breaches of duty. One such alleged breach was related to an allegation that Blyth had underestimated the amount of steel bar reinforcement required. This had been reflected in the employer's requirements and therefore in the contractor's proposals. It was alleged by Carillion that they had been involved in extra costs including the provision of additional reinforcement, which they sought to claim from Blyth. A central feature of this case concerned the meaning and effect of the novation agreement and its relationship with the deed of appointment and the design and build contract.

The central question before the court was whether Carillion could claim against Blyth for alleged pre-novation breaches, and if so on what basis in terms of the loss which could be claimed.

The novation agreement was, so far as can be gathered from the judgment, in a fairly typical form. For example, clauses 4 and 5 were as follows:

> '4. The liability of the Consultant under the Appointment whether accruing before or after the date of this Novation shall be to the Contractor and the Consultant agrees to perform the Appointment and to be bound by the terms of the Appointment in all respects as if the Contractor had always been named as a party to the Appointment in place of the Employer.
> 5. Without prejudice to the generality of clause 3 of this Novation the Consultant agrees that any services performed under the Appointment by the Consultant or payments made pursuant to the Appointment by the Employer to the Consultant before the date of this Novation will be treated as services performed for or payments made by the Contractor and the Consultant agrees to be liable to the Contractor in respect of all such services and in respect of any breach of the Appointment occurring before the date of this Novation as if the Contractor had always been named as a party to the Appointment in place of the Employer.'

Carillion argued their case in two ways (set out below). The first and primary contention was eventually abandoned during the hearing:

(1) Blyth owed duties to Carillion as if Carillion had always been named as a party to the deed of appointment in place of THI and that therefore the appointment was to be read as if references to the employer were to the contractor. In effect a kind of rewrite of the terms of appointment replacing THI with Carillion;
(2) If (i) was not successful then it was argued that even though pre-novation duties might still be owed to THI, Carillion as a party to the novation agreement were entitled to claim in respect of any breaches of the appointment by Blyth and in so doing could claim its own loss. In other words, whatever description, e.g. a novation, may have been given to the agreement, it was essentially a three-sided relationship the effect of which was that Carillion engaged Blyth to perform services and give advice to THI on behalf of Carillion and that therefore, if there was a breach of the terms of the appointment, it entitled Carillion to claim its own losses.

The first contention referred to above, the rewriting of the deed of appointment approach, produced some absurd results if the employer's name was simply removed and the contractor's name inserted. The judge, Lord Eassie, gave a couple of examples from the schedule of services to the deed of appointment:

- 'Assist the Architect in advising [the Contractor] as to the technical suitability for carrying out the project of firms tendering for the main contract;
- Assist the Architect and Quantity Surveyor in advising [the Contractor] as to the relative merits of tender prices and estimates received for carrying out the project.'

Clearly, these pre-construction services were wholly inappropriate in terms of a duty being owed at the outset by Blyth in favour of Carillion.

The situation was even more absurd if the recitals were considered, e.g. recital B would read:

> 'The [Contractor] proposes to enter into a building contract with a main contractor...'

More fundamental, however, was what the judge saw as a conflict of interest for Blyth depending upon whether the party for whom the service was carried out was the employer or the contractor. In other words, the same duty owed to both may require a different performance depending upon whether it was carried out for the benefit of the employer or the contractor. The judge held that it was inherently unlikely that the novation agreement intended to recast a given duty owed (and already performed) to the employer as becoming a duty to the contractor.

[*Author's note:* There is a tension between the best interests of the employer and the best interests of the contractor. Both are likely to be commercial in their approach. Both would legitimately wish to maximise profitability. For the employer this would involve obtaining the highest quality building for the lowest possible price. For the contractor the objective would be to make the greatest amount of profit by providing a building which would meet the Employer's Requirements for the lowest possible outlay. These are both legitimate objectives but they may result, in any particular situation, in a professional approaching the performance of his duties in different ways.]

Carillion abandoned its primary approach and concentrated its efforts on its second contention: that the effect of the novation agreement was not that the pre-novation services provided by Blyth ceased to be owed to THI, becoming instead owed to Carillion; rather, they continued to be owed to THI but Carillion became a creditor of those obligations and could claim their own losses as a result of a breach of those obligations by Blyth.

It may be pointed out that the novation did not provide for the extinction of THI's rights and obligations. These were to remain in force. In addition, there was a provision in clause 3.5 of the novation agreement providing that if Blyth were at any time entitled as between themselves and Carillion to treat their appointment as having been repudiated by Carillion, THI was entitled to give notice that it wished to enter into a further novation agreement to reverse the terms of the original novation agreement, thus enabling THI to remain in control of the project process.

Based on Carillion's second contention, the key issue was whether Carillion could claim its own losses in respect of any pre-novation breach by Blyth which caused Carillion loss.

Looking at clause 4 of the novation agreement, it distinguished between liabilities before and after novation so if Carillion were to be entitled to claim in respect of pre-novation breaches, it had to be by way of some sort of assignment to them of the employer's claim. As to clause 5 of the novation agreement, though it stated that pre-novation duties owed to THI were to be treated as owed to Carillion, those duties, and any losses following from their breach, had to be determined by the duty owed originally to THI and not to Carillion who might have a different interest.

In summary, Carillion argued that on the assumption that the service or advice provided to THI contained errors amounting to breach of the appointment, then, in the pre-novation situation, THI would be contractually entitled to have the errors corrected or the consequences made good. After the novation, Carillion was the party entitled to call for their correction and for the making good of their consequences. Accordingly, argued Carillion, they were entitled to claim for their own losses. Carillion supported this argument partially on the basis that THI had itself suffered no loss as a result of Blyth's alleged breaches as, under the design and build contract, Carillion had to accept responsibility for these mistakes and had to put them right. This argument was rejected by the judge who made the point that THI may have had a sound claim against both Blyth and Carillion, and just because THI could choose to claim against one or the other did not mean that the one not claimed against had no liability. This point could be highlighted by supposing that one or the other was insolvent. This would clearly not prevent THI from claiming against the other.

In the result the judge reached the conclusion that Carillion could not claim for its own losses, though it could, by way of assignment, have an interest in losses which may have been suffered by the employer based on alleged breaches of duty by Blyth prior to the date of novation in relation to duties then owed to THI.

Comment

The potential effect of this decision has caused some concern. It appears to suggest that typically worded novation agreements do not enable the building contractor to claim its own losses, namely the cost of correcting a professional's mistakes and/or paying damages to the employer for breach by the contractor of his contractual obligations under the design and build contract. Instead, at best the contractor appears to be able to claim from the professional concerned in respect of the breach of duty, losses based upon the loss which the employer may have suffered. In the *Blyth* case itself, for instance, if the consultants had not been at fault, as alleged, in underestimating the amount of reinforcement bars required, the inclusion of the correct quantity in the employer's requirements may well have increased the contractor's price, but in a competitive tendering situation not necessarily by an amount equal to the actual extra cost to the contractor of providing these. This position is certainly less than satisfactory from a contractor's point of view.

Possible solutions to the problem

In practice, the effect of this decision has resulted in contractors seeking an independent warranty from the novated professionals acknowledging that at all times the professional owes a duty to exercise reasonable skill and care etc. in favour of the contractor and in contemplation of the contractor being directly affected by any breach of such duty in terms of liability under the design and build contract. It will also provide an express warranty by the professional to the contractor that he has exercised reasonable skill and care in the performance of his duties under the appointment and acknowledges that the contractor has relied upon the information and advice provided by the consultant to the employer under its appointment. Such a warranty can either be in a separate document or contained within the novation agreement itself.

Even though the provision of an independent warranty acknowledging a direct duty of care owed by the professional to the contractor may meet the contractor's needs, it still leaves the professional in a potentially difficult and invidious position. It means that in connection with the exercise of pre-novation duties, he owes a duty of care both to the employer and to the contractor. The professional is therefore serving two masters with different interests which, on occasions, might produce a conflict for the professional in deciding what action he should take to fulfil any particular duty.

Finally, bearing in mind that novation agreements are, in effect, predicated on the basis that there is a notional rewriting of the original appointment, namely, the deleting of references to the employer and their replacement with a reference to the contractor, it is important that any services already provided by the professional to the employer which are inappropriate to the contractor's responsibilities under the design and build contract, should be excluded from the rewriting effect, and only those duties or responsibilities of the professional to the employer which coincide with the duties, responsibilities or contractual risks of the contractor to the employer under the design and build contract should be subject to the rewriting exercise.

Fitness for purpose versus skill and care

4-74 Where a novation is contemplated, another issue in practice will be the comparison between the duty of care owed by the professional to the contractor following a novation, and the contractual obligations of the contractor to the employer under the design and build contract. The different implied terms arising out of a contract for services and one for a service coupled with the production of a physical product, such as a building, have been discussed earlier (see paragraph 2-68 *et seq.*). The result can sometimes be that under the design and build contract the contractor warrants that the building will be reasonably fit for its intended purpose, whereas the professional under the terms of his novated appointment with the contractor will owe a duty only to exercise reasonable skill and care. If this is the position, clearly the contractor can find himself having a liability to the employer, e.g. for a non-negligent design defect, with no remedy against the professional responsible for it.

4-75 Clause 18 of the MPF dealing with pre-appointed consultants provides that they are to be novated to the contractor, and the contractor will execute a Model Form of Novation in respect of all such pre-appointed consultants. The Model Form of Novation agreement is not provided as a standard form and is to be attached to the requirements. Under the MPF, a 'Pre-Appointed Consultant' is defined by clause 39.2 as:

> A consultant identified in the Requirements as having been appointed by the Employer with the intention that the appointment be novated to the Contractor in accordance with clause 18 (*Pre-Appointed Consutants*).

4-76 Clause 18 in relation to pre-appointed consultants will be considered in detail below (paragraph 4-80) after briefly considering the other aspect of clause 18, the selection by the employer of named specialists to become sub-contractors to the contractor.

Employer selecting named specialists

4-77 Clause 18 provides that a named specialist can be chosen by the employer to be appointed by the Contractor in connection with the preparation of any design or the execution of any works identified for that purpose in the requirements. It covers therefore both the provision of services by professionals and also the provision of services, e.g. design and/or construction work, by a specialist contractor. A 'Named Specialist' is defined in clause 39.2 as:

> A sub-contractor or consultant that is either identified by name in the Requirements or that is to be selected by the Contractor from a list of specialists contained in the Requirements.

Many construction contracts give the employer this facility. Whether or not the employer's privilege in being able to select a named specialist carries with it a partial responsibility for their performance or failure to perform should be clearly spelt out in any construction contract.

The allocation of responsibility or risk of non-performance by selected specialists is a matter upon which there is often considerable debate. Different contracts deal with this in different ways. Reference is often made to the employer nominating, naming or pre-selecting specialists. Before considering in detail the provisions of clause 18 in relation to the appointment of such named specialists, it is worth putting the system of employer selection into a general context.

NO EMPLOYER SELECTION

4-78 The general principle of English law is that a main contractor will be responsible to the employer for the defaults and breaches of contract of a sub-contractor in respect of any work carried out by them which the contractor puts forward in the fulfilment of his own contractual obligations to the employer. This is the position with domestic sub-contractors chosen by the contractor. Sub-contracting of contractual obligations, i.e. vicarious performance of the actual work, is permitted under the general law unless the obligations are personal in nature (see next paragraph). In many instances, so long as the original party remains responsible for any failure to properly and fully perform the burden or obligation, it does not matter who actually carries it out. For example, a main contractor may of his own choice arrange for a sub-contractor to carry out certain of the work. Subject to what the express terms of the contract may say, the main contractor will be permitted to do this while remaining fully liable to the employer for any failure of the sub-contractor to adequately fulfil his contractual obligations where this leads to a similar failure of the main contractor vis-à-vis the employer. The employer will be able to claim from the main contractor who must in turn claim from the sub-contractor. The relationship between the main contractor and the sub-contractor is of no concern to the employer. This arrangement is known in the construction industry as domestic sub-contracting to distinguish it from nominated or named sub-contracting.

If the employer can demonstrate that he is placing reliance on the particular skill or expertise of a given contractor, then even a building contract may be regarded as of a personal nature, so preventing the sub-contracting of any of the work. A good example of this is to be found in the case of *Southway Group Ltd* v. *Wolff* (1991). This case concerned a contract between a developer and the owners of a warehouse under which the developer undertook to have the warehouse refurbished prior to its resale. The developer was closely involved in many decisions concerning the refurbishment, such as the number of windows, the location of staircases and entrance and the internal sub-division of walls, such that it was clear that the expertise of the developer, and in particular its principal shareholder, was being heavily relied on. It was held that this made the contract personal to the extent that it could not be vicariously performed. By analogy, this case shows that the personality of the contractor with regard to control and co-ordination of a construction project may be sufficiently important to prevent that part of the work being vicariously performed by sub-contractors, even if physical work can be sub-contracted.

The MPF does not seek to control the use by the contractor of domestic sub-contractors.

EMPLOYER SELECTION

4-79 Where a building contract provides that the employer may select a sub-contractor whom the contractor must appoint, the building contract concerned, when compared with the situation in relation to domestic sub-contracting, may well alter the contractual responsibility of the contractor for any failure to perform on the part of the selected sub-contractor.

There are different types of employer selection available in the standard forms of construction contract ranging from what might be called full nomination, e.g. in JCT 98 to named sub-contractors in WCD 98 and in IFC 98. There are in addition many ad hoc employer selection provisions usually targeted at improving the employer's position compared with standard forms, at the risk of the contractor, particularly where the sub-contractor fails to complete. It is not proposed to delve into these various options in this book. It is sufficient to point out that the allocation of risk between employer and contractor in relation to a failure to perform by an employer selected sub-contractor (whether nominated or named and whether selected before or after the building contract is entered into), including where the failure is as a result of the sub-contractor's insolvency, varies considerably from one contract to another.

The way in which clause 18 deals with pre-appointed consultants and named specialists will now be considered.

Summary of clause 18

4-80 In summary, clause 18 provides as follows:

- The reference in clause 18 to pre-appointed consultants only applies where the appendix states that it is to apply. If no such selection is made in the appendix, the provisions of clause 18 in relation to pre-appointed consultants do not apply.
- Turning to the appointment of a named specialist, if the requirements provide that a named specialist is to be appointed by the contractor to prepare designs or execute work identified in the requirements, the contractor is obliged to appoint them. The requirements may state a single name or provide a list of names from which the contractor may choose a single name.

- Having entered into the MPF, the contractor must immediately take steps to execute the Model Form of Novation for all pre-appointed consultants. He must also appoint any named specialists and notify the employer of their identity and provide a copy of the contract with them (except for financial details).
- The contractor is to be solely responsible for the services provided by pre-appointed consultants (whether carried out before or after the MPF is entered into) and the work of any named specialists. This is however subject to clause 5.1 which declares that the contractor is not responsible for the contents of the requirements or the adequacy of the design in them.
- If the contractor fails to execute the Model Form of Novation with pre-appointed consultants or fails to appoint a named specialist, the employer will have no obligation to pay the contractor in respect of their services.
- The contractor is not to amend his contract with any pre-appointed consultant or Named Specialist or waive any compliance with its terms or estop himself from enforcing any of the contractual obligations in such contracts without the employer's prior consent, which is not to be unreasonably delayed or withheld. The contractor is not to terminate his contract with the pre-appointed consultant or named specialist without similar consent being obtained from the employer.
- If there is a termination or proposed termination, the contractor must notify the employer of any proposed replacement who, where possible, will be selected from any relevant list in the requirements. The employer can raise reasonable objection to such a proposed appointment, in which case a further selection will have to be made by the contractor.
- Where there is a need for a replacement, the contractor remains entirely responsible for any delay or additional cost as a consequence of the appointment of a replacement.

We now turn to consider clause 18 in more detail.

Application in respect of pre-appointed consultants governed by appendix

18.1 The provisions of clause 18 relating to Pre-Appointed Consultants apply only where so stated in the Appendix

4-81 With the exception of clause 18.3, the remainder of the sub-clauses of clause 18 are all relevant to pre-appointed consultants. However, in order for those clauses to be relevant, there must be an entry in the appendix against clause 18.1 stating that the provisions of clause 18 are to apply. If no selection is made in the appendix the provisions of clause 18 in relation to pre-appointed consultants do not apply. In practice, the novation of pre-appointed consultants from employer to contractor, which this clause covers, is very common and it is unusual to find design and build contracts where this is not done. It can be anticipated therefore that almost always this appendix entry will be selected.

Execution of Model Form of Novation Agreement

18.2 Immediately upon entering into the Contract the Parties shall take all steps necessary to execute a Model Form in respect of all Pre-Appointed Consultants.

4-82 On the assumption that the appendix entry indicates that clause 18 is to apply, the employer and contractor must immediately, on entering into the MPF, take all necessary steps to execute a model form in respect of all pre-appointed consultants. 'Model Form' is defined in clause 39.2 as:

> Where applicable, the Model Form of Novation Agreement that forms part of the Requirements.

This is attached to the conditions and forms part of the MPF. There is no standard form of novation for use with the MPF.

The model form is to be entered into with all pre-appointed consultants. A 'Pre-Appointed Consultant' is defined in clause 39.2 as:

> A Consultant identified in the Requirements as having been appointed by the Employer with the intention that the appointment be novated to the Contractor in accordance with clause 18 (*Pre- Appointed Consultants*).

Accordingly, the employer will identify in the requirements those consultants it has engaged who are to be novated to the contractor and in respect of whose services, whether performed for the employer or contractor, the contractor will be solely responsible, subject to clause 5.1 (clause 18.4).

Appointment of named specialist

> 18.3 The Contractor shall appoint a Named Specialist to prepare any designs or execute any works that are identified by the Requirements as having to be undertaken by a Named Specialist. The Contractor shall immediately notify the Employer of the identity of any Named Specialist upon their appointment and shall supply to the Employer a copy of the contract entered into by the Contractor with the Named Specialist (other than the financial details contained in it).

4-83 Clause 18.3 deals with the appointment of a named specialist. A named specialist can either be a sub-contractor or a consultant. Any such person must either be identified by name in the requirements or contained in a list of specialists contained in the requirements from whom the contractor is allowed to choose. That named specialist will prepare the designs or execute the works identified in the requirements as having to be undertaken by a named specialist.

The contractor has an obligation to appoint them and must immediately notify the employer of the appointment, including the identity of the named specialist. A copy of the contract with the named specialist (except for any financial details contained in it) is to be provided to the employer. The employer has no role to play in terms of determining the price which the contractor pays for such designs or works and it is a matter for the contractor alone as to what financial arrangements have been made between himself and the named specialist.

The employer has no control over the terms of the appointment or sub-contract. Some employers take the view that either they should dictate the form of contract to be used or should at least have the right to approve it, even if such approval is not to be unreasonably delayed or withheld. Employer amendments to this clause can therefore be anticipated.

Pricing considerations

4-84 The contractor is likely to have a limited tender period in which to attempt to price the work of named specialists. In practice it is unlikely that the requirements will provide in detail the designs or work to be executed by a named specialist. In a perfect world, the detail would be sufficient for the contractor to accurately price the work after consultation with the named specialist. The contractor would ideally not only need to obtain a firm price from the named specialist against an adequate description of the work required, but also be able to agree such matters as appropriate conditions of contract, attendances and programme details. Unlike some other JCT contracts, e.g. JCT 98, there is no detailed procedural mechanism through which this exercise takes place.

The pricing document for the MPF will contain the contract sum analysis and pricing information. However, whether it will, in connection with named specialists, provide an adequate basis for valuing variations/changes in the work of named specialists, for example, is doubtful. One can therefore anticipate problems both in establishing an initial realistic price to form part of the contract sum analysis and also in the valuation of variations/changes which will often have to be based upon a fair valuation under clause 20.6 of the MPF, with little in the way of prices and principles contained in the pricing document.

Where a price is agreed between the contractor and the proposed named specialist during the tender period, or even later but before the tender is accepted, the contractor will need to ensure that the named specialist is in some way legally bound to subsequently enter into a contract containing the agreed price. The contractor may do this by obtaining an offer from the proposed named specialist which, supported by consideration, is to remain open for a sufficient period after the main contract is entered into, to enable the sub-contract to be executed, or alternatively to enter into a sub-contract subject to a condition precedent that it is not to come into force unless and until the main contract is formed. In practice there is a risk that the contractor will submit a main contract tender without having been able to properly address such matters with the obvious risks which flow from that. In addition, there is the risk that a proposed named specialist will consider himself to be in a very strong bargaining position and either drive up the price or hang back in agreeing a price knowing that the contractor's only definite option will be to decline to tender. A successful bid in such circumstances could lead to the contractor being held to ransom. Any inability on the part of the contractor to reach any necessary agreement with the proposed named specialist after the main contract is entered into is likely to place the contractor in an extremely difficult position, especially where the named specialist is a single name in the requirements. The contractor would not have the option to engage anyone else to do the work. Even if he had the work carried out by someone else, clause 18.5 makes it clear that the employer would not be liable to pay the contractor in respect of it. None of the safeguards for the contractor provided in other JCT standard forms to cover such situations (e.g. WCD 98, clause S4.2) are to be found in the MPF. Those for whom this contract is intended are likely to be fully aware of the position, which in any event only reflects what many ad hoc amendments to the standard forms seek to achieve.

Contractor solely responsible for pre-appointed consultants and named specialists

18.4 Subject to clause 5.1 the Contractor shall be solely responsible under the Contract for the services provided by any Pre-Appointed Consultant whether before or after the date of the Contract and for the works undertaken by any Named Specialist or replacement specialist appointed in accordance with clause 18.7.

The contractor's responsibility for the services of pre-appointed consultants

4-85 The requirements should set out all the employer's requirements for the project which the contractor has to meet through his proposals. One would not have thought therefore that the pre-appointed consultants' services for which the contractor is to be responsible should include anything which does not coincide with those services to be provided by the contractor under the contract. Examples could be initial viability studies, the drawing up of tender lists, the evaluation of tenders, and possibly initial planning considerations and site investigation activities. It depends what services the requirements require from the contractor and also what risks the contractor is taking under the MPF, e.g. the risk of bad ground. However, and unfortunately, this is not made clear contractually. There appears to be a risk for the contractor here that he could unwittingly become responsible for all of the pre-appointed consultants' services whether the requirements separately identify them or not. Taken literally it is obvious that on occasions this would produce an absurdity, e.g. making the contractor responsible for advising the employer on which tender to accept. In such a case the court might construe clause 18.4 as impliedly making the contractor responsible for only those services expressly stated in or implicit from the requirements. Contractors need to be assiduous in checking the scope of the pre-appointed consultants' services to determine exactly what they cover, and to clarify if necessary those for which the contractor is to be responsible.

4-86 If the contractor agrees to a fitness for purpose obligation under an amended clause 5.3 (see paragraph 2-101), the consultant's appointment may well contain only a requirement that the consultant exercises reasonable care and skill and does not give a warranty of fitness for purpose. Unless the contractor can include a similar express warranty from the consultant in the novation arrangements, which is highly unlikely, the contractor could be accepting a liability without being able to pass it on to the professional designer concerned.

EXCEPTION IN RESPECT OF THE CONTENTS OF THE REQUIREMENTS OR THE ADEQUACY OF THE DESIGN IN THE REQUIREMENTS

4-87 Clause 18.4 is expressed to be subject to clause 5.1. This provides that the contractor is not responsible for the contents of the requirements or the adequacy of the design within the requirements. This is clearly a significant exception, though not as wide as it could have been. Clause 5.1 is commented on earlier (see paragraph 2-79 *et seq.*). Further, it is worth restating that if the employer is in a strong commercial bargaining position, he may well transfer responsibility for the whole of the design to the contractor upon his appointment. This may be achieved by ensuring that whatever the employer requires and wishes to stipulate for is in fact kept out of the requirements and instead is included only in the proposals from the contractor; alternatively the employer may amend the MPF to expressly place responsibility for the requirements onto the contractor.

The contractor's responsibility for the work of named specialists

4-88 This provides that the contractor is to be solely responsible for the works undertaken by any named specialist (or any replacement specialist appointed in accordance with clause 18.8). It is made clear that even though the named specialist is selected by the employer or chosen by the contractor from a list of names provided by the employer, the risks associated with inadequate performance or non-performance lie squarely with the contractor. Any design or work executed by any such named consultant or sub-contractor is to be treated as if it was the design or work directly of the contractor. The drafting of this clause could usefully have referred expressly to the services provided by, as well as the work undertaken by, named specialists (see paragraph 4-92 for further comment on this).

4-89 Those contractors entering into the MPF are likely to fully appreciate these risks when choosing to tender. The possibility of the employer or any of his professionals owing the contractor a duty of care in the tort of negligence in relation to the choice of named specialists for inclusion in the requirements would be fanciful in such circumstances. However, any of the professionals engaged by the employer to advise him as to the inclusion of a named specialist may well owe a duty of care to the employer with the possibility that the appointment will be novated to the contractor, thereby owing the same duty to the contractor and providing the contractor with a possible remedy.

4-90 The contractor may need to take particular care in respect of design liability if clause 5.3 is amended (see footnote to that clause), so that the contractor warrants that the project when completed will be suitable for the purpose or purposes stated in the requirements. If a named specialist carries out design work, it is imperative that the contractor ensures that a similar fitness for purpose obligation applies under his contract with the specialist. If the named specialist will not accept anything other than a reasonable skill and care obligation in respect of any design, the contractor's only option, if he is not prepared to take on the additional risk, will be to decline to tender for the main contract itself. There is no provision in the MPF for the contractor to raise any reasonable objection to entering into a contract with a named specialist. The position is the same where the contractor takes on a fitness for purpose obligation in relation to any performance specification or in the selection of suitable materials (see clauses 5.2.2 and 5.4).

EXCEPTION IN RESPECT OF THE CONTENTS OF THE REQUIREMENTS OR THE ADEQUACY OF THE DESIGN IN THE REQUIREMENTS

4-91 This has been commented on in relation to pre-appointed consultants and the same issues arise here (see paragraph 4-87).

Consequences if contractor fails to enter into a contract

18.5 If the Contractor either fails to execute a Model Form in the manner provided by clause 18.2 or fails to appoint a Named Specialist in the manner provided by clause 18.3 the Employer shall not be liable to pay the Contractor in respect of services that were to be provided by that Pre-Appointed Consultant or works that were to be undertaken by the Named Specialist.

4-92 As mentioned also in relation to clause 18.4 (see paragraph 4-88), the reference at the end of this clause to 'works' to be undertaken by the named specialist is too limiting.

The named specialist will more often than not be providing a design service along with installation. The earlier reference in this clause to 'services' of the pre-appointed consultants makes it more difficult to argue that 'works' also covers services. On balance however, and adopting a purposive construction, it probably will cover design services etc. but the matter should have been put beyond doubt.

4-93 Clause 18.5 is potentially a devastating clause from the contractor's point of view. It provides that if the contractor fails to execute the model form with a pre-appointed consultant or fails to appoint a named specialist, the employer will not be liable to pay the contractor in respect of any of the services that were to be provided by that consultant or works to be undertaken by that specialist. But is this a penalty clause?

Is it a penalty?

4-94 What is to happen if the consultant or specialist (particularly if uniquely named in the requirements) refuses to enter into the model form or sub-contract as the case may be? If in such a case the contractor must either look elsewhere for the provision of such services or work, or undertake it himself, the employer is not liable to pay for it. But can he argue that he is not permitted by the MPF to arrange for those services or that work to be carried out by others or to undertake it himself? To do so would be contrary to the requirements which will provide for those services or that work to be carried out by the pre-appointed consultant or named specialist. The contractor would therefore be in breach of contract. Assuming this to be so, the next question is whether the contractor is in breach of clause 18.2 or 18.3 if he has tried everything reasonably possible to comply. Both clauses, though differently worded, are peremptory in nature and neither is conditioned by any reference to the use of reasonable or even best endeavours. The contractor is probably in breach irrespective of the absence of any culpability on his part, unless the employer has in some way prevented the contractor from fulfilling his obligations under these clauses, and this is highly unlikely.

Following this line of reasoning, the employer would have to arrange for such services or work to be carried out and would no doubt seek to recover any additional costs from the contractor. It may be, depending on the circumstances surrounding the contractor's breach, that the contractor could demand the opportunity to arrange it or carry it out himself on the basis that he could do it at less cost and thereby mitigate the loss, but under clause 18.5 he would not be entitled to any payment at all for this. If the employer did arrange for such services or work to be carried out by others, he could surely only recover from the contractor the extra costs, if any, over and above the sum which would have been payable to the contractor had the work been carried out by the pre-appointed consultant or named specialist. An argument by the employer that this would effectively negate the provision in clause 18.5 that the employer is not liable to pay the contractor for such services or work, would be tantamount to arguing for a penalty. Indeed it brings out the important point that clause 18.5 could well contain an unenforceable penalty provision.

An alternative option for the employer, to issue an instruction under clause 2.1 altering the requirements to require the contractor to carry out the work himself or through others of his choice, arguing that it did not give rise to a change because it was brought about by the contractor's negligence or default (see the definition of 'Change' in clause 39.2), would be unlikely to work. Firstly because, despite the wide wording of clause 2.1 as to the scope of instructions which the employer may issue, there has to be some limit, and to require the contractor to be responsible for arranging for others to carry out such services or work (or to do it himself) would be to rewrite the contract; and secondly, even if this was not the case, there would be an issue as to whether the contractor, even if in breach of clause 18.2 or 18.3, could argue that he was not in 'default' in the sense in which that word is used in the definition of change (see paragraph 5-19).

Contractor's suggested good practice

4-95 Whether the contractor, if he fails to enter into the model form with the pre-appointed consultant or the sub-contract with the named specialist, would be liable to bear the whole of the cost of the pre-appointed consultant's services or named subcontractor's work or just the additional cost of engaging others or carrying out the work himself, the contractor could easily find himself in a very weak bargaining position with pre-appointed consultants or named specialists. The contractor should seek to protect himself by obtaining during the tender period and before committing himself to the main contract, a contractually binding commitment from the pre-appointed consultants to enter into the Model Form of Novation Agreement. In obtaining this commitment it should be made clear to the pre-appointed consultant what would be the likely consequences and foreseeable losses to the contractor in the event that in breach of that obligation the consultant fails to enter into the model form. Similarly in relation to named specialists, a binding legal obligation should be obtained from them, preferably in the form of a binding sub-contract with a condition precedent that it is only to come into force upon the contractor entering into the main contract. In relation to named specialists, if their work is not already fully described and is not capable of being adequately priced during the tender period, the contractor is faced with a problem of achieving legal certainty in seeking to reach a legally binding sub-contract (subject to a condition precedent) with them. In addition, there are of course the problems associated with unreasonably short tender periods together with the costs associated with seeking to obtain such reasonable protection.

4-96 Unlike many other standard forms dealing with employer selection of specialists (e.g. compare WCD 98, clause S4.2.2), there is no protection for the contractor even where he has been unable to enter into a sub-contract for good reason.

4-97 In strictly literal terms, clause 18.5 applies where, although the pre-appointed consultants have entered into the model form, they have failed to do so in the manner provided by clause 18.2. Literally, the employer would not be liable to pay the contractor in respect of services provided by them. For example, under clause 18.2 if the contractor failed to *immediately* take all steps necessary to execute the model form but nevertheless still entered into it, it is inconceivable that where any breach of clauses 18.2 is technical or procedural in nature rather than substantive, the courts would allow the employer to rely upon clause 18.5 as a basis for rejecting liability to pay.

Contractor not to prejudice employer under form of novation or specialist sub-contract

18.6 The Contractor shall not without the prior written consent of the Employer amend its contract with any Pre-Appointed Consultant or Named Specialist or waive strict compliance by a Pre-Appointed Consultant or Named Specialist with the performance of its obligations under such contract or estop itself from enforcing such obligations, such consent not to be unreasonably delayed or withheld. Where the Contractor amends or waives or estops itself from enforcing its contract with any Pre-Appointed Consultant or Named Specialist other than in accordance with this clause then the Employer shall not be liable to pay the Contractor in respect of services that are no longer being provided by the Pre-Appointed Consultant or Named Specialist.

4-98 By clause 18.6, the contractor is not permitted to amend any of the terms of the professional appointment with a pre-appointed consultant or the sub-contract with a named specialist, or to waive strict compliance with performance of the obligations under them or estop himself from enforcing any obligations owed to him by any pre-appointed consultant or named specialist without the prior written consent of the employer. Any such consent is not to be unreasonably delayed or withheld.

Literally, any amendment, waiver of strict compliance or estoppel requires consent, presumably even if, for example, the amendment relates to just a revised fee arrangement or sub-contract price, although clearly, if the effect of the amendment, waiver or estoppel is unlikely to be significant in terms of jeopardising the proper and adequate performance of contractual obligations, the employer will be bound to consent. The reference to seeking prior consent in relation to the creation of an estoppel is very odd. The likelihood of intentionally creating an estoppel is remote indeed. Invariably an estoppel is created inadvertently by some action on the part of one party which is relied upon to his detriment by the other party in such circumstances that it would be wholly inequitable to thereafter seek to enforce some contractual obligation which had effectively been treated as if it would not be enforced. The more important area will be where the effect of any amendment or waiver will be that either the service or work will not be provided at all or where it will be provided by others. In such a case the issue of overall adequacy of performance will become relevant and the employer is entitled to be reasonably satisfied that he will not be prejudiced.

If the contractor does, without consent, amend the appointment or sub-contract or waive performance of obligations under them, or create an estoppel to the extent that this means that the services are no longer being provided by the consultant or named specialist, for example, if they are performed by someone else, clause 18.6 provides that the employer is not liable to pay for them. Surely this clause should have referred to work undertaken by named specialists as well as services (see for example clauses 18.4 and 18.5). This is probably an oversight in the drafting but it could pose problems. Clause 18 is riddled with confusion between services being provided and work being undertaken.

Clause 18.6 is clearly designed as a very strong incentive for the contractor to obtain any necessary consent rather than to unilaterally reallocate any of the services or work which are the subject of the professional appointment or named specialist sub-contract. The position is similar in respect of any unauthorised waiver, but note the error in drafting which refers to the contractor waiving the contract rather than waiving strict compliance with the performance of obligations under it. It must clearly be the latter which is intended in order to make proper sense of the last sentence in its relationship with the first sentence. However, there is a real possibility that this amounts to a penalty clause, in the same way as clause 18.5 might (see paragraph 4-94).

No termination without consent

> 18.7 The Contractor shall not terminate its contract with any Pre-Appointed Consultant or Named Specialist without the prior written consent of the Employer, such consent not to be unreasonably delayed or withheld. When the Contractor terminates its contract with a Pre-Appointed Consultant or Named Specialist it shall immediately notify the Employer of the termination.

4-99 By clause 18.7, the contractor is not to terminate the appointment with any pre-appointed consultant or the sub-contract with any named specialist without the prior written consent of the employer. That consent is not to be unreasonably delayed or withheld. Whenever the contractor does terminate the appointment or sub-contract he must immediately notify the employer of the termination. That notification will be a communication and accordingly under clause 38.1 it must be in writing or by some other agreed means, e.g. electronic communication. The most likely reason for the contractor wishing to terminate any such contract is the breach, possibly a repudiatory breach, by the pre-appointed consultant or named specialist of the appointment or sub-contract respectively. This clause means that where there is conduct amounting to a repudiation on the part of the pre-appointed consultant or named specialist, entitling the contractor, if he wishes, to accept that repudiation, thereby bringing the contract to an end and discharging his own obligations under it, he must not do so without the prior consent of the employer. There will be very few instances where in circumstances such as this the employer could reasonably withhold his consent. More likely, the purpose of the requirement for the contractor to obtain consent is to enable the employer to see if there is any way of salvaging the situation. If the employer places significant value on the continued use of the pre-appointed consultant or named specialist concerned, the employer is likely to become embroiled in any arguments between the contractor and pre-appointed consultant or named specialist as to the allocation of blame and responsibility. Where the contractor has strong legal grounds for terminating, the employer may have to negotiate some relaxation of the effects of clause 18.10 (contractor responsible for delay and additional costs as a consequences of replacement), if he is keen for the pre-appointed consultant or named specialist to be retained.

On occasions the contractor may have little choice in connection with terminating the contract. The pre-appointed consultant or named specialist may be unwilling or unable to further perform its obligations, e.g. where insolvency intervenes. In such a case the contractor may be faced with a *fait accompli*. The requirement for the employer's consent in such a case will be purely academic. If the contractor does not seek it, he will nevertheless still have to find a replacement (see clause 18.8) and will still suffer any delay and meet any additional costs incurred (clause 18.10).

As it is the contractor who takes all of the time and financial risks consequent upon the appointment of any replacement (see clause 18.10), the contractor is only likely to seek the termination in the most extreme circumstances.

'The Contractor shall not terminate its contract'

4-100 There could be a problem with the drafting of this clause. It refers to the termination of the contract but not to the termination of the pre-appointed consultant's or named specialist's employment under the contract. In the MPF itself, the termination provisions (see clauses 31 to 34 inclusive) refer not to the termination of the contract but to the termination of the contractor's employment under it. These are arguably different things. The main purpose of the MPF termination provisions referring to termination of the contractor's employment rather than the contract itself, is to ensure that any contractual terms, e.g. those dealing with the financial consequences following the termination of employment, can operate satisfactorily on the basis that the contract itself remains alive. The termination of the contract itself will generally, apart from any dispute resolution provisions, bring the contract to an end making it more difficult to operate such provisions. However, the termination itself will not affect accrued rights and liabilities under the contract. It may well be that the appointment or the sub-contract enables the pre-appointed consultant's or named specialist's employment to be terminated without a termination of the contract itself. This would appear not to require the written consent of the employer. The provisions regarding replacement under clause 18.8 which follow 'after the termination of the contract' would not therefore strictly apply and the employer would lose his control over the replacement process. While it might have been sensible to have referred to the termination of the contract or the pre-appointed consultant's or named specialist's employment under it, the probability is that the reference to the termination of the contract will cover both situations, given a purposive approach to the construction of these provisions. However, the fact that clause 18 refers to the termination of the contract whereas clauses 32 to 34 inclusive only ever refer to the termination of the contractor's employment under the contract (with the presumption therefore that they are intended to mean different things), is not helpful to such a construction.

Appointment of replacement

18.8 Either before or as soon as possible after the termination of the contract of a Pre-Appointed Consultant or Named Specialist the Contractor shall notify the Employer of its proposed replacement consultant or specialist. Where possible the replacement specialist shall be selected from any list contained within the Requirements. The proposed replacement consultant or specialist shall be appointed by the Contractor unless the Employer raises reasonable objection within 7 days of the notification. Where a reasonable objection is raised by the Employer the Contractor shall propose a further replacement for consideration by the Employer in accordance with this provision.

4-101 By clause 18.8, either before or as soon as possible after any termination the contractor must notify the employer of his proposed replacement. The clause goes on to provide that where the replacement is of a specialist rather than a consultant, the specialist shall be selected from any list which is contained in the requirements. No doubt contractors would rather have had this obligation conditioned by references to where it was reasonable to so select rather than 'where possible'. It may well be that the prices of those on the list will be higher than the contractor may be able to secure elsewhere but this hardly seems to render impossible the appointment of a replacement specialist from the list. There are likely to be arguments about this. Certainly, if any of the listed specialists seek to hold the contractor to ransom, any court is likely to construe the reference to 'where possible' as protecting the contractor in such a situation; likewise if the price differential between listed specialists and a non-listed specialist who can satisfy the contractor's contractual obligations in relation to such work, is significant. What is 'possible' will in practice take its meaning from the overall context and despite its apparent absolute quality, there is little doubt that it will be qualified by reference to what is reasonable in the circumstances.

Employer's objection

4-102 Once the contractor has notified the employer of the proposed replacement consultant or specialist, the employer has 7 days in which to raise any reasonable objection, failing which the contractor must make the appointment. If the objection relates to the employer's view that there will be a sub-standard performance, as the contractor will remain responsible for this under the contract, it is rarely that any such objection would be valid. It is perhaps only where there is a realistic concern on the part of the employer either that liquidated damages for any delay due to inferior performance (such as the time taken to put matters right) is an inadequate remedy in all the circumstances, or perhaps where the employer has genuine concerns that the contractor will not have the resources to fulfil all of his obligations if things do go wrong, that a reasonable objection could be raised.

If the employer is unhappy with the form of appointment or sub-contract, it is difficult to see how clause 18.8 would entitle the employer to object to the replacement, particularly in relation to a specialist as the employer has no say in the terms of the sub-contract with the original named specialist. Some employers are keen to keep control over such forms of appointment and sub-contract and it is likely that clause 18 will be amended to reflect this.

In the event of a reasonable objection being raised by the employer, the contractor must propose a further replacement for consideration. The issue of what is a reasonable objection can of course, if necessary, be referred to adjudication.

4-103 Any replacement consultant or specialist falls outside the definition of 'Pre-Appointed Consultant' and 'Named Specialist' (see clause 39.2). Neither definition includes replacements. Accordingly, only clauses 18.9 and 18.10 will apply to the replacement or any further replacement. At this point the employer loses control over the services or work which have been reserved in the requirements to be carried out by consultants or named specialists. Some employers will not be content with this state of affairs and are likely to make amendments to ensure that all replacements are subject to the machinery of clause 18.

Contractor to supply copy of contract

> 18.9 The Contractor shall immediately supply to the Employer a copy of the contract entered into with any replacement consultant or specialist (other than the financial details contained in it).

4-104 Clause 18.9 provides that the contractor is to immediately supply the employer with a copy of the contract entered into with any replacement consultant or specialist. This is no doubt a good housekeeping clause so far as the employer is concerned. A copy of the contract might be useful in the event of any further problems arising between the contractor and the replacement consultant or specialist.

Contractor's responsibilities unaffected by appointment of replacement

> 18.10 The Contractor's liabilities and obligations under the Contract shall not be affected by the appointment of a replacement consultant or specialist and the Contractor shall be entirely responsible for any delay or additional cost incurred as a consequence of the appointment of a replacement consultant or specialist.

4-105 Clause 18.10 provides that where there has been a replacement, this will not affect the contractor's liabilities or obligations under the contract.

4-106 The contractor is to be entirely responsible for delays or additional costs incurred as a consequence of the appointment of a replacement consultant or specialist. This would, for example, include the difference in price between the original and replacement consultant or specialist, in relation to the outstanding work. However, a significant element of delay or additional cost is likely to be incurred not as a consequence of the appointment of a replacement consultant or specialist but more as a result of the problems which have caused the termination to arise and the delay and disruption caused whilst a replacement is found. The wording of this clause is therefore unfortunate. There is little doubt that contractors will contend that the words used exclude any reference to the delay and disruption caused to the contractor by the original pre-appointed consultant or named specialist. Employers will contend that in any event clause 18.4 is sufficient to make the contractor responsible for delays or additional costs leading up to or caused as a result of any termination. Contractors may point out that the responsibility of the contractor appears to be aimed at the quality of the services provided or the work undertaken rather than obligations as to the time of performance. Bearing in mind that clause 18 gives the employer significant control over the appointment by the contractor of certain consultants and specialists in what some courts might not consider to be a particularly even-handed way, it would have been better, from the employer's point of view, if the contractor's responsibility for delays and disruption caused by such consultants and specialists had been made clearer.

The wording is also unfortunate in referring to the delay and costs only as a consequence of the appointment of a replacement as these words are hardly appropriate to make the contractor responsible in the event that the replacement consultants or specialists fail to perform adequately. Such non-performance would not in causation terms be a consequence of their appointment. The delay or additional cost incurred would be the result of the failure to perform, not the appointment, even though the appointment was a necessary pre-condition to such failure. The appointment would not be the cause of the failure.

4-107 A further problem in connection with the drafting of clause 18.10 in connection with delay and additional costs, is that it can be argued that named specialists are to prepare any designs or execute any works 'identified by the Requirements as having to be undertaken by a Named Specialist' (see clause 18.3). Following the appointment of a replacement who is not on any list of names in the requirements, such designs or works will not thereafter be required to be carried by a named specialist. This could be regarded as an 'alteration by the Employer of a restriction or obligation set out in the Requirements and/or Proposals as to the manner in which the Contractor is to execute the Project, or the imposition of additional restrictions or obligation', and therefore as a change (see clause 39.2). It might be arguable therefore that at any rate from the termination onwards, there will have been a change entitling the contractor to an adjustment to the completion date and a valuation of the change which could include the costs of prolongation or disruption between the termination and the replacement. Much will depend upon whether the proviso in the definition of change applies to any given situation, namely 'Provided always that the alteration ... is not required as the result of any negligence or default on the part of the Contractor'. Clearly if the problems are as a result of the contractor's negligence then it will not be a change. In terms of whether the fact that the contractor is responsible for the works undertaken by named specialists (see clause 18.4) means that if they fail to perform the contractor is in 'default' depends upon what that word means in this context. In any event there is a counter argument that as clause 18 envisages that work reserved in the requirements for named specialists may cease to be so, no change is required.

Contractor's responsibilities unaffected by variations to services etc.

> 18.11 No variation or alteration in the services to be provided by the Pre-Appointed Consultant or Named Specialist to the Contractor and no waiver or forgiveness or other action or inaction by the Contractor shall serve to alter or diminish the Contractor's liability to the Employer in respect of the services provided under the Contract.

4-108 Clause 18.11 makes it clear that the contractor's obligations to the employer under the contract in respect of the services to be provided by pre-appointed consultants or named specialists will not be altered or diminished by any variation or alteration in those services as between the pre-appointed consultant or named specialist and the contractor. Additionally, any waiver or forgiveness or any action or inaction of the contractor shall again not alter or diminish the contractor's liability to the employer in respect of the services provided by the contractor under the MPF.

This clause is to some extent related to clause 18.6 (though unlike clause 18.6, it fails to refer to estoppel), which deals with amendments to the appointment with the pre-appointed consultant or sub-contract with the named specialist as well as any waiver of strict compliance with the performance by them of their obligations.

While it may well be that without such a provision the contractor would nevertheless remain responsible under the MPF for providing the services etc. under the contract and referred to in the requirements, nevertheless it is probably worth expressly confirming the position, which is what clause 18.11 purports to do. Unfortunately, it refers only to services provided by the named specialist whereas in other parts of clause 18 reference is only made to work undertaken by the named specialist (e.g. clauses 18.4 and 18.5). This is more likely to be an oversight in the drafting rather than any intention of distinguishing between services provided and work executed, but the results of it are unpredictable.

Cost savings and value improvements – clause 19

Introduction

4-109 Clause 19 deals with cost savings and value improvements suggested by the contractor. Before engaging a contractor, the employer will generally have had the benefit of advice from professionals. However, contractors often have a good deal to offer in terms of practical suggestions as to what can be achieved and how, often being able to offer practical and economic ways of achieving a desired result.

Even in relation to standard forms which do not expressly provide for such matters, in practice, agreements for cost savings or value improvements are often reached between the contractor and employer on an ad hoc basis. However, such discussions can sometimes be surrounded by uncertainty and concern on the part of the contractor that his good idea will be utilised by the employer without the contractor receiving an adequate reward. Accordingly, it is sensible for the contract to expressly encourage the taking of such initiatives by the contractor, with the incentive of both a share of any savings or benefits and also some protection of the contractor's position if agreement cannot be reached between the parties. This is what clause 19 seeks to do by inviting the contractor to suggest amendments to the requirements or proposals which, if adopted, would give the employer a financial benefit in terms of cost, quality or time, with the contractor benefiting by being entitled to an agreed percentage share of the employer's benefit expressed in financial terms.

4-110 **Summary of clause 19**

- The contractor may suggest amendments to the requirements or proposals. If these are then contained in an instruction from the employer requiring a change, which would result in a financial benefit, the employer is liable to pay the contractor on practical completion a proportion of the agreed financial benefit. This proportion is set out in an appendix item for clause 19.
- The contractor is to provide details of suggested amendments together with a quotation in the same form as the quotation which the contractor produces in connection with other change instructions (clauses 20.3 and 20.4).
- If the employer wishes to implement the amendments suggested, the parties then negotiate with a view to agreeing the contractor's quotation and the level of financial benefit to the employer. That agreement is then confirmed by the employer issuing an instruction which identifies the agreed value of the change, any agreed adjustment to the completion and the agreed financial benefit to the employer.
- Once the contractor has suggested any amendment pursuant to clause 19, the employer may only instruct those amendments in accordance with the clause 19 procedures.
- However, if the employer and contractor cannot agree, then no instruction for the change can be issued whether under clause 19 or under clause 20. However, once practical completion of the project has been achieved, the employer may utilise other contractors to implement the suggested amendments.

Clause 19 is now considered in more detail.

Scope of savings and improvements

19.1 The Contractor is encouraged to suggest amendments to the Requirements and/or the Proposals which, if instructed as a Change, would result in a financial benefit to the Employer. The benefit may arise in the form of:

.1 a reduction in the cost of the Project;
.2 a reduction in the life cycle costs associated with the Project;
.3 the achievement of Practical Completion at a date earlier than the Completion Date; and/or
.4 any other financial benefit to the Employer.

4-111 The introductory words in clause 19.1 are not drafted in traditional common law contract style. They state that 'The Contractor is encouraged to suggest amendments'. Traditionally, a contract is drafted as a series of contractual obligations and promises. This is different. It is in effect a statement from the employer of what he would like to see happen. It has more of a civil (European) law flavour. Though unusual, it does no harm and indeed reflects the modern development in contracts which increasingly deal with both general objectives and the underlying spirit or culture, e.g. certain partnering-type arrangements.

The suggestion to amend the requirements and/or proposals, will, if adopted by the employer, generally give rise to a change instruction. The benefit to the employer can arise in the following ways:

- A reduction in the cost of the project;
- A reduction in the life cycle costs associated with the project;
- The achievement of practical completion on a date earlier than the completion date; and/or
- Any other financial benefit to the employer.

Reduction in cost of project

4-112 The easiest benefit on which to place a financial value will of course be any reduction in the cost of the design and construction of the project, and most suggestions will no doubt be of this kind.

The contractor may be encouraged to offer means by which a reduction in the cost of the project can be achieved where the employer needs to engage in a cost-cutting exercise. However, there may be a problem here for the employer if he forms the view that the contractor's share of the financial benefit (as stated in the appendix) is too great. Once the contractor has put forward the suggestion and quotation etc., if the employer chooses not to proceed and issue the instruction, he is unable to implement the change until after practical completion of the project (clause 19.5). Often satisfactory negotiation will be possible in these circumstances. Alternatively, the employer may be reluctant to even discuss such matters with the contractor and instead will seek advice from his professional advisors who are unlikely to have similar benefit provisions as part of the terms of their appointment.

Reduction in life cycle costs associated with the project

4-113 The issue of life cycle costs, i.e. the costs associated with the operation and maintenance of the buildings forming the project and any plant and equipment which also forms part of it, is increasingly regarded as an important, even fundamental, part of project planning and objectives. Some examples would be energy savings in terms of heat conservation in colder weather and adequate ventilation (without air conditioning) in warmer weather, and the use of low maintenance materials and products. The financial benefits of such matters will be somewhat more difficult to measure but by no means impossible, and given a sensible approach should not cause a problem between employer and contractor.

The reduction in life cycle costs may only be achievable with an associated increase in the capital cost of the project and this will need to be taken into account in evaluating any financial benefit.

The contractor has the obligation under clause 19.2 to provide the initial calculation of the financial benefit to the employer. This may by no means be an easy thing to do in this situation. If unsure, the contractor may be tempted to put in a figure on the high side, particularly as having done so the employer, if no agreement is reached, will not be able to proceed unilaterally before practical completion of the project. However, the figure should nevertheless be within the range of what the contractor, acting reasonably, would believe to be realistic.

Achievement of practical completion earlier than the completion date

4-114 Any financial benefit related to achieving practical completion earlier than the completion date will be based upon the value to the employer of being able to utilise the completed project. This will not necessarily be a calculation which the contractor can easily make under clause 19.2. It is very much therefore a question of trust and agreement. It may be that the level of liquidated damages is some indication of the saving to the employer in achieving early practical completion, though this is not necessarily the case, e.g. the liquidated damages figure may well be less than the real costs of not having the project completed.

The means by which the contractor can achieve an earlier practical completion may involve extra costs in terms of the way in which work is resequenced or accelerated, and clearly this would form part of the calculation of any financial benefit to the employer.

If the contractor forms a view during the tender period, which could well happen, as to how the project can be completed more quickly, he may wish, instead of waiting for the contract to be entered into before suggesting an amendment to the requirements or proposals, to discuss it beforehand with the possibility of negotiating a bonus under clause 14 (see paragraph 3-114 *et seq.*). If the contractor does this, he will need to carefully consider how to protect his position so that the employer does not adopt the amendment while awarding the contract to someone else.

If it is the employer who wishes to take the initiative in seeking an earlier practical completion, he can do so by utilising clause 13 dealing with acceleration (see paragraph 3-106 *et seq.*).

Any other financial benefit to the employer

4-115 This could cover many possibilities. Some of them may actually increase the capital cost but perhaps increase the overall value of the project, e.g. by making it more versatile or flexible in use. Suggested amendments relating to aesthetic design or choice of materials will be difficult to evaluate financially even if the employer is satisfied that they are beneficial. It is unlikely in practice, though by no means impossible, that clause 19 will be used to generate benefits in terms of improving the aesthetic quality of the project. Other more likely examples could be in relation to the functional operation of the project, e.g. more efficient use of space available in order to increase the investment value of the project in terms of selling or leasing.

Details of amendments and quotation

> 19.2 The Contractor shall provide details of its suggested amendments to the Requirements and/or Proposals together with a quotation as provided in clause 20.4 and its calculation of the benefit it believes the Employer will obtain, expressed in financial terms.

4-116 The contractor is to provide details of any suggested amendments to the requirements and/or Proposals, together with a quotation. The quotation is to be provided in accordance with clause 20.4 in relation to changes (see paragraph 5-36 *et seq.*). The contractor must also calculate the benefit which he believes the employer will obtain, expressed in financial terms. As already discussed, this may not be easy for the contractor to do, for example in relation to the financial value to the employer of early practical completion and therefore earlier use of the project, or the financial value attached to any amendments having an aesthetic quality. The contractor's belief will be judged objectively rather than subjectively, so that it must be based upon what a contractor acting reasonably would believe to be an appropriate calculation. If the calculation is not based on an objectively judged reasonable belief on the contractor's part, the employer may argue that he can proceed unilaterally by issuing a routine change instruction under clause 20, as the prohibition in clause 19.5 on having the work carried out before practical completion only applies where the contractor has suggested amendments to the requirements and/or proposals 'in this manner', i.e. in the manner required by clause 19. This is an issue which could be adjudicated or litigated.

4-117 The extent (or limit) of the 'details' of the amendment to be provided by the contractor could be crucial where the employer has in mind to proceed unilaterally with a routine change which has similarities but is different from that proposed by the contractor. It is a matter of consideration in each case whether a routine change can be seen as an attempt to make use of the contractor's suggested amendment without the employer sharing any financial benefits accruing from it. It is a difficult area. The contractor may well wish to provide a broad description, maybe even going not much further than the expression of an idea, so that the employer cannot simply proceed with a similar but different approach to that contained in a detailed description provided by the contractor. This is certainly an area where trust and fair dealing are paramount requirements.

Negotiation of the financial benefit

> 19.3 Where the Employer wishes to implement an amendment suggested by the Contractor the parties shall negotiate with a view to agreeing the Contractor's quotation and the financial benefit to the Employer. Such agreement shall be confirmed by the Employer by an instruction identifying the agreed value of the Change, any agreed adjustment to the Completion Date and the agreed financial benefit to the Employer.

4-118 Following the contractor's submission of his quotation and assessment of any financial benefit to the employer, if the employer wishes to implement the suggested amendment then the parties are required to 'negotiate with a view to agreeing' the quotation and financial benefit. Probably, the parties cannot be contractually obliged to negotiate although certainly they can be obliged to attempt to negotiate. As it is assumed that agreement must be reached, there is no express provision in clause 19 as to what happens if agreement is not in fact reached.

However, clause 19.5 implicitly provides that if agreement is not reached, the employer's hands are tied until practical completion, both in terms of not being able to instruct a change under clause 20 in respect of such suggested amendments, or in arranging for another contractor to carry out the work.

If the parties, despite negotiating in good faith, cannot reach an agreement, the only fall-back position appears to be that in clause 19.5. It would not be possible to take the issue of the genuine failure to agree to an adjudicator or court. It is not for an adjudicator or judge to impose upon the parties his or her own version of what a reasonable agreement would have been.

When agreement is reached it is to be confirmed by the employer in terms of an instruction which identifies:

- The agreed value of the change itself;
- Any agreed adjustment to the completion date; and
- The agreed financial benefit to the employer.

Payment of contractor's share of financial benefit

> 19.4 Upon Practical Completion of the Project the Employer shall be liable to pay the Contractor the proportion identified in the Appendix of the agreed financial benefit.

4-119 This provides that the contractor is to receive his share of the financial benefit to the employer in the proportion identified in the appendix item for clause 19. If no percentage is stated in the appendix the default position is 50%. Employers need therefore to be careful to give consideration to the operation of clause 19 before entering into the contract. If they simply ignore it, leaving the appendix item blank while leaving clause 19 as part of the MPF conditions, they may find it difficult later to make sensible changes where the contractor has outmanoeuvred them by putting forward numerous suggested amendments covering a wide range of matters. In such a case, either the contractor would then obtain the 50% proportion of any financial benefit to the employer, or the employer would be forced to wait until after practical completion before having the work carried out. This could have significant cost implications for the employer.

Where the matter proceeds pursuant to clause 19, the contractor is not entitled to payment from the employer until practical completion of the project has been reached. The contractor will of course be entitled to the cost of actually implementing the change instruction in the normal way. The payment in respect of the contractor's share of the financial benefit will follow practical completion in accordance with clause 22.2 (and clause 22.5.4) which provides that after practical completion of the project the employer must issue further interim payment advices at intervals of not less than one month, except that their issue will not be required where the amount identified as due is less than that stated in the appendix item against clause 22. Where no figure is inserted the minimum amount is to be £10 000.

Restrictions on employer's use of contractor's suggested amendments

> 19.5 Any amendments suggested by the Contractor shall be clearly identified as being an amendment suggested under the provisions of clause 19. Where the Contractor has sug-

> gested amendments in this manner the Employer may only instruct those amendments in accordance with the procedures set out in clause 19, provided always that nothing shall prevent the Employer from utilising other contractors to implement suggested amendments after Practical Completion of the Project.

4-120 Where the contractor is seeking to operate clause 19, any suggested amendment which he makes must be clearly identified as being made under the provisions of clause 19. Any failure to do so could result in the employer being able to adopt the amendment and instruct it as a routine change, outside the provisions of clause 19.

4-121 Once the contractor has suggested amendments under the provisions of clause 19, the employer can only instruct those amendments to be incorporated in accordance with the clause 19 procedures. Once, therefore, the contractor has lodged his suggested amendment pursuant to clause 19, the employer is locked into the clause 19 procedures; the employer may well find this, on occasions, highly inconvenient, particularly if the contractor has decided to come up with a significant list of potential financial benefits to the employer, many of which the employer might have himself come up with at some later date. There could well be situations where the employer, faced with such difficulties, will seek to issue a routine change which appears to require an alteration to the requirements and/or proposals similar but somewhat different to the contractor's suggested amendments.

4-122 If the employer, for whatever reason, decides not to instruct the amendments to be implemented, clause 19.5 makes it clear that this is not to prevent the employer from engaging other contractors to implement the suggested amendments after practical completion of the project.

Design matters

4-123 Clause 19.5, in providing that nothing is to prevent the employer from utilising other contractors to implement the contractor's suggested amendments once practical completion of the project has been achieved, does not address the question of the copyright which the contractor may have in any design which forms part of his suggested amendment. Is the wording of clause 19.5 sufficient to implicitly give the employer a licence to use such designs over which the contractor may have copyright? Clause 7, which deals with copyright, gives the employer an irrevocable licence to copy and use 'Design Documents' which clause 39.2 defines as: 'Drawings, specifications, details ... required to be prepared by the Contractor for the purpose of explaining and amplifying the Requirements and/or Proposals which are necessary to enable the Contractor to execute the project or which are required by any provision in the Requirements ...'

Clearly this definition does not encompass designs forming part of suggested amendments by the contractor, which are not pursued under clause 19. There is therefore no express licence. While it is not difficult to contend that an implied licence is given where it is in effect necessary in order for the employer to obtain proper beneficial use of the project, this is not the case here. The employer will obtain the originally anticipated benefit of the project following practical completion without any subsequent alterations being made to the project. In addition, the contractor would obtain no share of any benefit to the employer of incorporating the contractor's suggested amendment after practical completion, and it is inherently unlikely that the contractor would simply gratuitously give a licence to the employer to use the contractor's designs.

However, clause 19.5 does of course give the employer the right to use the contractor's suggested amendments and this right would be useless if it involved any design on the contractor's part without at least an implied right for the employer to use that design.

This issue is far from easy to determine. Contractors would be well advised to confirm when entering into the contract that in such circumstances no copyright licence is given.

Chapter 5
Valuation and payment

Content

5-01 Clauses 20 to 25 inclusive fall under the general heading of 'Valuation and Payment'. They cover the following topics:

- Changes
- Loss and/or expense
- Payments
- Withholding
- Interest
- VAT.

In addition, it will be convenient to deal with the pricing document in this chapter.

Changes – clause 20

Background

5-02 All standard forms of construction contract will make provision for the introduction of changes, often also called variations, to the work to be undertaken by the contractor. This can be by way of the addition, substitution or omission of work. Many standard forms go far beyond this and include provision for the introduction of or changes to such matters as access to the site or specific parts of the site; limitations of working space; limitations of working hours; and the sequence in which work is carried out. Many such contracts will also contain instances where matters which are not strictly speaking changes at all are nevertheless stated to be treated as if they are, essentially in order that the contractor will receive an appropriate valuation and an adjustment to the completion date as a result of the happening of the identified events. As we shall see later (paragraph 5-15), the MPF, in clause 20 and the definition of 'Change' in clause 39.2, includes a considerable number of such matters.

5-03 It is important that the employer, when issuing instructions requiring a change, does so strictly in accordance with the terms of the contract giving him that authority. A failure to do so can cause severe difficulties with an outside possibility of the contractor being disentitled to payment for such unauthorised, though ordered, changes, and the possibility of any employer's representative running the risk of personal liability in such circumstances for losses suffered by the contractor.

5-04 In the absence of an express power to make changes, there is no implied right for the employer to have changes carried out by the contractor. Any contract documents describing the work or conditions under which it is to be carried out, e.g. employer's requirements, contractor's proposals including drawings, specification, etc., would in fact be frozen at the date that the contract was entered into, and if the employer required any departure from these, it could only be achieved by a separately negotiated agreement between himself and the contractor.

5-05 Most change clauses contain a power not only to add or vary work but also to omit it. It would not be a proper exercise of this power to omit work in circumstances where the employer intended to have the omitted work carried out by a third party: see *Amec Building Ltd* v. *Cadmus Investment Co Ltd* (1996).

Work indispensably necessary

5-06 As a general principle, an obligation to do specified work includes an obligation to do all necessary ancillary work. Items or work processes which are not fully described, but which are nevertheless indispensably necessary to achieve completion in accordance with the contract, will not amount to a change or variation and will have to be carried out within the original contract sum. Even where the work required is set out in considerable detail, e.g. a bill of quantities prepared in accordance with a standard method of measurement, it may not descend to every detail of construction and the same point is at least arguable.

Limits to changes and variations

5-07 Although most standard forms of construction contract, including the MPF, contain no express restrictions as to the permitted size or extent of any change, there must be some point at which a so-called change would so fundamentally alter the contract that it would not be regarded as a change within the meaning of the contract: see *Sir Lindsay Parkinson & Co* v. *Commissioners of Works* (1950). Depending on the nature of the contract, it will often be the case that it would require changes to be extreme in content or number before the courts would declare them as falling outside the scope of the change clause – see for example *McAlpine Humberoak Ltd* v. *McDermott International Inc (No. 1)* (1992), considered earlier (see paragraph 2-10).

Changes and variations after practical completion

5-08 Although it is common practice to issue a certificate or statement of practical completion when some relatively unimportant items still remain to be completed, and indeed the MPF expressly permits this (see the definition of 'Practical Completion' in clause 39.2), it is generally thought that, unless a contract makes it an express term, there is no power to order changes after practical completion. Certainly so far as the MPF is concerned, its overall structure suggests that there is no such power. Firstly, a change instruction after practical completion of the project might, for example, require significant additional work, resulting in the project being rendered insufficiently complete to warrant a statement of practical completion, i.e. a change in such circumstances could appear to undo an existing practical completion statement. Secondly, the provisions dealing with adjustment to the completion date and loss and expense are geared either to delays in completion, i.e. delays to practical completion, or disturbance to regular progress respectively, and it is difficult to see how the progress of the works can be affected once they are practically complete.

It clearly makes good sense that the works should not be capable of being altered by a change after they have been completed. However, the consequence of this is that once practical completion is achieved, even though there are some minor work outstanding items, the employer will not be able to issue change instructions relating to them. As the outstanding work, if any, will often relate to final finishes which are a frequent area for changes, it could in strict contractual terms, though perhaps not often in practice, give rise to problems.

The use of clause 11 permitting the taking over of part of the project prior to practical completion of the whole, treating just the part taken over as having achieved practical completion, may be a solution in appropriate circumstances (see paragraph 3-40 *et seq*.).

Summary of clause 20

5-09 Essentially clause 20 provides as follows:

- A change is defined in clause 39.2 to include:
  - Alterations to the requirements or proposals which alter design, quality or quantity;
  - Alterations of restrictions or obligations set out in the requirements or proposals as to the manner in which the contractor is to execute the project, including additional restrictions or obligations;
  - Any matter that the contract requires to be treated as giving rise to a change.

  There is a proviso that to be a change the alteration or matter must not have been required as a result of the negligence or default of the contractor.
- A change can come about by means of an instruction for a change or the occurrence of an event which is required under the contract to be treated as giving rise to a change.
- There is a pre-instruction procedure enabling the employer, before deciding whether or not to issue a change instruction, to provide details of the proposed change requesting the contractor to submit a quotation in which he will value the change, including any requirement in respect of reimbursement of any loss or expense, and identifying any required adjustment to the completion date.
- Alternatively the employer can issue an instruction without seeking a prior quotation, in which case the contractor must respond by providing details of the proposed valuation and information supporting it to enable the employer to make a fair valuation. Any adjustment to the completion date is dealt with separately under clause 12. The position is similar where the employer does not accept the contractor's quotation.
- The same valuation procedure described immediately above applies also where the contract requires an event to be treated as a change.
- A fair valuation must have regard to matters such as the nature and timing of the change and the effect on other parts of the project, and it must have regard to the prices and principles in the pricing document as applicable. It is also to include any recoverable loss and expense.
- There is a valuation review procedure following practical completion.

The MPF has no detailed valuation rules such as those contained in WCD 98, clause 12.5. The MPF simply provides for a fair valuation to include relevant matters such as prices and principles in the pricing document, the effect of the change on other parts of the project, and the nature and timing of the change. Likewise, even though in clause 12.4.2 of WCD 98 there is a contractor's price statement arrangement similar to the MPF post instruction valuation procedure, the MPF procedure is far simpler.

The MPF has no procedure whereby an oral instruction, including an oral instruction requiring a change, can be perfected by the employer or contractor subsequently confirming it in writing, c.f. WCD 98, clause 4.3.2. Hence, the contractor will be taking a significant risk in complying with an oral instruction for a change. He may be able to argue that an oral instruction can still give rise to a change as defined in clause 39.2 (see commentary on clause 2.1 in paragraph 2-29). Even if this is not possible, he may still be able to obtain payment based on a separate contractual agreement outside the MPF, or possibly on the basis of a restitutionary claim against the employer for unjust enrichment or, more rarely, on the basis that the employer is estopped from denying the validity of the oral instruction.

It is worth noting that the MPF has no provisional sum procedure and accordingly there are no provisions dealing with instructions for the expenditure of provisional sums or their valuation.

Figure 5.1 shows the procedure for changes.

We now turn to consider clause 20 in more detail.

The definition of a 'Change'

Clause 39.2 defines a 'Change' as:

- Any alteration in the Requirements and/or Proposals that gives rise to an alteration in the design, quality or quantity of anything that is required to be executed in accordance with the Contract; or
- any alteration by the Employer of any restriction or obligation set out in the Requirements and/or Proposals as to the manner in which the Contractor is to execute the Project, or the imposition of additional restrictions or obligations; or
- any matter that the Contract requires to be treated as giving rise to a Change.

Provided always that the alteration or matter referred to above is not required as a result of any negligence or default on the part of the Contractor.

Comments on definition

ALTERATIONS IN DESIGN, QUALITY OR QUANTITY

5-10 The first paragraph of this definition dealing with alterations to the requirements or proposals which give rise to alterations in design, quality or quantity, is relatively self explanatory. It can be contrasted with the more detailed provision covering similar ground in WCD 98, clause 12.1.1. The first paragraph of the clause 39.2 definition may well be an attempt to summarise clause 12.1.1. However, it should be noted that, unlike in clause 12.1.1, there is no express reference to the addition, omission or substitution of any work. Additions and omissions are almost certainly covered by the reference to alterations in design, quality or quantity. However, the substitution of work is not quite as clear-cut. It would be covered where the substitution had an effect on the design, quality or quantity, but would appear not to

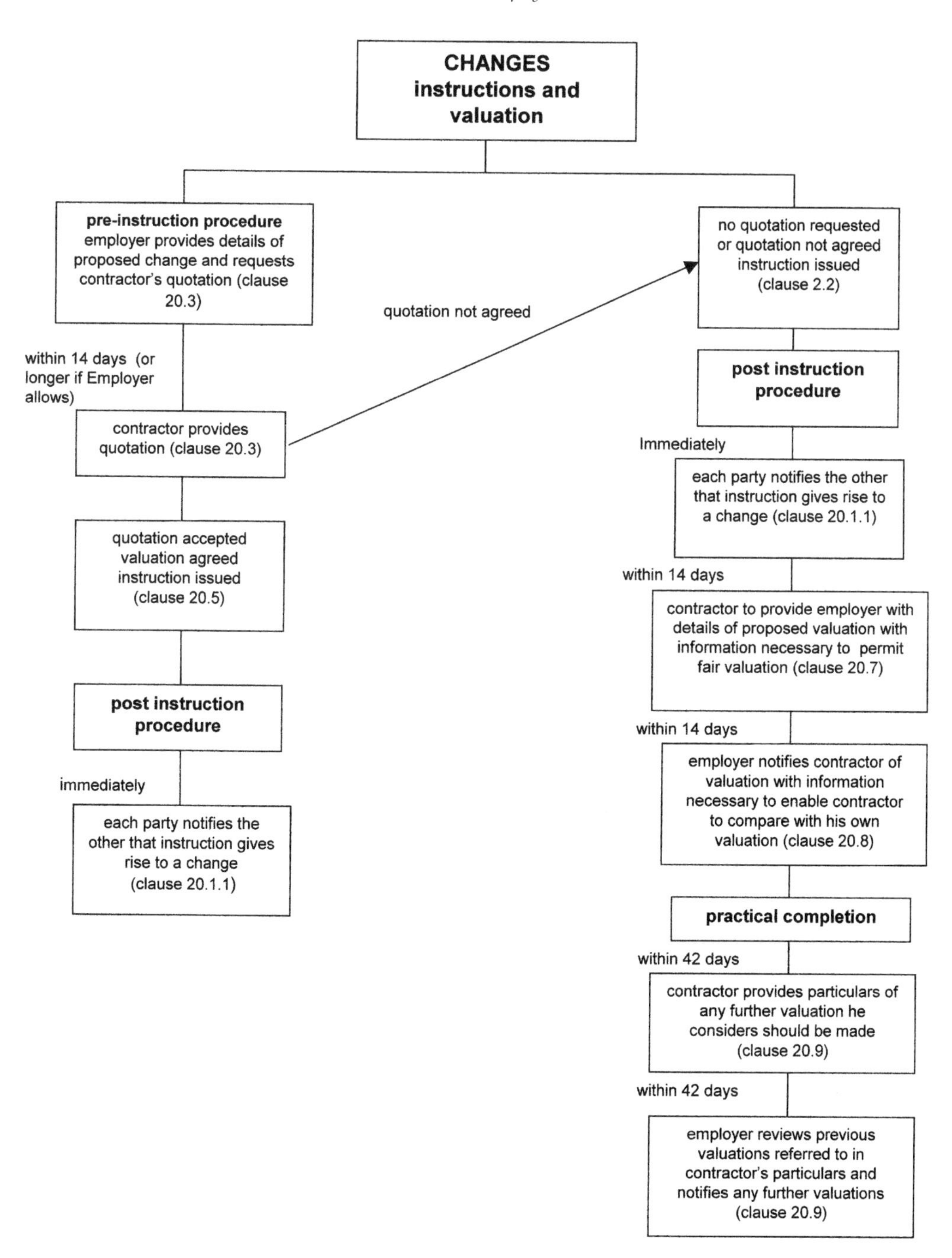

Fig. 5.1 Changes flow chart.

readily encompass a replacement of goods or materials with others of equal quality which have no effect on design or quantity. A straight substitution of materials or goods may very often be on the basis that they are equivalent. Oddly therefore, on this basis, the employer could not instruct such an alteration and it would therefore have to be negotiated outside the terms of the contract. This cannot have been intended. The answer may that the alteration by means of changing goods or materials for equivalent ones is a sufficient exercise of choice to be treated as a design issue and therefore an alteration in design, even if they are identical in terms of appearance and performance.

5-11 A further point to note is that it appears that 'Design Documents' (see definition in clause 39.2) which explain or amplify the requirements or proposals, and which the employer would like to alter, cannot be altered unless the alteration to them would also have the effect of altering the requirements or proposals as drawn. In other words, if the contractor's explanation or amplification of the requirements or proposals contained in the design documents is not inconsistent with the requirements or proposals but nevertheless the employer would like them altered in some way, strictly speaking this could only be done by negotiation outside the terms of the contract. This is an example of a bigger point. The requirements and proposals taken together may leave the contractor with considerable discretion as to how any particular design requirement is to be achieved. He may adopt a design solution which, while being consistent with those documents, is not to the employer's liking. The employer may want to instruct a 'Change'. However, that 'Change' may well not alter the requirements or the proposals. Strictly therefore, and somewhat surprisingly, this will not be a change as defined. However, in practice it invariably will be treated as a change.

5-12 Another point of comparison with clause 12 of WCD 98 is that, unlike the clause 39.2 definition in the MPF, the clause 12 definition refers only to changes in the employer's requirements with no mention being made of the contractor's proposals. This could be significant. Under the WCD 98 form, the contractor's proposals are his statement of how he intends to meet the employer's requirements. They have contractual force and he can be held to them. It is his part of the bargain and they can only be indirectly changed by the employer changing the employer's requirements. Having considered and accepted the contractor's tender, the employer cannot thereafter seek to change the way in which the contractor will meet the employer's requirements. Under the MPF he can do this by means of a change instruction. This gives the employer very considerable power to interfere with, and possibly undermine, the contractor's basic tender offer, and even though any instruction will fall to be valued and the completion date adjusted as appropriate, contractors may still feel uneasy about this possibility.

5-13 Mention has already been made concerning the size or nature of any alteration and the inherent limits which will be placed on the scope of the definition (see paragraph 5-07).

ALTERATION OF ANY RESTRICTION OR OBLIGATION

5-14 The second paragraph of the definition deals with alterations in connection with any restrictions or obligations in the requirements or proposals as to the manner in which the contractor is to execute the project, together with the imposition of additional restrictions or obligations. This is extremely wide. Again it can be compared with WCD 98, this time clause 12.1.2. The MPF appears again to summarise and shorten that which is set out in more detail in WCD 98. Clause 12.1.2 refers to obligations or restrictions with regard to specified matters:

- Access to the site or the use of any specific parts of it;
- Limitations of working space;
- Limitations of working hours;
- The execution or completion of the work in any specific order.

No doubt all of these are adequately covered in the MPF definition. The MPF definition may, however, go considerably further, referring as it does in general terms to restrictions or obligations set out in either the requirements or proposals 'as to the manner in which the Contractor is to execute the Project'. The meaning of these words can give rise to considerable debate and conjecture. For example, does it extend to any of the following:

- Changing the identity of a named specialist or altering the list of named specialists (see clause 18);
- Altering the design programme (see clause 6.2);
- Changing a stipulation identified by the requirements as being essential for practical completion to take place (see the definitions of 'Practical Completion' and 'Project' in clause 39.2 and the description of the project in the appendix); for example requiring operating manuals or additional or different operating manuals.
- Changing the boundaries of the 'Site' (see definition in clause 39.2 together with the definition of 'Requirements');
- Changing what is within a 'Section' (see definition in clause 39.2).

Depending upon any given situation, any of the above could have implications for the manner in which the contractor is to execute the project. If the employer does seek to instruct any such alterations as a change, while the contractor may of course be entitled to an adjustment to the completion date and a valuation to include reimbursement of loss and expense, nevertheless, he may not wish to be placed in this position. Firstly, many of these matters may be regarded as fundamental contractual obligations which, under other standard forms of construction contract, could not be the subject of such alteration without the contractor's consent. In other words, they could only be achieved by renegotiating terms of the contract. Secondly, the question of what should be an appropriate adjustment to the completion date, or an appropriate fair valuation, could give rise to considerable differences of opinion.

It is unfortunate that the definition of 'Change' is so wide in this particular respect. It may, however, be conditioned by the inherent limits to its meaning, i.e. that it cannot turn the contract into something completely different to that originally envisaged. Even so care needs to be taken.

MATTERS THAT THE CONTRACT REQUIRES TO BE TREATED AS GIVING RISE TO A CHANGE

5-15 The MPF makes liberal use of the facility of declaring to be a change many issues which ordinarily would not perhaps be regarded as changing the project. It is a convenient device for treating the occurrence of certain events in a way which enables them to be dealt with under the contract in terms of valuation and adjustments to the completion date. The matters covered in this way are as follows:

- Clause 4.2 dealing with discrepancies within the requirements;
- Clause 4.5 dealing with alterations to statutory requirements after the base date;
- Clause 8.2 dealing with the contractor's response on encountering adverse ground conditions or man-made obstructions under clause 8.1;
- Clause 13.2 where the employer issues an instruction in relation to acceleration;
- Clause 16.1 in relation to opening up work, materials or goods for inspection or test;
- Clause 27.8 in certain circumstances where the contractor carries out additional works following loss due to terrorism.

Certain clauses in the contract specifically state that an instruction is not to be treated as giving rise to a change, in which case the contractor is not entitled to any additional payment or any adjustment to the completion date (see clause 2.2 and paragraphs 2-32 and 3.76).

PROVISO

5-16 At the end of the definition of change is a proviso that anything which would fall within any of the three sections of the definition, is only to be included where the alteration or matter referred to is not required as a result of any negligence or default on the part of the contractor. The choice of the word 'required' in relation to the last section, namely a 'matter that the Contract requires to be treated as giving rise to a Change' is confusing. Surely the matter itself cannot be required to be treated as a change as a result of the contractor's negligence or default, nor does the MPF anywhere directly provide that a matter caused by the contractor's negligence or default is to be treated as giving rise to a change. Presumably, the proviso should have referred to alterations not being required, and to matters not resulting from the negligence or default of the contractor. Hopefully the choice of words will not defeat the intention behind this proviso.

5-17 The intention behind this proviso is clear. There should be no possibility that the contractor could claim to have a valuation made and an adjustment to the completion date where the reason for the instruction for a change or the reason behind the event or matter that the contract requires to be treated as giving rise to a change, was attributable to the contractor's negligence or default.

5-18 In practice there could well be much argument about whether there has been any negligence or default, but that may be inevitable.

5-19 There may also be question as to what the word 'default' means in this context. The proviso refers to any negligence or default on the part of the contractor. Clearly therefore default must mean something other than negligence. Set against a contractual background, it might seem obvious that it would include any breach by the contractor of any of his contractual obligations, whether culpable or not. Generally speaking, breaches of contract are not dependent on whether the contract breaker can be regarded as morally blameworthy or at fault. With few exceptions a failure to fulfil a contractual promise is a breach of contract whatever the excuse may be.

However, in a number of cases, the absence, fault or blameworthiness of the contractor appears to have been regarded as relevant where the fault or blameworthiness lies further down the contractual chain, such as with a sub-contractor. See, for example, *Greater Nottingham Co-operative Society Ltd* v. *Cementation Piling & Foundations Ltd* (1989); *Scott Lithgow* v. *Secretary of State for Defence* (1989) and *John Jarvis* v. *Rockdale Housing Association* (1987). On this basis and looking at the definition of 'Contractor' in clause 39.2, namely 'the Party to the Contract named as such', there may just be an opportunity here for the contractor to argue that he must be personally at fault and that he is not automatically in default where the fault lies with a sub-contractor or even a consultant. Having said this, a relevant point (e.g. see the *Jarvis* case), is whether some clauses in the contract refer only to defaults of the contractor while others expressly refer to the defaults of the contractor, his servants, agents or sub-contractors. Under the MPF, the only reference to sub-contractors is to named specialists and clause 18.4 makes the contractor solely responsible for the works undertaken by them. The proviso will therefore most probably extend to defaults of any sub-contractor with the effect that any instruction resulting from it will not be treated as giving rise to a change.

Notification

> 20.1 Either party shall immediately notify the other:
>
> ...

5-20 This introductory line to clause 20.1 is odd. It requires that 'Either party shall ...'. In other words it must mean that both parties must notify the other except that if and when one does the other need not notify. It is possible to have a provision that either party *may* notify but not that either party *shall* notify. If no one notifies, which party is in breach. Presumably it is both, so both should be referred to even if there is a suitable qualification, e.g. 'Unless and until one party has notified the other, both parties shall immediately ...'.

The notification under clause 20.1 sets in motion the valuation procedure (see clause 20.7).

Instructions giving rise to change

> 20.1.1 ... whenever it considers that an instruction gives rise to a Change; and/or

5-21 By clause 20.1.1, where it is the instruction which gives rise to a change, the notification is required immediately either party 'considers' that an instruction gives rise to a change. In its context, this appears to be more of a subjective than objective obligation. In other words, the obligation to notify arises only when the particular employer or particular contractor actually thinks that this is the position. On this basis a party is not in breach for failing to notify provided he actually did not consider the instruction as giving rise to a change, even if his failure to appreciate this was a failure to recognise the obvious.

5-22 The obligation is to notify 'immediately' either party considers that an instruction gives rise to a change. This must mean straight away.

5-23 An instruction requiring an alteration to the requirements or proposals which satisfies the definition of 'Change' in clause 39.2 will very often be self-evident. However, it is still necessary to give notice. Wherever possible, the employer should identify the instruction as giving rise to a change either on the face of the instruction itself, and/or in any covering communication. In any event, as a notification will be a communication within clause 38.1, it must be in writing, so the contract does not permit, for example, the issue of an instruction in writing which does not identify the instruction as giving rise to a change, accompanied by an oral statement that it is considered to be a change.

In practice, if an instruction giving rise to a change is issued by the employer without a notification under this clause, the onus will fall upon the contractor to notify as he will wish to ensure an appropriate valuation and adjustment to the completion date.

5-24 Apart from being a good housekeeping point so far as audit trails and record keeping is concerned, the main purpose of the obligation to notify must be to bring into the open as soon as possible any potential disagreement between the parties as to whether an instruction does give rise to a change. This is particularly relevant having regard to the proviso in the clause 39.2 definition of 'Change' (see paragraph 5-16). However, if neither notifies the other, what are the consequences?

Failure to notify

5-25 A failure to notify does not have the effect that an instruction which in fact gives rise to a change, is to be treated as if it did not give rise to a change. What it does do, however, is prevent the valuation procedure from operating until notification has taken place (see for example clause 20.7 requiring the contractor to provide the employer with details of the proposed valuation etc. and which starts the procedure in motion). Clearly therefore if nothing is ever notified, the contractor would not actually get a valuation or adjustment to the completion date.

It should be appreciated that while the obligation to notify would appear to cover both instructions issued in the form of accepting the contractor's pre-instruction quotation under clause 20.5, as well as instructions where there is no such quotation or the quotation is not accepted, in relation to the former, as the valuation and time consequences will have already been agreed, the failure to notify does not appear to affect the contractor's rights at all. It is of course in any event self-evident in such circumstances that any such instruction gives rise to a change and it may be that the instruction accepting the quotation for a proposed change is a fulfilment by the employer of the obligation to notify.

5-26 The most likely scenario surrounding a failure to notify is the absence by the contractor of immediate notification followed by a belated notification. This could, of course, equally be a breach by the employer, though not if the employer does not consider that the instruction gives rise to a change. Having notified belatedly, the contractor will still be entitled to seek the valuation and adjustment to the completion date. It is not easy to see how the contractor's breach in notifying late can affect this right, though it may possibly affect the amount of the valuation or the extent of any adjustment to the completion date if the employer can establish that an earlier notification would have altered the situation. For example, had the notification been immediate, an employer who was not in breach of the duty to notify as at the time of issue he did not consider the instruction as giving rise to a change, might have had an opportunity to issue a further instruction altering the earlier one in whole or in part which might have had an effect on both costs and time. That opportunity may have been lost, e.g. due to the stage of the project reached by the time of the contractor's belated notification. However, if the instruction does give rise to a change, it may be that the employer is at fault in not having recognised that fact, even though not in breach of the notification provision if he did not, subjectively, appreciate that it was a change. For the employer to argue that, although as a matter of fact he did instruct a change, he did not appreciate that fact, is hardly a contractual basis upon which to refuse to value etc. the instruction actually issued.

Occurrence of event treated as giving rise to a change

> 20.1.2 ... of the occurrence of any event that under the Contract is required [to] be treated as giving rise to a Change.

The events which the contract requires to be treated as a change have been listed earlier (see paragraph 5-15).

5-27 Under clause 20.1.2, dealing with the occurrence of events that the contract requires to be treated as giving rise to a change, the duty on the parties to notify is not dependent upon whether or not they subjectively consider that the occurrence of the event did fall within the contractual requirement that it be treated as a change. It would have been sensible for the definition (in clause 39.2) itself to have referred to the '*occurrence of any event* that the Contract requires to be treated as giving rise to a Change' instead of referring to 'any matter' or else to have begun clause 20.1.2 by referring to 'any matter arising' rather than the 'occurrence of any event'.

5-28 As the notification in respect of clause 20.1.2 is not dependent upon what either party, subjectively, considers to be the position, the failure to give notice will put both parties in breach. In these circumstances, a belated notification on the part of the contractor can hardly be complained of by the employer who is himself in breach. So, for example, if a contractor belatedly notifies that his proposals for dealing with ground conditions or man-made obstructions under clause 8.1 are, by clause 8.2, to be treated as giving rise to a change, the employer will be in difficulty in arguing that had the contractor notified him earlier, he would have issued instructions to amend what the contractor proposed to something less costly or less time consuming. The contractor can point out that the employer is equally in breach in not notifying the contractor that he considered the ground conditions or man-made obstructions to be such that they could not have reasonably been foreseen by an experienced and competent contractor, however unlikely it is in practice that an employer would ever do this.

5-29 It is quite possible that it cannot be determined whether or not an instruction is required to be treated as giving rise to a change until some time after the instruction has been issued. This is the case, for example, under clause 16.1 where the employer can instruct the contractor to open up for inspection or test any work, materials or goods. If that test is not already provided for by the contract, the instruction is to be treated as giving rise to a change unless the inspection or test discloses that the work, materials or goods are not in accordance with the contract, in which case the instruction shall not be treated as giving rise to a change. In such a case notification must await such results.

Value agreed or a fair valuation

> 20.2 Other than in respect of Changes instructed in accordance with the provisions of clauses 13 (*Acceleration*) or 19 (*Cost savings and value improvements*), the consequences of any Change shall be determined in accordance with the provisions of clause 20 so that either:

.1 the value of the Change and any adjustment to the Completion Date is agreed in accordance with clause 20.5 prior to an instruction being issued; or

.2 a fair valuation of the Change is made in accordance with clause 20.6 and any adjustment to the Completion Date is notified in accordance with clause 12 (*Extension of time*).

5-30 This is a neat and effective clause which deals with the financial and time consequences of changes.

It expressly excludes from the financial and time consequences of changes, instructions issued in accordance with the provisions of clause 13 (Acceleration - see paragraph 3-106) or clause 19 (Cost savings and value improvements - see paragraph 4-109). These are both examples of pre-instruction agreements between the parties as to costs and time. All other changes are subject to clause 20. What is not altogether clear is whether clause 13 or 19 changes require notification under clause 20.1. Perhaps not, but even if they do, the failure to notify would not appear in any way to disentitle the contractor from receiving any agreed additional costs, benefits or adjustments to the completion date.

5-31 Clause 20.2 introduces the two methods by which changes are valued:

- Under clause 20.2.1 on the basis of the value of the change and any adjustment to the completion date being agreed prior to an instruction being issued; or
- Under clause 20.2.2 on the basis of a fair valuation of the change being made following the instruction, with any adjustment to the completion date being dealt with under clause 12 (extension of time).

Employer's request for a quotation

20.3 Prior to instructing any Change the Employer may provide details of the proposed Change and request the Contractor to submit a quotation in respect of the Change. The Contractor shall provide the quotation within 14 days of the request, or within such longer period as the Employer states in its request.

5-32 Clause 20.3 introduces the pre-instruction procedure under which the employer has the option to seek a quotation from the contractor covering his valuation of the proposed change, together with his assessment of any adjustment to the completion date. Even without a contractual mechanism for this, it is often done in practice. The call for the contractor to respond to the employer's request for a quotation is framed as a contractual obligation. In truth, if the parties cannot amicably reach an agreement there is very little that the employer can do about it. To satisfy his contractual obligation, an unwilling contractor is likely to provide the minimum detail possible to meet the requirements for the quotation under clause 20.4 and, while obliged to value the proposed change in accordance with the provisions for a fair valuation set out in clause 20.6, the valuation is likely nevertheless to reflect the contractor's unwillingness to go along with the request. Even if the contractor fails to comply with the requirements to provide a quotation, there is generally not a great deal that the employer can do in practice. In particular, the contractor will not of course be failing to comply with an instruction as all that is being provided by the employer is a request. Accordingly the employer's rights under clause 2.3, where the contractor fails to comply with an instruction, do not apply. The option open to the employer if he does not wish to abandon the proposed change altogether will be to instruct it as a change and then seek to argue that had the contractor provided a quotation it would have in some way produced a lower valuation, or a lesser adjustment to the completion date. There are obvious difficulties with this. The employer may argue that had he known, from a quotation, the likely valuation and adjustment to the completion date which in fact follows the routine instruction for a change, he would have elected not to proceed at all with it, and thereby seek in some way to claim a loss. This is such a speculative area as to render such a course of action highly unlikely. Such a procedure would almost certainly, therefore, work equally well outside the contractual provisions. If catered for within the contract, it would have been sensible to limit the contractor's obligation to considering whether or not to provide a quotation rather than being obliged to do so, particularly as in any event the contractor cannot recover the costs associated with preparing the quotation whether it is accepted or not (see paragraph 5-41).

5-33 The quotation procedure naturally enough does not apply to events which the contract requires to be treated as a change as, for the most part, these just happen.

5-34 The quotation procedure can, of course, apply to an omission of work as well as an addition or substitution. In such circumstances the contractor's quotation may well be in terms of an appropriate reduction in the cost of the project.

Employer to provide details of proposed change

5-35 Before the contractor can be obliged to provide a quotation, the employer must provide 'details' of the proposed change. The level of detail required will depend upon the circumstances. It could amount to a straightforward performance specification. Alternatively, the employer may have detailed designs prepared. As the employer is seeking to agree a valuation and adjustment to the completion date before issuing an instruction, it is obviously in his interests to provide adequate information, and any shortcoming in this regard is likely to meet with no quotation or a loaded quotation from the contractor if the contractor is in any difficulty in determining a fair valuation under clause 20.6, or in ascertaining the likely effects of any proposed instruction on the completion date.

In providing a quotation with the information required by clause 20.4, the contractor is having to take a risk. He should at least be able to take a calculated rather than uncalculated risk. The less detail the contractor has, the more he will need to provide something for the unmeasurable risks both in terms of the valuation and the adjustment required to the completion date. This is likely to result either in the quotation not being accepted or in the employer paying more and allowing more time than may in the event be necessary.

Contractor to provide quotation within 14 days (or such longer period stated in the employer's request)

Fourteen days is a very tight timetable for the contractor to provide a quotation containing all the matters set out in clause 20.4. Hopefully the employer will recognise this and provide a longer period wherever this is appropriate, or alternatively will agree to further time if the contractor is having difficulty in meeting it. Failure to approach the matter in such a way on the employer's part is likely to lead to a loaded quotation. Even if this amounts to a breach of clause 20 on the part of the contractor, this will in reality take the matter no further from the employer's point of view.

Contents of the quotation and period of validity

20.4 The quotation provided by the Contractor shall:
.1 give a valuation of the Change calculated in accordance with the principles set out in clause 20.6;
.2 identify any adjustment to the Completion Date that will be required as a consequence of the Change;
.3 be in sufficient detail for the Employer to assess the amounts and periods required and, in particular, shall state separately any amounts included in respect of loss and/or expense;
.4 identify the period, being not less than 14 days, for which the quotation remains open for acceptance.

5-36 Clause 20.4 deals in four sub-clauses with the required contents of the contractor's quotation and with its period of validity.

Valuing the change in accordance with clause 20.6

5-37 The contractor's quotation must value the change instruction on the basis set out in clause 20.6, namely having regard to:

- The nature and timing of the change;
- The effect of the change on other parts of the project;
- The prices and principles set out in the pricing document, so far as applicable;
- Any loss or expense that will be incurred as a consequence of the change, unless contributed to by some other cause which is neither a change nor one of the loss and/or expense matters set out in clause 21.2.

These specific requirements are considered later when dealing with clause 20.6 itself (see paragraph 5-48 *et seq.*).

5-38 It will very often be impossible for the contractor to produce anything like a realistic figure for a quotation within a short period of time and in advance of implementing the change. There could be a considerable degree of risk to one or other of the parties if a quotation is accepted. Contractors are of course well used to submitting lump sum quotations under tight time constraints in respect of specific pieces of work, but in this instance they must take account of its effect on other parts of the project and this may be difficult to fully appreciate at the time. Even more problematic could be the requirement to include any loss and expense that will be incurred. It will often be extremely difficult to accurately forecast the extent to which the regular progress of the project is likely to be affected as a consequence of the change. It is likely therefore that in the majority of instances, the amount actually payable in respect of an accepted quotation will be considerably above or below the amount which would have been payable had the same exercise been undertaken once the effects on the project of the change had been fully worked through. The provisions of clause 20.9 enabling the contractor to seek a further valuation within 42 days of practical completion in respect of any change, do not apply to a clause 20.4 quotation.

While the contractor is to include in his quotation any loss and expense incurred as a consequence of the change (see clause 20.6.4), its exclusion from a fair valuation under clause 20.6.4 where any element of the loss or expense was contributed to by any cause that is either not a change or not a clause 21.2 listed matter, seems entirely inappropriate to an exercise which is based on looking into the future. The point behind this proviso to clause 20.6.4 is explained later (see paragraph 5-60). What can be said here is that it involves a consideration of what actually happens. Having agreed a sum in advance for loss and expense, the employer would not be able to subsequently challenge a quotation on the basis that as it turned out the proviso would have applied in the absence of an agreed quotation.

Adjustment to the completion date

5-39 The contractor's quotation is to identify the adjustment to the completion date which the contractor requires as a consequence of the change. Again, to provide an accurate assessment of what the actual delay might be would be very difficult in the limited time available and before the consequences of the change upon the project completion date are worked through. Once a quotation is accepted, the chances of the actual delaying effects being in the order of that agreed will not be very great in respect of the majority of changes. The contractor does not have the opportunity to seek a review of the adjustment with the benefit of hindsight, as would be the case if the adjustment to the completion date was to be undertaken following an instruction for a change rather than a quotation (see clauses 20.2.2, 12.5 and 12.6). The contractor is bound to be driven to err on the side of overestimating the effect of the change on the completion date, particularly when he is under a contractual obligation to provide a quotation rather than it being the result of the exercise of his free choice to provide one.

Contractor to provide sufficient detail to enable the employer to assess amounts and periods

5-40 The contractor's quotation must be in sufficient detail to enable the employer to properly assess the contractor's valuation, including any amount in respect of loss and expense which must be separately stated. It must also be in sufficient detail to enable the employer to assess the extent of the adjustment to the completion date being sought. If the employer considers that the contractor has provided insufficient detail, there appears little in practice that the employer can effectively do other than to consider issuing an instruction for the change or abandoning the proposal so far as the contractor is concerned. The employer's difficulties in this respect have been considered earlier (see paragraph 5-32). If the employer considers that the contractor has provided a quotation in the spirit of clause 20 but has in fact not provided enough information, the employer can request the contractor to submit a revised quotation for consideration (see clause 20.5).

5-41 As the contractor will not be entitled to the costs of preparing the quotation (such costs do not appear to fit anywhere within the provisions of clause 20.6 dealing with the calculation of a fair valuation), whether the quotation is accepted or not, there exists a significant disincentive for the contractor to supply 'sufficient' details if this would be a time consuming exercise.

Period for which the quotation is to remain open for acceptance by the employer

5-42 The contractor is required to indicate the period for which the quotation is to remain open for acceptance by the employer. This is to be not less than 14 days.

No doubt if the quotation is in respect of a proposed change which is extensive and which has required a particularly detailed and technical response in the contractor's quotation, it will be sensible for the period for which the quotation remains open to be significantly longer than 14 days. The time taken for the employer to respond will often be determined not only by the amount of detail which the contractor supplies but also by the current state of the project in relation to any implementation of the proposed change.

Acceptance of quotation

> 20.5 The Employer may accept the quotation or request the Contractor to submit a revised quotation. When the Employer accepts a quotation it shall issue an instruction identifying the quotation that is being accepted, the agreed value and any agreed adjustment to the Completion Date.

5-43 If the employer decides to accept the contractor's quotation, he must issue an instruction (in writing) which must identify the quotation being accepted and must also state the agreed value and any agreed adjustment to the completion date. Once accepted, it is not possible for the contractor to seek a reconsideration of the valuation under clause 20.9.

5-44 Rather than accept the quotation, the employer may request the contractor to submit a revised quotation. There are no procedural requirements stated for this. Presumably any such quotation will be subject to the same requirements as for the original quotation set out in clauses 20.3 and 20.4. In practice this would be appropriate in situations where the employer decides, e.g. because of the level of the quotation, to amend the proposed change. This could equally be done by rejecting the quotation and requesting a new one based on the revised proposed change. If the employer wishes to negotiate any aspect of the contractor's quotation, any resulting agreement will probably be contained in a revised quotation which would then be accepted under this clause.

5-45 There is no provision entitling the employer to select only part of the quotation for acceptance, though the parties can always reach agreement to do so.

Fair valuation

> 20.6 Where agreement is not reached under clause 20.5, a fair valuation of any Change shall be made by the Employer. Such fair valuation shall have regard to the following:
> .1 the nature and timing of the Change;
> .2 the effect of the Change on other parts of the Project;
> .3 the prices and principles set out in the Pricing Document, so far as applicable and,
> .4 any loss and/or expense that will be incurred as a consequence of the Change, provided always that the fair valuation shall not include any element of loss and/or expense if that element was contributed to by a cause other than a Change or a matter set out in clause 21.2.

5-46 Clause 20.6 provides that where the employer has not accepted a contractor's quotation, then the employer will make a fair valuation of any change having regard to the principles contained in sub-clauses 20.6.1 to 20.6.4 inclusive. However, it should be appreciated that the requirement for a fair valuation extends to a number of situations. Fair valuations are to be carried out in the following situations:

- By the contractor when he provides a quotation for consideration by the employer pursuant to clause 20.4 (see clause 20.4.1);
- By the employer where the contractor's quotation is not accepted and the employer nevertheless wishes to proceed with the issue of an instruction requiring a change (see clause 20.2.2 and the opening words to clause 20.6);
- By the contractor where the employer issues an instruction requiring a change which has not previously been subject to the quotation procedure under clause 20.3 (see clauses 20.2.2 and 20.7);
- By the contractor where the contract requires an event to be treated as a change (see clauses 20.1.2, 20.2 and 20.7);
- By the employer following the contractor's valuation required under clause 20.7 (see clause 20.8).

Put simply, the principles set out in the sub-clauses to clause 20.6, to which regard must be had in reaching a fair valuation, have a role to play in the valuation of all changes other than those instructed under clause 13 (acceleration) and clause 19 (costs savings and value improvements) (see clause 20.2 introductory paragraph).

5-47 There is no general provision inviting the parties to agree to value a change otherwise than in accordance with these fair valuation principles as there is in some other JCT contracts, e.g. WCD 98, clause 12.4.1, which refers to the application of the contractual valuation mechanisms 'unless otherwise agreed by the Employer and the Contractor'. Even so, there is nothing of course to prevent the parties from stepping outside the valuation provisions in clause 20.6 if they wish. In addition, the fact that the sub-clauses of clause 20.6 are matters to which a fair valuation 'shall have regard' leaves some room for manoeuvre as to 'have regard' does not mean that it should slavishly be followed.

Summary

5-48 Any such fair valuation is to have regard to the following matters:

- The nature and timing of the change;
- The effect of a change on other parts of the project;
- The prices and principles set out in the pricing document, so far as applicable; and
- Any loss or expense that will be incurred as a consequence of the change, excluding any elements of loss and expense contributed to by a cause other than:
    - A change; or
    - Any breach or act of prevention by the employer or his representatives or advisors; or
    - Interference with the contractor's regular progress by 'Others' on the site authorised by the employer; or
    - A valid suspension by the contractor for non-payment.

THE NATURE AND TIMING OF THE CHANGE

5-49 Clearly the nature of a change and the timing of its implementation may have a bearing upon how it should be fairly valued. Even if prices and principles set out in the pricing document (see clause 20.6.3) are generally applicable, it may still be appropriate to make some allowance to take the nature and timing of the change into account. An obvious example would be where the change required excavation work to be carried out during winter months, making excavation a slower more difficult operation, perhaps requiring more powerful plant and equipment. Much will depend upon the precise principles set out in the pricing document.

THE EFFECT OF THE CHANGE ON OTHER PARTS OF THE PROJECT

5-50 It is inevitable that in the case of certain changes, there will be an effect on other parts of the project.

5-51 It is not difficult to envisage an instruction which affects the whole operation of the contract, e.g. an instruction 'as to the manner in which the Contractor is to execute the Project' (clause 39.2 definition). The employer or his advisors would therefore be well advised to consider the effect that such instructions may have on the valuation of other work before issuing the instruction, particularly if the instruction could result in a complete revaluation of all the work remaining to be done if the conditions under which it is executed change substantially. The contractor's obligation to provide details of his assessment of the valuation (see clause 20.7) is a major safeguard for the employer in this regard.

As any valuation carried out by the contractor under clause 20.7 will generally be on the basis of its *anticipated* rather than actual effect on other parts of the project, the contractor may well be very grateful for the valuation review procedure following practical completion (see clause 20.9).

The situations in which this sub-clause could be relevant are almost infinite. This is particularly so if the change impacts upon the design which, as a result, requires modification.

APPLICATION OF THE PRICES AND PRINCIPLES SET OUT IN THE PRICING DOCUMENT

5-52 It might have been thought appropriate to begin the sub-clauses with this provision as it is likely in practice to be the starting point in valuing many changes. Where the subject matter of the change is similar to that covered in the pricing document, that will almost invariably at least be the starting point.

5-53 The prices and principles set out in the pricing document will apply only so far as applicable. If there is work which is of an identical nature and where the pricing of the work is not affected by the nature and timing of the change, then the provisions of the pricing document are likely to produce a fair valuation. To the extent that the prices and principles of the pricing document include, as in reality they must, provision for preliminaries, overheads and profit, the fair valuation will take this into account.

5-54 To the extent that the prices and principles of the pricing document are not applicable, the fair valuation should nevertheless include a fair assessment in respect of a contribution to preliminaries, overheads and profit. The overall pricing structure of the contract will be a relevant factor in determining what that profit level might be - see for example *Sanjay Lachhani and Another* v. *Destination Canada (UK) Ltd* (1996).

5-55 So far as contribution to profit is concerned, in *Weldon Plant Ltd* v. *The Commission for New Towns* (2000), Weldon entered into a contract with the Commission using the ICE Conditions 6th Edition. A variation was ordered under clause 51 and disputes arose as to whether Weldon should be paid overheads and profits on costs incurred by reason of the variation. The appropriate variation rule was that requiring a fair valuation to be made. It was held that a 'fair valuation' will normally mean cost plus a reasonable percentage for profit with a deduction for any proven inefficiency by the contractor, but if there is proof of a general market rate for comparable work it may be taken into consideration or applied completely. In this case, Judge Humphrey Lloyd QC said:

> '... it would not be fair if the valuation did not include an element on account of such contribution. It would mean that such a contribution would have to be found elsewhere, presumably from the contractor's margin for profit or risk. In my view a valuation which in effect required the contractor to bear the contribution itself would not be a fair valuation, in accordance with principles of clause 52 (1) which are intended to secure that the contractor should not lose as a result of having to execute a variation (except, as I have stated, to the extent its costs etc. are of its making).'

5-56 The 'fair' valuation is not to be carried out on a general or global basis. Each item of work should be particularised and priced: *Crittall Windows Ltd* v. *T.J. Evers Ltd* (1996).

Pricing document containing pricing errors

5-57 As a general principle, if the contractor has made a mistake in his rate or price he will nevertheless generally be bound by it both for the original work and in applying the same rate or price in respect of a change. The fact that the contractor has priced in error is not a reason making the rate or price inapplicable for the purposes of valuing the change. However, the pricing document will have to be carefully considered to see if it in any way alters this general principle. The application of this principle could be to the contractor's advantage or disadvantage: *Dudley Corporation* v. *Parsons & Morrin* (1967).

An example where the application of this general rule gave the contractor a windfall is to be found in a case under the ICE Conditions 6th Edition: *Henry Boot Construction Ltd* v. *Alstrom Combined Cycles Ltd* (2000). In this case the mistake in pricing took the form of overpricing by the contractor in the accepted tender so that applying the rate to the varied work provided a 'windfall' for the contractor of approximately £2 500 000, whereas otherwise a fair valuation would have produced a figure of approximately £575 000.

At first instance Judge Lloyd QC considered the relevant clauses of the ICE Conditions dealing with variations and changes in quantity – clauses 52(1), 52(2), 56(2) and 57 – and held that:

- The Civil Engineering Standard Method of Measurement (CESMM) cannot be used to prevent a genuine (if mistaken) contract price from becoming a price for the purpose of a clause 52 (valuation of a variation).
- The price in the tender was sacrosanct, immutable and not subject to correction (clause 55(2)). A mistake in price or rate bound both sides.
- In valuing variations, the fact that in 'fair value' terms a price or rate may be too high or too low is immaterial and the words in clause 52(1) 'executed under similar conditions' do not refer to economical or financial conditions or con-

siderations. The intrinsic profitability of a price or rate is irrelevant in applying the valuation principles, or when dealing with increases or decreases in quantities under clause 56(2).

On appeal to the Court of Appeal, the first instance decision and reasoning was upheld and confirmed.

LOSS AND EXPENSE THAT WILL BE INCURRED AS A CONSEQUENCE OF THE CHANGE

5-58 As part of calculating a fair valuation, the MPF expressly requires that associated loss and expense will be included as part of the valuation. However, excluded from the valuation should be any element of the loss and expense which was contributed to by a cause other than a change or one of the matters in respect of which the contract allows recovery of loss and expense (see clause 21.2 and paragraph 5-60 below). The recovery by the contractor of loss and expense generally is covered by clause 21 and this is discussed in detail later (see paragraph 5-67 *et seq.*). However, where the loss and expense that will be caused to or incurred by the contractor is as a consequence of the change, then it forms part of the fair valuation under clause 20.6. Unlike under clause 21 (see clause 21.3), clause 20.6.4 does not refer to the need for the loss and expense to be the result of the regular progress of the project being materially affected. The phrase 'loss and/or expense' is not a defined term in the MPF and it is possible that its application in determining a fair valuation is not restricted to situations involving the regular progress being materially affected, e.g. it could include the payment of damages under necessarily cancelled supply contracts or sub-contracts: see *Tinghamgrange Ltd* v. *Dew Group Ltd and North West Water Ltd* (1995).

'... that will be incurred ...'

5-59 The loss and expense which is to be included as part of the fair valuation is that which 'will be incurred as a consequence of the Change'. Although at first sight this may appear to look forward to loss and expense which has not yet been incurred at the time of the valuation, especially as, under clause 20.7, the contractor is to provide details of the proposed fair valuation within 14 days of any instruction giving rise to a change, in its overall context it can certainly be argued that even if the loss and expense has already been suffered or incurred when the valuation is carried out, it will nevertheless still be loss and expense 'that will be' incurred as a consequence of the change. In other words the phrase 'that will be' starts from the time when the change is effective rather than at the time of the valuation.

Loss and expense valuation to exclude any element contributed to by a cause other than a change or a loss and expense reimbursable matter set out in clause 21.2

5-60 The purpose of this exclusion appears to be to keep out of the calculation of loss and expense any element operating concurrently with or overlapping with the change and which is either attributable to the contractor's own default, or to some more neutral event for which the contractor is to take the risk so far as loss and expense is concerned, even if, under Clause 12, the contractor is entitled to an adjustment of the completion date, e.g. *force majeure* or the occurrence of a specified peril or the exercise after the base date by the UK Government of any statutory power directly affecting the execution of the project, etc. In other words, one operating cause may be the consequence of the change instruction and another equally effective cause may be a matter which is neither a change instruction nor one of those events listed under clause 21.2. In such a situation, the exclusion seeks to make it clear that the contractor is not to be entitled to reimbursement of loss and expense (see also clause 21.8). This exclusion in relation to such events is consistent with the approach contained in the Society of Construction Law Protocol for Determining Extensions of Time and Compensation for Delay and Disruption, published October 2002 (corrected 2003). Clause 1.10 of the Protocol provides:

> 'If the Contractor incurs additional costs that are caused by both Employer Delay and Contractor Delay, then the Contractor should only recover compensation if it is able to separate the additional costs caused by the Employer Delay from those caused by the Contractor Delay.'

This is the converse of the situation where the contractor is seeking an adjustment to the completion date as a result of concurrent delays (see paragraph 3-103). Put another way, if the contractor would have suffered or incurred the loss and expense in any event as a result of a delay for which he is responsible (including for this purposes a neutral delay), the contractor will not be entitled to recover the additional costs. In such a situation, clause 20.6.4 seeks to make it clear that the contractor is not to be entitled to reimbursement of loss and expense except to the extent that he is able to isolate an element of loss and expense as being solely attributable to consequences of a clause 21.2 matter or a change (see also clause 21.8 and the comments on it - paragraph 5-100). The JCT Guidance Note for the MPF provides an example of this: a change instruction issued which would have caused delay and prolongation to site establishment costs but for the fact that the contractor is already in delay having failed to submit design documents in accordance with the design programme. The contractor would get an extension of time but would not be able to recover the prolongation to the site establishment costs incurred. However, he would be able to recover in respect of the costs of disruption to trades as a consequence of complying with the change instruction. All of the above begs the question of whether in overlapping 'causes' there can be more than one actual cause, i.e. the first in time being the true cause. The relationship between what the Protocol provides (and therefore what the MPF also provides) and legal issues of causation need to be considered (see the discussion of this when considering clause 12.7.3 - see paragraph 3-103).

It might be questioned whether at the time the contractor is making his assessment of loss and expense, namely within 14 days of the change being identified, it is even possible for him to exclude from his assessment of loss and expense any element which will be contributed to by his own default or some neutral event, when it cannot possibly be anticipated that this will happen at some future time. In other words, the contractor's calculation cannot make an allowance for this unless, exceptionally, the valuation is made after some loss and expense has already been suffered or incurred, but which is nevertheless caught by the 'will be' wording (see paragraph 5-59). In the great majority of cases, therefore, this exclusion provision should be irrelevant. However, in practice it may be that the clause 20 procedures do not operate at the time or in the way intended so that the change has been implemented or has begun to be implemented and loss and expense has been suffered or incurred before the clause 20.7 proposed valuation has been submitted by the contractor. In this case the exclusion would apply to prevent any element of loss and expense which was contributed to by a non-claimable cause from being included in the fair valuation.

Proposed valuation

20.7 Within 14 days of a Change being identified by either party the Contractor shall provide to the Employer details of the proposed valuation of the Change together with such information as is reasonably necessary to permit a fair valuation to be made.

Time limit

5-61 The time limit of 14 days begins with a change being identified under clause 20.1 which requires either party to immediately notify the other if they consider that there has been an instruction for a change or that an event has occurred which the contract requires to be treated as a change. This notification requirement has already been discussed (see paragraph 5-21 *et seq.*) and it is possible that the notification will be given some time after the instruction has been issued or the event has occurred, without this being the result of any default on the part of either party. This could mean that by the time a valid notification identifying the change is given, implementation of the change may have begun or its effects may already have impacted upon the project. This is another reason why the reference to loss and expense 'that will be incurred' should be interpreted so as to cover loss and expense already suffered or incurred at the time of the valuation (see paragraph 5-59).

Late notification by the contractor is unlikely to cause prejudice to the employer in carrying out his valuation under clause 20.8.

If the contractor is late in providing details of the proposed valuation together with reasonably necessary information, the main consequence will be that the contractor will be paid later than otherwise would be the case. It is inconceivable that the provision of details and further information within the 14-day period is a condition precedent to the operation of the valuation procedures so as to prevent the contractor receiving any payment at all if it is supplied late.

Contractor to provide information reasonably necessary to permit a fair valuation

5-62 There is bound to be room for argument as to what is reasonably necessary information to permit a fair valuation to be made by the employer under clause 20.8. In general terms, it is whatever amount of information is available and can reasonably be provided which will enable the employer to verify or otherwise the contractor's valuation, and independently carry out a fair valuation. It seems certain that the contractor must identify all of the elements contained in clauses 20.6.1 to 20.6.4 inclusive. However, this does not take the question of how much information is reasonably necessary any further. This can only be considered on a case-by-case basis.

Employer's notification of valuation

20.8 Within 14 days of receipt of the information referred to by clause 20.7 the Employer shall notify the Contractor of its valuation of the Change, that valuation being calculated by reference to the information provided by the Contractor. The valuation shall be in sufficient detail to permit the Contractor to identify any differences between it and the Contractor's proposed valuation.

5-63 Within 14 days of receipt of the details and information required by clause 20.7, the employer must notify the contractor of his fair valuation. This is to be calculated by reference to the information provided by the contractor and is to be in sufficient detail to enable the contractor to identify any differences between the employer's valuation and his own proposed valuation under clause 20.7.

Although the calculation of the employer's valuation is to be by reference only to the information provided by the contractor, clearly the employer may be in possession of information which is also relevant to a fair valuation and which ought therefore to be taken into account with details of it provided to the contractor as part of the employer's obligation to provide sufficient details to permit the contractor to identify differences between the two valuations. It might have been better to require the calculation of the employer's valuation to 'have regard to' the information provided by the contractor rather than for the calculation of the valuation to be 'by reference to the information provided by the Contractor'. This apparent requirement to link the employer's valuation directly to that carried out by the contractor in his proposed valuation is almost certainly intended to ensure that the employer's response deals with the contractor's proposed valuation item-by-item and figure-by-figure so that it is clear where any differences lie. In other words, it is probably not intended to mean that the employer is restricted to considering only the information provided by the contractor in carrying out his own valuation; rather it means that the employer's response must mirror the contractor's approach. This would not prevent the employer therefore reaching a different conclusion based upon information available to him separately, with that information being provided to the contractor to explain any differences. It might have been sensible therefore to have provided firstly that the employer should have regard to the information provided by the contractor, and then to state that the employer's calculation of the valuation should be by reference to the contractor's valuation so that any differences are clearly identified.

If the contractor is unhappy with the employer's valuation, he can of course refer it to adjudication at any time or alternatively await the outcome of the review procedure under clause 20.9. If he does decide to refer it to adjudication, the result of that adjudication will not affect the contractor's right to seek a review under clause 20.9.

Post practical completion review

> 20.9 No later than 42 days after Practical Completion of the Project the Contractor shall provide particulars of any further valuation it considers should be made in respect of any Change. Within 42 days of the receipt of those particulars the Employer shall undertake a review of its previous valuations of each Change to which those particulars relate and notify the Contractor of such further valuation as it considers appropriate.

5-64 Clause 20.9 provides a review procedure in connection with the valuation of variations. Not later than 42 days after practical completion of the project the contractor is obliged to provide particulars of any further valuation which he considers should be made in respect of any change. Within 42 days of receipt of those particulars the employer must undertake a review of any of his previous valuations to which those particulars relate and notify any further valuation he considers appropriate.

The original valuation carried out by the employer under clause 20.8 will almost always be before or shortly after the change instruction is implemented by the contractor, or begins to take effect (for instance where the change affects restrictions or obligations as to the manner in which the contractor is to execute the project). Accordingly, it is sensible to provide for the contractor to seek a review if events turn out in such a way as to render the employer's valuation, even if based upon the contractor's proposed valuation, inadequate having regard to actual events. It enables the contractor to revalue with the benefit of hindsight and to request the employer to change his original valuation accordingly. This exercise is likely to produce a far more accurate valuation than that carried out originally.

If the employer considers as a result of the review that a further valuation is appropriate, the contractor is to be notified within the 42-day period. The requirement therefore appears to be for the contractor and employer to go through a similar exercise, in whole or in part as appropriate, to that required under clauses 20.7 and 20.8, using the fair valuation criteria set out in clause 20.6.

Any disagreement can be referred to adjudication.

5-65 If the contractor misses the 42-day deadline, it may well not be possible to seek an adjudication of, or to litigate, the original valuation if to do so is dependent on the further particulars not available to the employer at the date of the original valuation. There would be no dispute as to the original valuation based on the information then provided (see for example *Edmund Nuttall Ltd* v. *R. G. Carter Ltd* (2002)). It would be otherwise if the contractor was in dispute over the original valuation based on the details and information already supplied under clause 20.7. The contractor's failure to meet the 42-day deadline could therefore have very serious consequences.

It should be noted that this review procedure does not apply to a valuation obtained by the quotation procedure (see clauses 20.3 to 20.5).

5-66 Clause 20.9 does of course include not only instructions which give rise to a change but also the occurrence of any event that under the contract is required to be treated as giving rise to a change.

Loss and/or expense – clause 21

Background

5-67 The MPF, as with other JCT standard forms (e.g. WCD 98, clause 26, and JCT 98, clause 26) provides for the contractor to be reimbursed in respect of loss and/or expense incurred as a result of regular progress being materially affected by various matters. The matters for which the employer is liable to the contractor for reimbursement and loss and expense are listed. Unlike the other JCT contracts just mentioned, the MPF has a short list of only three matters. It is kept short principally by the inclusion of a general provision referring to any breach or act of prevention on the part of the employer or his representatives or advisors. This reduces the need to include in the list specific examples of breaches or acts of prevention on the part of the employer, such as work by other direct contractors on the site of the project; delay on the part of the employer or his advisors in supplying any necessary information to the contractor; or insufficient access to the site. In addition, many matters which might otherwise feature in a list are classified under the MPF as giving rise to a change, and loss and expense incurred as a consequence of changes are expressly excluded from clause 21 and included instead as part of the valuation of the relevant change (see clause 20.6.4).

Recovery of loss and expense incurred is dependent upon the regular progress of the project being materially affected. These words are very similar to those found in the equivalent clauses of other JCT contracts, including those mentioned above. However, in other JCT contracts, reference is made to 'direct loss and/or expense', whereas in the MPF reference is made to 'loss and/or expense'. The absence of the word 'direct' is discussed below (paragraph 5-70).

The matters listed in clause 21 (see clause 21.2) are made up of:

- A general provision dealing with breaches or acts of prevention by the employer, his representatives or advisors already mentioned above;
- Interference with the contractor's regular progress of the project by other persons on the site whose presence has been authorised by the employer; and
- The valid exercise by the contractor of his right to suspend performance of his obligations under section 112 of the Housing Grants, Construction and Regeneration Act 1996 for non-payment of sums due.

These are discussed in detail later (see paragraph 5-80 *et seq.*).

Alternative claims for breach of contract

5-68 Often, many of the matters listed in clause 21.2 will not only give rise to the possibility of a claim by the contractor for recovery of loss and expense, but may also amount to a breach of contract by the employer. Though the MPF does not expressly preserve the contractor's right to claim damages for breach of contract as an alternative to claiming reimbursement of loss and expense (c.f., clause 26.4 of WCD 98, which does), silence on the point is enough to preserve the contractor's common law right to do so.

The general legal principles on which damages are awarded for breach of contract are considered later when dealing with termination by the contractor of his own employment under clause 33 of the contract (see paragraph 8-21 *et seq.*). The application of these general principles is likely to overlap to a considerable degree with any discussion of the particular principles on which recovery can be claimed for loss and expense incurred under the express wording of the contract itself.

Where the cause of the disturbance of regular progress is a breach of contract by the employer, the contractor may therefore claim damages at common law for breach of contract and thereby include in his claim any matters which the employer may seek to reject as part of a claim for reimbursement of loss and expense. For example, in the case of *London Borough of Merton* v. *Stanley Hugh Leach* (1985), Mr Justice Vinelott in considering clause 24 of JCT 63 said (at 32 BLR 107-8):

> 'Moreover there is a clear indication in the contract that the draftsman contemplated that the contractor might have parallel rights to claim compensation under the express terms of the contract and to pursue claims for damages. That arises under clause 24(2) which I have already read and which, of course, expressly provides that the provisions of the conditions are to be without prejudice to other rights and remedies of the contractor. The effect of clause 24(2) (as I understand it) is this. Clause 24(1) specifies grounds upon which the con-

tractor is entitled to make a claim for reimbursement of direct loss or expense for which he would not otherwise be reimbursed by a payment made under the other provisions of the contract. The grounds specified may or may not result from a breach by the architect of his duties under the contract; a claim by the contractor under sub-paragraph (a) will normally, though not perhaps invariably, arise from a failure by the architect to answer with due diligence a proper application by the contractor for instructions, drawings and the like, while a claim by the contractor under sub-clause (b) following a proper instruction requiring the opening up of works under clause 6(3) normally (though not perhaps invariably) will not involve any breach by the architect of any obligation under the contract. In either case the contractor can call on the architect to ascertain the direct loss or expense suffered and to add the loss or expense when ascertained to the contract sum. The contractor will then receive reimbursement promptly and without the expense and delay of a claim for damages. But the contractor is not bound to make an application under clause 24(1). He may prefer to wait until completion of the work and join the claim for damages for breach of the obligation to provide instructions, drawings and the like in good time with other claims for damages for breach of obligations under the contract. Alternatively he can, as I see it, make a claim under clause 24(1) in order to obtain prompt reimbursement and later claim damages for breach of contract, bringing the amount awarded under clause 24(1) into account.'

Scope of recovery

5-69 Before considering the listed matters and the machinery of clause 21, it is appropriate to outline the scope of the recovery and to at least mention some typical heads of claim likely to be included in a contractor's assessment of, or claim for reimbursement of, loss and expense under clause 21.4.

LOSS AND/OR EXPENSE

5-70 The governing words limiting the extent of recovery are 'loss and/or expense'. Other JCT contracts, e.g. WCD 98, clause 26, refer to '*indirect* loss and/or expense'. Under these other JCT contracts the right of recovery does not therefore extend to *indirect*, or what may be called consequential, loss or expense. The case of *Croudace Construction Ltd* v. *Cawoods Concrete Products Ltd* (1978) gives some assistance in drawing this distinction.

FACTS

The plaintiffs were main contractors for the erection of a school. They entered into a sub-contract with the defendants for the supply and delivery of masonry blocks. The sub-contract contained a clause which included the following words:

> '... if any materials or goods supplied to us should be defective or not of the correct quality or specification ordered our liability shall be limited to free replacement of any materials or goods shown to be unsatisfactory. We are not under any circumstances to be liable for any consequential loss or damage caused or arising by reason of late supply or any fault, failure or defect in any materials or goods supplied by us or by reason of the same not being of the quality or specification ordered or by reason of any other matters whatsoever.'

The main contractor sued the sub-contractor for losses alleged to have arisen because of late delivery and defects in the materials and goods supplied, seeking as part of the claim to recover loss of productivity and additional costs of delay in executing the main contract works and also the cost to them of meeting other sub-contractors' claims which were brought about by the delays of the defendant sub-contractor. A preliminary issue was ordered to be tried as to:

> 'Whether on a proper construction of such Contract or Contracts, including the Defendants' Standard Conditions of Trading, the Plaintiffs are entitled to recover damages under any, and if so, which, of the Heads of Damage which they had pleaded.'

HELD

Consequential loss or damage meant loss or damage which did not result directly and naturally from the breach of contract complained of. The Court of Appeal held that the meaning of the words 'consequential loss or damage' had already been decided in the case of *Millar's Machinery Co Ltd* v. *David Way & Son* (1935) and that they were bound by that decision. They also agreed with its reasoning. The plaintiffs could therefore recover the losses which they had pleaded.

It is clear therefore that 'consequential' will be treated as meaning indirect and will be interpreted quite restrictively as including only such heads of loss or expense which do not flow naturally from the breach without other intervening cause and independently of special circumstances. The *Croudace* case has been approved and followed by the Court of Appeal in the cases of *British Sugar Plc* v. *NEI Power Projects Ltd and Another* (1997) and *Deepak Fertilisers & Petrochemical Corporation* v. *Davy McKee (London) Ltd and Another* (1999).

Further, in the case of *Wraight Ltd* v. *P.H. & T. (Holdings) Ltd* (1968), Mr Justice Megaw said:

> 'In my judgment, there are no grounds for giving the words "direct loss and/or damage caused to the contractor by the determination" any other meaning than that which they have, for example, in a case of breach of contract.'

It is therefore apparent that the words 'direct loss and/or damage' mean the same as damages at common law for breach of contract.

The MPF, in clause 33.4.2 refers to 'loss and/or damage' as being recoverable by the contractor in the event of his terminating his own employment for certain stated reasons. It is a general principle in construing the meaning of words in a contract to presume that the same words have the same meaning and that different words are intended to have a different meaning: in other words, that the words 'loss and/or expense' mean something different to the words 'loss and/or damage'. If this is so, then the former words appear to be more restrictive in their scope than do the latter. Despite the present trend, therefore, of allowing under 'loss and/or expense' such items as the loss of profit which could have been earned on another contract had the disruption and delay not occurred, it may not yet be beyond doubt that such claims, whilst recoverable where the word 'damage' is used (as in the case of *Wraight* v. *P.H. & T. (Holdings) Ltd* above), will not be recoverable where the word 'expense' is used instead. However, it is equally arguable that it is the common interpretation of the word 'loss' which appears in both phrases, and which therefore can be interpreted in both places as being consistent with common law damages for breach of contract.

This interpretation of the words 'direct loss and/or expense' would bring them very close to the general common law position as to the measure of damages for breach of contract.

In the MPF, there is no reference to 'direct' in clause 21. In addition, unlike other JCT contracts, the contractor's express right to claim (following his termination of his own employment) under clause 33.4.2 refers to 'loss and/or damage' and not to '*direct* loss and/or damage' (e.g. WCD 98, clause 28.4.4.4). The likelihood is therefore that the scope of any claim for reimbursement of loss and expense under clause 21 will be identical in scope to a claim for damages for breach of contract arising from the same facts.

REMOTENESS OF DAMAGE

5-71 At common law it is well settled that, while the governing purpose of damages is to put the party whose rights have been violated in the same position, so far as money can do so, as if his rights had been observed, this overall position is qualified in that the aggrieved party is only entitled to recover such part of the loss actually resulting from the breach as was at the time of the contract reasonably foreseeable as liable to result from it. This will depend firstly on imputed knowledge and secondly on actual knowledge.

So far as imputed knowledge is concerned, a reasonable person is taken to know that in the ordinary course of things certain losses are liable to result from a breach of the contract. This is known as the first limb of the rule in the leading case of *Hadley* v. *Baxendale* (1854) (dealt with in more detail in Chapter 8, paragraph 8-24). Secondly, the contracting parties may have particular knowledge which would lead them to the conclusion that a breach of contract would result in losses being suffered over and above those which might be thought to flow naturally from the breach in the ordinary course of things (the second limb): *Victoria Laundry (Windsor) Ltd* v. *Newman Industries* (1949). In a claim for damages for breach of contract, both types of damages are recoverable although in relation to the second, i.e. that depending on the actual knowledge of the parties, specific evidence must be adduced to demonstrate this. The measure of damages for breach of contract at common law is dealt with briefly later (see paragraph 8-21 *et seq*.).

There is no reason why the same principles should not apply to the operation of the express recovery provisions under clause 21.

There are those who might argue that as we are dealing here with an express contractual right to make a claim rather than with a claim for damages for breach of contract, there is no reason to apply the rules as to remoteness of damage which apply in the latter situation. If this is the case, then provided there is a sufficiently direct relationship between the loss and expense caused by disturbance to regular progress and any of the matters listed in clause 21.2, the sum attributable to this would be recoverable without being subject to the application of the remoteness of damage principle. However, as yet, the courts have not adopted such an approach. Note also possible relevance of remoteness to prolongation costs between the contractor's programmed completion date and the contractual completion date dealt with under the next heading.

RESTRICTION ON PERIOD OF RECOVERY FOR PROLONGATION COSTS

5-72 The contractor may have programmed to complete before the contractual completion date. His pricing will be based on a construction period shorter than the contractual period. If the regular progress of the works is materially affected by a clause 21.2 matter which causes the contractor to overrun his programme, even though not the contract period, he will suffer site-wide preliminary costs, e.g. hutting, supervision, security and general site services which he otherwise would not have incurred. Can he recover these as loss and expense?

It has been argued that these costs are irrecoverable in the sense that they are not additional costs at all. It is argued that in tendering against a known contract period it must be assumed that the contractor has priced his preliminaries on the assumption that he may be on site for the whole contract period – see for example the case of *J.F. Finnegan Ltd* v. *Sheffield City Council* (1988) and in particular the editorial commentary on it at 43 BLR 126.

It is submitted that it cannot be 'deemed' that the contractor's tender includes for all time-related costs for the entire contract period. This would often give the employer the best of both worlds. On the one hand he might be looking towards the lowest tender, which in practice will generally be at least partly due to the contractor shortening the construction period. On the other hand, if the employer is responsible for causing the contractor to remain on site beyond that shorter period, he ought not to be able to contend that nothing is due for the period before the contract period has expired.

A possible approach, it is submitted, is to be found in the ordinary principles of remoteness of damage. What could be said at the date of entering into the contract as to the likely result of such disturbance to progress of the kind set out in clause 21.2. The knowledge of the employer at the time of entering into the contract is crucial. For instance, if the contractor has provided, with his tender, a realistic programme, showing a construction period shorter than the contract period, it is likely to be foreseeable that prolongation costs will be suffered by an overrun of that shorter period; whereas, if the information available at the time of entering into the contract indicates that the contractor intends or expects to be on site for the whole contract period, clearly no prolongation costs will be suffered during that period.

CAUSATION

5-73 There must be a sufficiently causal link between the loss and expense for which reimbursement is claimed and the matter under clause 21.2 upon which reliance is placed as having caused the disruption to progress and consequentially the loss and expense. If the cause is not sufficiently direct then the claim will fail. The line is sometimes difficult to draw, but the principle must be kept firmly in mind.

THE DUTY TO MITIGATE

5-74 At common law, following a breach of contract, it is the duty of the aggrieved party to take reasonable steps to mitigate his loss. Clause 21.3 of MPF requires the contractor to take all reasonable steps to reduce the loss and/or expense to be incurred. In any event, there would be a duty to mitigate the loss caused by one of the matters referred to in clause 21.2 either in relation to the extent of disturbance of regular progress or the financial consequences of it. There is a general duty on the contractor to take reasonable steps to mitigate. This may be by reducing the extent to which regular progress is disturbed by the matter in question, or alternatively by limiting the loss and expense which flows from it. If the contractor fails to mitigate the former, then it can be argued that, to that extent, it is not the matter referred to in clause 21.2 which caused the disturbance of regular progress, and if he fails in relation to the latter, then it can be argued that, to that extent, not all of the loss and expense is attributable to the disturbance of regular progress as part of it is the result directly of the contractor's failure to mitigate his loss. There are of course limits to the steps which the aggrieved party must take in order to mitigate his loss. It must be a reasonable step to take in all the circumstances. It will certainly not include the expenditure of substantial sums of money.

While the employer may in appropriate circumstances be able to challenge part of the contractor's claim on the basis of his failure to mitigate, it should also be borne in mind that if the contractor does take reasonable steps to mitigate his loss, and thereby incurs expenditure, then even if the effect of this is to inadvertently aggravate the loss, this is nevertheless recoverable by him.

GLOBAL OR ROLLED-UP CLAIMS

5-75 Loss and expense, in order to be reimbursable, must have been caused by one of the clause 21.3 matters. The link between cause and effect must be established. It is sometimes not practicable to relate loss or expense to one specific instance of one specific matter under clause 21.3. If this can be done, then it should be. However, there could be a series of events, the interaction of which prevents this approach from working. In the case of *Crosby* v. *Portland Urban District Council* (1967), a case which concerned the ICE Conditions of Contract, Mr. Justice Donaldson said (at 5 BLR 135-6):

> 'Since, however, the extent of the extra cost incurred depends upon an extremely complex interaction between the consequences of the various denials, suspensions and variations, it may well be difficult or even impossible to make an accurate apportionment of the total extra cost between the several causative events.
>
> . . . so long as the Arbitrator . . . ensures that there is no duplication, I can see no reason why he should not recognise the realities of the situation and make individual awards in respect of those parts of individual items of the claim which can be dealt with in isolation and a supplementary award in respect of the remainder of these claims as a composite whole.'

In suitable cases, therefore, a global or rolled up claim may be permissible. It is, however, to be regarded as the exception rather than the rule. Nevertheless, in practice, by far the majority of contractors' claims for reimbursement of direct loss and expense are framed, at any rate initially, on this global basis.

In the case of *London Borough of Merton* v. *Stanley Hugh Leach* (1985) this issue was considered again in relation to JCT 63 by Mr Justice Vinelott who said (at 32 BLR 102):

> 'In *Crosby* the arbitrator rolled up several heads of claim arising under different heads and indeed claims for which the contract provided different bases of assessment. The question accordingly is whether I should follow that decision. I need hardly say that I would be reluctant to differ from a judge of Donaldson J's experience in matters of this kind unless I was convinced that the question had not been fully argued before him or that he had overlooked some material provisions of the contract or some relevant authority. Far from being so convinced, I find his reasoning compelling. The position in the instant case is, I think as follows. If application is made (under clause 11(6) or 24(1) or under both sub-

> clauses) for reimbursement of direct loss or expense attributable to more than one head of claim and at the time when the loss or expense comes to be ascertained it is impracticable to disentangle or disintegrate the part directly attributable to each head of claim, then, provided of course that the contractor has not unreasonably delayed in making the claim and so has himself created the difficulty, the architect must ascertain the global loss directly attributable to the two causes, disregarding, as in *Crosby*, any loss or expense which would have been recoverable if the claim had been made under one head in isolation and which would not have been recoverable under the other head taken in isolation. To this extent the law supplements the contractual machinery which no longer works in a way in which it was intended to work so as to ensure that the contractor is not unfairly deprived of the benefit which the parties clearly intend he should have.
>
> ...a rolled up award can only be made in a case where the loss or expense attributable to each head of claim cannot in reality be separated and ... where apart from that practical impossibility the conditions which have to be satisfied before an award can be made have been satisfied in relation to each head of claim.'

This issue of global claims is closely linked to the requirement to properly plead the claim with adequate particulars. In terms of the facts, there should be an analysis of cause and effect which establishes the contractual entitlement to make a claim. Provided this is done, it may then, in terms of the amount of the claim, be possible to claim globally where the loss or expense suffered arises from more than one of the causes particularised.

A real danger for contractors in advancing a composite financial claim is that it could fail completely if any significant part of the disruption or delay is not established and the court finds no basis for awarding less than the whole amount claimed. Even so, a court may not be willing to dismiss such a claim before trial if there is any chance that the claimant could still adduce evidence of a causal connection at least between some individual loss and the event which caused it. In the Scottish case of *John Doyle Construction Ltd* v. *Laing Management (Scotland) Limited* [2002], Lord Macfadyen held while the general approach stated above is correct, nevertheless it would not always be appropriate to strike out the case at an early stage. It may be appropriate to give the claiming party at least an opportunity of demonstrating by evidence that some of the loss could nevertheless be specifically related to a cause or causes; or that even taking into account a non-qualifying cause, the whole of the loss was caused by the remaining events. Lord Macfadyen said:

> 'The rigour of that analysis is in my view mitigated by two considerations. The first of these is that while, in the circumstances outlined, the global claim as such will fail, it does not follow that no claim will succeed. The fact that a pursuer has been driven (or chosen) to advance a global claim because of the difficulty of relating to each causative event an individual sum of loss and expense does not mean that after evidence has been led it will remain impossible to attribute individual sums of loss or expense to individual causation of events. The point is illustrated in certain of the American cases. The global claim may fail, but there may be in the evidence a sufficient basis to find a causal connection is between individual losses and individual events, or to make a rational apportionment of part of the global loss to the causative events for which the defender has been held responsible.

> The second factor mitigating the rigour of the logic of global claims is that causation must be treated as a commonsense matter... That is particularly important, in my view, where averments are made attributing, for example, the same period of delay to more than one cause.'

This case also provides a useful resume of judicial and academic thinking on the topic of global claims.

Also relevant is the Australian case of *John Holland Property Ltd* v. *Hunter Valley Earthmoving Co Pty Ltd* (2002). This case is another example of a court not being prepared to strike out a global claim at an early stage on the basis that evidence may at trial be presented which establishes a right to part if not all of the money claimed. In a striking out application alleging that the claim was a global claim and should not therefore stand, Judge McClellan said:

> 'The fate of any strike out application may often depend upon the capacity of a plaintiff to provide necessary particulars of its claim. But as these decisions make plain [the judge was referring to a number of earlier Australian cases], a plaintiff who has a claim will not be denied the opportunity to prosecute that claim only because there may be difficulty in identifying with precision each individual element of the claim. Whether the claim can be sustained will depend upon the evidence in relation to it. If that evidence allows a conclusion that the plaintiff has a quantifiable loss, then it is open to the tribunal determining the matter to bring in a verdict for the plaintiff for the sum which it is satisfied is appropriate. It is not material that the claim is described as a "global claim" or given any other label.'

In addition, the following cases dealing with this issue will repay study:

Wharf Properties v. *Eric Cumine Associates* (1991)
ICI v. *Bovis* (1992)
Mid Glamorgan County Council v. *J. Devonald Williams & Partners* (1992)
John Holland Construction & Engineering Pty Ltd v. *Kvaerner R.J. Brown Pty Ltd and Another* (1996)
Bernhard's Rugby Landscapes Ltd v. *Stockley Park Consortium Ltd* (1997)
British Airways Pension Trustees Ltd v. *Sir Robert McAlpine & Sons Ltd* (1994)
Amec Building Ltd v. *Cadmus Investment Co Ltd* (1996)
Inserco Ltd v. *Honeywell Control Systems* (1998).

Typical heads of claim in loss and expense claims

5-76 It is not proposed to discuss the various heads of claim typically found in claims for the recovery of loss and expense under construction contracts. There is a wealth of published information available on this topic. The reader should refer to the specialist works.

5-77 Typical heads of claim are likely to cover, as a minimum, the following matters:

- Loss of productivity;
- Site-overheads;
- Head office overheads;
- Interest and financing charges;
- Loss of profit;
- The costs of the claim. Generally not recoverable but it can perhaps be mentioned in passing that there is authority for the proposition that professional

fees paid to a claims consultant for work done as an expert witness in the preparation of a building case for arbitration are allowable in the taxation of legal costs: see *James Longley & Co Ltd* v. *South West Thames Regional Health Authority* (1983).

Summary of clause 21

5-78 Before considering clause 21 in detail, it can be summarised briefly as follows:

- The matters which can give rise to a claim for reimbursement of loss and expense incurred are:
  - Breach or act of prevention of the employer or his representatives or advisors;
  - Interference with the contractor's regular progress of the project by other persons on site authorised by the employer;
  - The contractor's valid suspension for non-payment of sums due.
- It should be noted that loss and expense attributable to a change is expressly excluded and is dealt with in the valuation of the change (see clause 20.6.4).
- The contractor is to notify the employer as soon as he becomes aware that the regular progress of the project is being or is likely to be materially affected as a result of any of the above listed matters, so as to cause loss and expense to be incurred.
- The contractor must take all practicable steps to reduce the loss and expense to be incurred.
- The contractor must provide an assessment both of the loss and expense that has been incurred and that which is likely to be incurred, together with such information as is reasonably necessary for the employer to make an ascertainment of the former.
- The contractor's assessment and accompanying information must be updated at monthly intervals until all information reasonably necessary has been provided by the contractor to allow the employer to ascertain the whole of the loss and expense.
- Upon receipt of the assessment and sufficient information, the employer must within 14 days ascertain the loss and expense incurred by reference to the information provided by the contractor and in sufficient detail to enable the contractor to identify any differences with his assessment.
- Within 42 days of practical completion the contractor can provide further documentation in support of a further ascertainment which he considers the employer should make in respect of notified matters. Within 42 days of receipt of the further documentation the employer must undertake a review of its previous ascertainment in respect of each matter for which further documentation has been provided, and notify the contractor of any further ascertainment.
- The employer is liable to pay the loss and expense as soon as it has been ascertained.
- No ascertainment of loss and expense is to include any element contributed to by any cause which is neither a change nor one of the matters contained in clause 21.2 (listed under the first bullet point above).

Figure 5.2 shows the timetable for a loss and expense reimbursement.

We now turn to a more detailed consideration of clause 21.

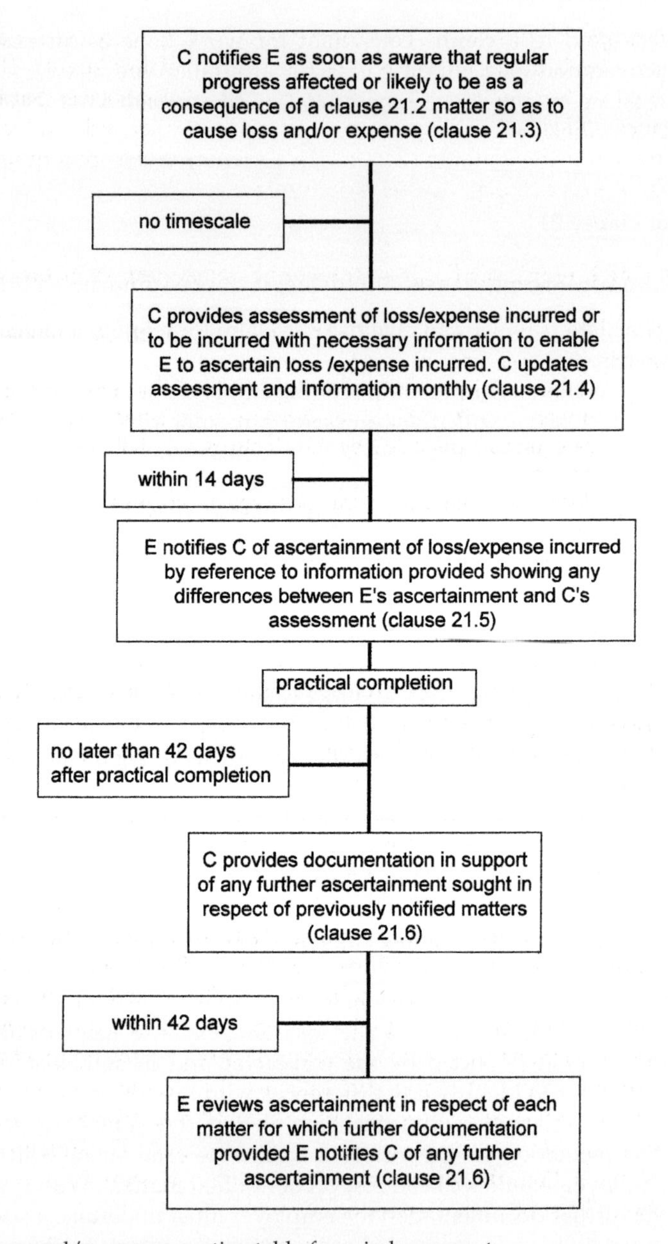

Fig. 5.2 Loss and/or expense – timetable for reimbursement.

Changes excluded

> 21.1 No Change or matter that is required by the Contract to be treated as giving rise to a Change shall, either individually or in conjunction with other Changes, give rise to an entitlement to be reimbursed loss and/or expense under clause 21.

5-79 Clause 21.1 expressly excludes loss and expense resulting from a change, including any matter that the contract requires to be treated as giving rise to a change (listed in paragraph 5-15). Instead, any consequent loss and expense is valued as part of the change itself under clause 20.6 (see especially clause 20.6.4). The main advantage offered by treating the loss and expense in this manner is that it enables all of the financial effects of a change to be dealt with together and in advance (even if subject to review if the valuation is not dealt with by means of a quotation under clause 20.5) of loss and expense being incurred.

Matters for which employer liable to the contractor in respect of loss and expense

> 21.2 Subject to clause 21.1 the only matters for which the employer will be liable to the Contractor in respect of loss and/or expense are:
> .1 a breach or act of prevention on the part of the Employer or its representative or advisors appointed pursuant to clause 15.2, other than any matters or actions that are expressly permitted by the Contract and that are stated not to give rise to a Change;
> .2 interference with the Contractor's regular progress of the Project by Others on the Site;
> .3 the valid exercise by the Contractor of its rights under section 112 of the HGCRA 1996.

Clause 21.2 lists the three matters in respect of which, if they cause loss and expense to be incurred, the contractor can claim reimbursement.

5-80 The matters set out in clauses 21.2.2 and 21.2.3 mirror exactly those events which can give an adjustment to the completion date under clauses 12.1.6 and 12.1.7. These have been commented on in detail when dealing with clause 12 (see paragraphs 3-70 and 3-71 to 3-73).

5-81 Clause 20.2.1 dealing with a breach or act of prevention on the part of the employer or his representatives or his advisors appointed pursuant to clause 15.2, is the same as that in clause 12.1.8. Although there is a proviso to clause 12.1 and an exception in clause 21.2.1 both seeking to exclude matters which the contract states expressly do not give rise to a change, the wording is somewhat different. In the last paragraph of clause 12.1, there is to be no adjustment to the completion date in respect of any matter where it is 'specifically stated by the contract that such a matter will not give rise to a Change'. In clause 21.2.1 the exception applies to any 'matter or actions that are expressly permitted by the contract and that are stated not to give rise to a Change'. So, in clause 21.2.1, the exception refers to permitted actions as well as to matters which are stated not to give rise to a change (rather than specifically stated not to give rise to a change). The reason for such different wording is difficult to understand and is indeed confusing.

The full list of such matters is:

- Clause 4.3 – instructions on discrepancies within the proposals;
- Clause 4.5 – alterations to statutory requirements announced before the base date;
- Clause 5.4 – the contractor proposing alterations to kinds or standards of materials or goods where those described in the contract are not procurable;
- Clause 6.9 – contractor not notifying employer of his disagreement with employer's comment on design documents under clause 6.8;
- Clause 8.1 – contractor's proposals to overcome problems on encountering ground conditions or man-made obstructions;

- Clause 16.1 – opening up for inspection or test where materials or goods are found not to be in accordance with the contract;
- Clause 16.3 – instructions issued under clause 16.2 as a consequence of work, materials or goods not being in accordance with the contract;
- Clause 27.1 – Contractor implementing remedial measures required by insurers due to non-compliance with the Joint Fire Code.

Each of the listed matters appears to fall squarely within both the proviso at the end of clause 12.1 and within the exception in clause 21.2.1, so identical wording would have been sensible.

Another point of difference is that the clause 12.1 proviso covers all of the listed events, whereas that in clause 21.2.1 refers only to breaches and acts of prevention by the employer and those for whom he is responsible. The proviso does not therefore qualify the other listed events, e.g. interference with the contractor's regular progress by persons on site authorised by the employer. So, for instance, an advisor to the employer interfering with regular progress whilst on site in connection with an inspection under clause 16.1, would not fall within clause 21.2.1, but would fall within clause 21.2.2. This makes no sense.

Notification issues

> 21.3 As soon as the Contractor becomes aware that the regular progress of the Project is or is likely to be materially affected as a consequence of any of the matters set out in clause 21.2 so as to cause loss and/or expense to be incurred it shall notify the Employer. The Contractor shall take all practicable steps to reduce the loss and/or expense to be incurred.

5-82 Clause 21.3 deals with the contractor's notification that the regular progress of the project is or is likely to be materially affected as a result of one of the clause 21.2 matters which has caused or will cause loss and expense to be incurred.

Timing of notification

5-83 The contractor is to notify (the notification is a communication and must therefore be in writing or in some other agreed manner – see clause 38.1) the employer as soon as he becomes aware that the following has happened:

- The regular progress of the project is being or is likely to be materially affected;
- As a consequence of one or more of the matters set out in clause 22.1;
- So as to cause loss and/or expense to be incurred.

All of these must therefore have happened. However, this means that if the contractor becomes aware that regular progress is likely to be affected, he should not wait until it is actually affected before notifying. Similarly, although the wording is not particularly apt, it must mean that if the loss and expense is anticipated rather than actual, there is still a requirement to notify. This is confirmed by the requirement in clause 21.4 for the contractor's assessment to include loss and expense which is 'to be incurred'.

Not only must the events referred to above have occurred, but the contractor must have become aware of them. Unlike, for example, clause 26.1.1 of WCD 98, there is no requirement for the notice ('application' under that clause) to be given as soon as it has become, *or should have become,* apparent that regular progress would be affected. The clause 26.1 wording has the unfortunate effect of disentitling the contractor from recovering loss and expense where he should have been aware but, in fact, was unaware that regular progress was likely to be affected. In other words, subjectively it may be impossible for him to make a written application but if he ought to have known the relevant facts then he loses his rights. Under clause 21.3 of the MPF, the requirement is more subjective. The reference to the contractor becoming aware does not include a requirement for notification where the contractor is unaware but ought to have been aware.

However, the MPF wording may, on occasions, favour the contractor who is lacking in diligence in terms of being aware of what is happening to a project, and yet is efficient in the skill of serving notices, while punishing the contractor who is always up to speed in the managing and running of a project but who is less diligent in the efficient service of notices.

Failure to give timely notification

5-84 If the contractor is late in notifying, he may find himself in some difficulty. Clause 21.4 requires the contractor to provide an assessment of the loss and expense 'notified in accordance with clause 21.3'. An employer may therefore seek to argue that a late notification is not a notification in accordance with clause 21.3 and that therefore the contractor is unable to provide a qualifying assessment of loss and expense. Two comments can be made in relation to this. Firstly, it is possible to construe the requirement for what amounts to immediate notification by the contractor as being procedural rather than fundamental to the operation of the rest of the clause dealing with assessment and ascertainment. Secondly, even if that is not the case, if the late notification is treated as valid by the employer who goes on to ascertain the loss and expense following receipt of the contractor's assessment, the employer is likely to be estopped from challenging the assessment on this ground. This will be particularly so if the employer has treated the notification as valid and the contractor has relied on this to his detriment.

Notification a condition precedent

5-85 The notification from the contractor appears to be a condition precedent. In other words, the contractor's assessment is dependent upon notification under this clause and if there is no valid assessment the employer has no obligation to ascertain. Accordingly, if there is no notification there will be no ascertainment. The failure to notify will not of course prevent the contractor from claiming damages for breach of contract where the clause 21.2 matter also amounts to a breach of contract by the employer.

Contents of the notice

5-86 Nothing is said about the contents of the notice itself though there is a requirement to provide in addition such information as is reasonably necessary to enable the employer to make the ascertainment (see clause 21.4). Accordingly, a liberal approach is likely to be adopted in relation to what the notification itself must provide by way of information. Clearly in order for the employer to appreciate what is being notified, the notification should state that the regular progress of the project is or is likely to be materially affected; the matter or matters in clause 21.2 which are being relied upon; and the fact that loss and expense is being incurred or is likely to be.

Disturbance to regular progress

5-87 If loss and expense which has been incurred is to be recoverable, the regular progress of the project must have been materially affected. So far as construction activity is concerned, regular progress in this sense probably refers to attending to the project on a regular basis with sufficient in the way of men, materials and plant to have the physical capacity to progress the work substantially in accordance with the contractual obligations (see *West Faulkner Associates* v. *London Borough of Newham* (1994) – per Lord Justice Brown). If therefore the contractor is prevented from proceeding in this manner, including in addition relevant off-site activities such as design work, then regular progress will have been affected.

The disturbance must be material, so if it is minimal and insignificant this will not warrant reimbursement of loss and expense.

5-88 Regular progress of the works can be materially affected without any overall delay occurring. It is not therefore necessary to establish overall delay or prolongation in order to found a claim for reimbursement of loss and expense. It could happen that the disturbance to progress is the result of out-of-sequence working brought about by one of the matters referred to in clause 21.2. In this way, for example, the labour force may be less effectively deployed. Certain skilled craftsmen may have to spend longer on a particular task than had been anticipated. The contractor will be involved in additional wage payments. There may be no overall delay to the completion date. Progress, while being disturbed, may not result in delay to completion, e.g. if the activity is not on a critical path.

5-89 As the claiming of loss and expense is tied to the regular progress of the works being materially affected, it is necessary to show that there has been a disturbance of the regular progress. Before there can be disturbance there must be some form of progress so that if any of the matters listed cause the contractor to be unable to commence progress on the works, it just might be argued that this is not covered, e.g. failure of the employer to give access to the site, in such circumstances that the contractor cannot even commence progress of the works. Although there is a certain logic in this argument, it is considered unlikely to succeed before an adjudicator or the courts. This possible uncertainty could have been avoided by including within clause 21.3 a reference to the *commencement* as well as regular progress of the works being materially affected (see for example clause 2.2.1 of the Conditions of the JCT Standard Form of Domestic Sub-Contract 2002 Edition). Further, the employer should appreciate that preventing the contractor from even commencing progress will often be a serious breach of contract so that the employer could well be faced with a substantial claim for damages for breach of contract at common law. If the breach of contract is serious enough, the contractor will also have the right to regard the employer's failure as a repudiation of his obligations under the contract and may then regard the contract as at an end, or terminate his own employment under clause 33 on the basis of the employer's 'Material Breach' (see clauses 33.1 and 39.2).

In a similar way, loss and expense caused by one of the listed matters after progress has apparently ended could cause difficulties to the contractor in seeking to claim under this clause. It may be that the works have been completed and are awaiting a statement of practical completion, but that before such a statement is issued certain tests have to be carried out, e.g. in relation to a heating and ventilation system. If the results of testing are delayed owing to one of the matters listed in clause 21.2, can it be argued that there has been no disturbance of regular progress as there is nothing left for the contractor to do except await the result? Again, the better view is that this argument would not succeed and that progress can be disturbed by any of the listed occurring between the date of access (see clause 9.1 and the appendix entry) and the date of practical completion.

It would be difficult to argue that progress can be affected after practical completion has been achieved, e.g. during the rectification period. If, for example, the employer disrupts the contractor in carrying out remedial works, making it more expensive than it otherwise would be, clause 21 appears not to apply, though the contractor may be able to argue that it is a breach of an implied term not to hinder or prevent the contractor from fulfilling his contractual obligations. The fact that there is a review procedure immediately following practical completion (see clause 21.6), intended to finalise loss and expense claims, supports this interpretation.

The regular progress referred to is that which 'is or is likely to be materially affected'. In comparison, clause 26.1 of WCD 98 refers to regular progress which 'has been and is likely to be materially affected'. There are probably no consequences to this difference in wording.

Causation

It must be one of the matters referred to in clause 21.2 which causes or is likely to cause the regular progress of the works to be materially affected. Causation is always therefore an important practical and legal consideration (see paragraphs 5-73 and 5-100).

Contractor to take all practical steps to reduce loss and expense

5-90 It is probably the case that even without a provision such as this, the contractor is under a duty to take reasonable steps to mitigate the extent of loss and expense to be incurred. Reference has already been made to the duty to mitigate in such circumstances (see paragraph 5-74). There is no equivalent express obligation on the contractor to take all practical steps to reduce the loss or expense in other JCT forms, including WCD 98 and JCT 98. Perhaps it is inserted here to put the matter beyond doubt as the duty to mitigate is relevant to claims for damages for breach of contract, and a claim for reimbursement of loss and expense incurred is not a claim for damages. However, as the courts have treated claims for loss and expense as equivalent for most purposes to claims for damages for breach of contract, including the application of principles of remoteness of damage (see paragraph 5-71), it is inconceivable that the duty to mitigate will not be applicable in the loss and expense context.

Does this requirement impose on the contractor an obligation over and above the equivalent common law duty to mitigate loss? In the loss and expense context the application of the common law duty to mitigate would almost certainly not require the contractor to expend significant resources even if this had the net effect of reducing the loss and expense incurred by more than the contractor's expenditure.

It would therefore require express wording to impose an obligation on the contractor to expend anything more than minimal resources. The last sentence of clause 21.3 does not do this. Nevertheless, it is suggested that the contractor probably does have an obligation to efficiently reschedule or redeploy his resources where these would have the effect of reducing the loss and expense incurred whether directly or by reducing the disruption to regular progress, provided it does not involve significant expenditure.

Contractor to provide assessment and supporting information

> 21.4 The Contractor shall provide to the Employer its assessment of the loss and/or expense incurred or to be incurred as a consequence of any matter notified in accordance with clause 21.3 together with such information as is reasonably necessary to enable the Employer to ascertain the loss and/or expense incurred. Such assessment and information shall be updated at monthly intervals until such time as the Contractor has provided all of the information that is reasonably necessary to allow the whole of the loss and/or expense that has been incurred, to be ascertained.

Assessment

5-91 Clause 21.4 obliges the contractor to provide an assessment of the loss and expense. This extends to loss and expense which has been incurred and also loss and expense which the contractor believes is to be incurred as a consequence of a clause 21.2 matter. There is no time limit stated for the provision of the contractor's assessment. It is of course very much in the contractor's own interest to provide the assessment and supporting information as rapidly as possible.

The contractor's assessment will no doubt include the items referred to earlier as typical heads for a loss and expense claim (see paragraph 5-76 *et seq.*). To the extent that the assessment is of loss and expense which has been incurred it should be relatively accurate, though it may not necessarily be so as the financial consequences of regular progress being affected may not be fully clear until some time after the loss and expense is incurred. This is particularly so where there are overlapping financial effects from different causes. So far as the assessment includes loss and expense to be incurred at some future time, this is clearly likely to be less accurate and is probably required to be provided to the employer for information purposes. It is clear from clause 21.5 that the employer is to ascertain only that loss and expense which has been incurred.

Supporting information

5-92 Where the loss and expense has actually been incurred, the contractor must provide the employer with such information as is reasonably necessary to enable the employer to ascertain such loss and expense. Obviously, there would be little or no information available in connection with loss and expense which is yet to be incurred in terms of evidence justifying the contractor's assessment and the employer's ascertainment.

Unlike, for example, clause 26.1.2 of WCD 98 which only requires the contractor to provide information in support of his application upon request by the employer, the information provided under clause 21.4 is required without the need for any request from the employer.

UPDATING

5-93 Both the assessment and the information are to be updated at monthly intervals until the contractor has provided all reasonably necessary information to enable the employer to ascertain the whole of the loss and expense that has been incurred. The contractor therefore must continue to provide information until the loss and expense caused by the clause 21.2 matter has ceased to change.

5-94 An initial reading of clause 21.4, and particularly the last sentence, might appear to suggest that the employer is not required to carry out any ascertainment until all the reasonably necessary information required to permit an ascertainment of 'the whole of the loss and/or expense' incurred has been provided by the contractor. If this were the case it would remove any obligation upon the employer to carry out any interim ascertainment and the contractor would have to wait until any loss and expense from any specific event had ceased to change before any ascertainment took place. This would be fundamentally unreasonable and would not be an appropriate way to construe the clause unless it was absolutely clear that it had no other meaning. Reading clause 21.4 in conjunction with clause 21.5 (particularly the words 'Upon receipt of *any* information') and clause 21.6 (review process), it is tolerably clear that the employer is required to carry out interim ascertainments for which the employer will then be liable to pay by including the sum in the next available payment advice. However, the drafting should have put this beyond argument.

Employer's obligation to ascertain

> 21.5 Upon receipt of any information referred to by clause 21.4 regarding loss and/or expense that has been incurred the Employer shall within 14 days notify the Contractor of its ascertainment of the loss and/or expense incurred, that ascertainment being made by reference to the information provided by the Contractor and being in sufficient detail to permit the Contractor to identify any differences between it and the Contractor's assessment of the loss and/or expense incurred.

5-95 Once the employer has received the contractor's assessment together with the information reasonably necessary to enable the employer to ascertain the loss and expense incurred, the employer has 14 days in which to carry out the ascertainment and notify the contractor of the ascertained loss and expense.

5-96 The ascertainment is to be made by reference to the information provided by the contractor. It might have been more appropriate for the ascertainment to be made by reference to the assessment rather than by reference to the information provided by the contractor. While the employer should of course have regard to the information provided by the contractor, to carry out the ascertainment by reference only to that might suggest that the employer could not take into account any information from any other source which he might happen to possess and which is relevant to the ascertainment. It is clearly appropriate for the ascertainment to be made by reference to the contractor's assessment so that the contractor can clearly see where any differences between the assessment and the ascertainment may lie. Indeed this is clearly intended by clause 21.5 which requires the ascertainment to be in sufficient detail for the contractor to identify any such differences. (See also the comments on the similar clause dealing with the valuation of variations, clause 20.8 – paragraph 5-63.)

If the contractor is unhappy with the employer's ascertainment, he can of course refer it to adjudication at any time or alternatively await the outcome of the review procedure under clause 21.6 provided he intends to supply further documentation in support of any further ascertainment. If he does decide to refer it to adjudication, the result of that adjudication will not affect the contractor's right to seek a review under clause 21.6.

What does ascertain mean?

5-97 The word 'ascertain' may be regarded as an unfortunate choice. Dictionary definitions refer to, for example, 'finding out for certain'. In the case of *Alfred McAlpine Homes North Ltd* v. *Property & Land Contractors Ltd* (1995) Judge Humphrey Lloyd QC said:

> 'Furthermore "to ascertain" means to "find out for certain" and it does not therefore connote as much use of judgment or the formation of an opinion as had "assess" or "evaluate" been used. It thus appears to preclude making general assessments as have at times to be done in quantifying damages recoverable for breach of contract.'

Such an approach has led some employers, and particularly auditors on their behalf, to take the view that if they cannot know for certain that every penny of the contractor's claim for loss and expense has been suffered or incurred, then nothing is due. In other words, if the contractor cannot prove every penny of his claim, he is entitled to nothing as it is not possible under the clause to make estimates. This is taking the matter too far. In *How Engineering Services Ltd* v. *Lindner Ceilings Floors Partitions Plc* (1999), Mr Justice Dyson dealing with the meaning of 'ascertainment' when applied to the recovery of loss and expense under JCT standard forms of building contract, referred to the earlier case of *Alfred McAlpine Homes North Ltd* v. *Property & Land Contractors Ltd* (1995) and the judgment of Judge Humphrey Lloyd QC and said:

> 'I do not understand Judge Lloyd to be saying that there is no room for the exercise of judgment in the process of ascertainment. I respectfully suggest that the phrase "find out for certain" might be misunderstood as implying that what is required is absolute certainty. The arbitrator is required to apply the civil standard of proof.
>
> In my view it is unhelpful to distinguish between the degree of judgment permissible in an ascertainment of loss from that which may properly be brought to bear in an assessment of damages. A judge or arbitrator who assesses damages for breach of contract will endeavour to calculate a figure as precisely as it is possible to do on the material before him or her. In some cases, the facts are clear, and there is only one possible answer. In others, the facts are less clear, and different tribunals would reach different conclusions. In such cases, there is more scope for the exercise of judgment. The result is always uncertain until the damages have been assessed. But once the damages have been assessed, the figure becomes certain: it has been ascertained. In my view, precisely the same

situation applies to an arbitrator who is engaged on the task of "ascertaining" loss or expense under one of the standard forms of building contract.'

It is likely that the more liberal approach reflected in the judgment of Mr Justice Dyson will prevail.

Review mechanism

21.6 No later than 42 days after Practical Completion of the Project the Contractor shall provide documentation in support of any further ascertainment it considers should be made in respect of any matter notified in accordance with clause 21.3. Within 42 days of receipt of such documentation the Employer shall undertake a review of its previous ascertainment in respect of each matter for which further documentation has now been provided and notify the Contractor of any further ascertainment that it considers appropriate.

5-98 Even though the ascertainment carried out by the employer under clause 21.5 is in respect of loss and expense which has been incurred at the date of the ascertainment, and even though the monthly updates under clause 21.4 are intended to enable the employer to ascertain 'the whole' of the loss and expense incurred, it is still sensible to have a review procedure under which the contractor is given the opportunity to seek a further ascertainment subject to providing documentation in support of it. From the employer's point of view it provides a cut-off point after which time the contractor cannot introduce fresh evidence in support of existing notifications.

The contractor must provide any further documentation not later than 42 days after practical completion of the project. Within 42 days of the receipt of that documentation the employer must undertake a review of his previous ascertainment in respect of each matter for which further documentation is provided, and also notify the contractor of any further ascertainment within that time.

The review mechanism only applies to notifications already made pursuant to clause 21.3. The review period should not therefore be seen as an opportunity for the contractor to make new notifications. Having said that, looking at the timescales it appears that, in theory at least, the contractor could give a first notification under clause 21.3 (provided it was as soon as he became aware that it was appropriate to do so) as late as up to 27 days after practical completion provided the notification was accompanied by the contractor's assessment and the reasonably necessary information required under clause 21.4. The employer would then have 14 days under clause 21.5 to carry out the ascertainment, leaving the contractor with 1 day in which to seek a review by providing further documentation.

If the contractor misses the 42-day deadline, it may well not be possible to seek an adjudication of, or to litigate, the original ascertainment if the contractor's case is dependent on the provision of further particulars not available to the employer at the date of the original ascertainment. There would be no 'dispute' as to the original valuation based on the information then provided. This issue has been discussed in relation to the similar provision dealing with the review of change valuations under clause 20.9 (see paragraph 5-65). It would be otherwise if the contractor was in dispute over the original valuation based on the details and information already supplied under clause 21.4. The contractor's failure to meet the 42-day deadline could therefore have very serious consequences.

Payment by employer following ascertainment

21.7 The Employer shall be liable to pay the Contractor any loss and/or expense that has been ascertained in accordance with clause 21.

5-99 Once the employer has carried out the original ascertainment under clause 21.5 or a revised ascertainment under clause 21.6, he becomes liable to pay the contractor the loss and expense ascertained. This sum should be included in the next payment advice and becomes due to the contractor on receipt by the employer of a VAT invoice for that amount. Payment is considered in more detail when dealing with that topic under clause 22.

Loss and expense ascertainment to exclude any element contributed to by a cause other than a change or some other loss and expense reimbursable matter under clause 21.2

21.8 No ascertainment of loss and/or expense under clause 21 shall include any element of loss and/or expense if that element was contributed to by a cause other than a Change or a matter set out in clause 21.2. Any loss and/or expense incurred as a consequence of a Change is to be included in a valuation made under clause 20 (*Changes*).

5-100 The purpose of this exclusion appears to be to keep out of the calculation of loss and expense any element operating concurrently with or overlapping with a matter under clause 21.2, and which is either attributable to the contractor's own default or to some neutral event for which the contractor is to take the risk so far as loss and expense is concerned. This is so even if, under clause 12, the contractor is entitled to an adjustment of the completion date e.g. *force majeure* or the occurrence of a specified peril or the exercise after the base date by the UK Government of any statutory power directly affecting the execution of the project etc. In other words, one operating cause may be the consequence of a clause 21.2 matter and another equally effective cause may be a matter which is neither a change nor another matter listed under clause 21.2. In such a situation the exclusion seeks to make it clear that the contractor is not to be entitled to reimbursement of loss and expense except to the extent that he is able to isolate an element of loss and expense as being solely attributable to consequences of a clause 21.2 matter or a change. This provision is not without its difficulties (see the similar provision in clause 20.6.4 relating to loss and expense flowing from a change and the discussion of it – paragraph 5-60; and a related issue in clause 12.7.3 relating to extension of time – see paragraph 3-103).

The final sentence of clause 21.8 should be read in conjunction with clause 21.1 and seems to have been inserted for the avoidance of any doubt, to make it clear that loss and expense associated with changes is to be included in the valuation under clause 20.

Payments – clause 22

Background

5-101 Clause 22 sets out the payment mechanism for the MPF. In practice it will of course be directly linked to the pricing document which contains the method for determining the manner in which the contractor is to receive payments in respect of the contract sum. It will also contain some form of contract sum analysis and other pricing information. Clause 22 and the pricing document envisage periodic payments which can be based on interim valuations, stage payments, progress payments or some other method attached to the pricing document. The pricing document also covers the possibility of advance payments with a supporting bond from the contractor.

5-102 The pricing document is considered in more detail at the end of this chapter (see paragraph 5-179 *et seq.*).

The MPF payment provisions are much shorter and simpler than in most other JCT contracts, including WCD 98 (see clause 30 which has about 5000 words compared with less than 900 words in the MPF covering the same ground in clauses 22, 23 and 24). The principal reason for this is that the MPF leaves much of the mechanics for calculating what is due to be dealt with in the pricing document and its attachments, e.g. in relation to stage payments or interim valuations. Other factors are the absence of any express retention provisions; the absence of any express provisions for payment for off-site goods and materials; and the provision for advance payments being located in the pricing document.

Before considering clause 22 further, it is appropriate to briefly set out the legal position in relation to payment of the contract sum and related matters.

Entire and severable contracts

5-103 In many everyday contracts, payment by one party is due only on the complete fulfilment by the other party of its contractual obligations. It may involve a contract for the supply of materials or goods or the performance of a service. Such contracts are known as entire contracts. A classic example of this is to be found in the very old case of *Cutter* v. *Powell* (1795).

FACTS

A sailor agreed the following terms in his employment contract:

> 'Ten days after the ship *Governor Pary* ... arrives at Liverpool I promise to pay to Mr T. Cutter the sum of 30 guineas, provided he proceeds, continues and does his duty as Second Mate in the said ship from here to the Port of Liverpool ...'

The sailor died before completion of the voyage and his personal representatives sought to recover a proportionate part of the agreed remuneration. It was held that they could not do so as this was an entire contract so that the sailor had to continue carrying out his duties until the ship arrived at Liverpool, and failure to do this, even though as a result of his death, disentitled his personal representatives to any part of the remuneration.

5-104 On the other hand, some contracts, by their terms or by their nature, permit final payment against an interim valuation of work done or on the completion of a stage of the work having been reached. These are known as severable contracts.

On a strict application of the law relating to entire contracts, a contractor carrying out work on the employer's land, which he fails to complete, will not be entitled to payment. This could of course provide an unexpected and possibly unwarranted benefit to the employer. The rigours of the operation of this principle are qualified by the law in a number of ways. Firstly, if in such a case the employer sues the contractor for breach of contract to recover the increased cost of having the work completed, he will, in the assessment of damages, have to give some credit against what he would have had to pay had the contract been properly performed. Secondly, the doctrine of substantial performance may aid the contractor. The essence of this doctrine is that, provided the contractor has substantially performed his obligations, he will be permitted to sue for the price, giving credit for the outstanding work left incomplete.

Most standard forms of building contract will contain express provisions for payment by instalments, though this does not, of itself, prevent them from being entire contracts as the provision for payment by instalments will usually be treated as being for payment on account of a final sum. The fundamental principle in relation to entire contracts, i.e. payment in full on completion in full, can still apply to the last instalment or the release of the final balance. However, if the contract expressly envisages that the contractor may be entitled to some payment notwithstanding his failure to complete, then this will mean that the contract in question is not an entire contract but is severable. See *Tern Construction Group Ltd (in Administrative Receivership)* v. *R. B. S. Garages Ltd* (1992), a case on JCT 80 (containing certain termination provisions not unlike those in the MPF clauses 32.4 to 32.8) in which Judge John Newey QC held that the contract was not an entire contract on the basis that it provided for the possibility that the contractor would be paid or credited with the value of work done even where he failed to complete the works in total. He said:

> 'In my judgment the contract made between the parties using the JCT Standard Form, with its elaborate and detailed provisions dealing with many matters, but most importantly employers going into partial possession, determination of the contractors' employment without determination of the contract and payment by instalments, was not simply a contract for the contractors to perform all or nearly all their obligations before the employers performed any of theirs, which can usefully be described as "entire".'

Even, however, in a contract which is not an entire contract, this need not mean that instalment payments are treated as several in nature. In other words, in JCT contracts (including in all probability the MPF, unless the detailed method of calculating payments under the pricing option selected in the pricing document provides otherwise) and similar contracts, the principle that instalment payments are treated as payments on account of a final sum which is finally adjusted by, in effect, a valuation of the whole of the works at completion, remains intact. A consequence of this is that instalment payments for work done or materials supplied create no estoppel against the employer if the work or materials are discovered subsequently to be defective. A later valuation can take this decrease in value into account. The mechanism by which interim instalments are paid is generally through the issue of interim payment certificates or advices.

Implied periodic payments

5-105 In contracts for work and materials which involve work being carried out over a significant period of time, the courts will readily imply a term, in the absence of an express term, that the presumed intention of the parties is for interim payments to be made – see for example the first instance decision in *Williams* v. *Roffey Bros &*

Nicholls (Contractors) Ltd (1989). The need for the implication of such a term in construction contracts (as defined) has now been largely removed by section 109 of the Housing Grants, Construction and Regeneration Act 1996 which provides that a party to a construction contract is entitled to payment by instalments unless the contract duration is of less than 45 days. If the parties have not in their contract stated the amount of, or the intervals at which, or circumstances in which, interim payments become due, the Scheme for Construction Contracts (England and Wales) Regulations 1998 will apply and part II of the Regulations provides in paragraphs 2 to 4 inclusive for payments based on 28-day cycles reflecting the value of work carried out including the value of materials on site.

All relevant JCT contracts, including the MPF, make express provision for payment by instalments so that these provisions in the statutory scheme will have no application to such contracts.

Where the contract expressly provides for payment by instalments, provided any conditions required to be fulfilled before an instalment becomes due have been met, a debt is created. This is likely to be so even if the relevant payment clause does not expressly state that the obligation to pay amounts to a debt: see *Re: Clemence Plc* (1992). If the contract requires a certificate to be issued in respect of an instalment this may be a condition precedent to the debt coming into existence. Under the MPF, the requirement for the contractor to provide a VAT invoice to the employer before a payment becomes due has the same effect (see clause 22.2).

The interim payment process

5-106 Most standard forms of building contract which do not provide for interim certificates to be issued by a named third party certifier such as an architect or contract administrator, allow for interim payments by means of a payment advice from the employer. This may relate to a stage which is reached in the work, in which case there is little or no act of valuation required. On the other hand, many contracts provide for interim payments against valuations of work carried out, and often (though not in the case of the MPF unless reflected in the value of stage payments or progress payments) goods and materials delivered to the site and sometimes those delivered off site but intended for incorporation into the works.

The contract sum

5-107 The contract sum is defined by clause 39.2 of the MPF as: 'The amount stated in the Appendix'.

It is exclusive of value added tax (see clause 25.1). The MPF is a lump sum contract. It is not adjustable in the way that a measure and value contract would be. Any additional sums which may become payable, e.g. in respect of changes or reimbursement of loss and expense, are dealt with separately. It is a lump sum contract in the sense that it is a contract to complete the whole work for a lump sum. The contractor must carry out and complete the project in accordance with the contract documents (see clause 1.1). While the contractor is entitled to interim payments in accordance with the conditions, these payments are on account of the finally adjusted contract sum. An interim payment is not therefore a final payment in respect of the work to which its value relates.

5-108 Generally, the parties will be bound by any errors or omissions incorporated in the contract sum unless sufficient grounds exist to persuade a court to grant the equitable remedy of rectification. This remedy is discretionary and the courts will have to be satisfied that the written contract fails in some way to express what was the clear intention of both parties. It does this firstly in order that the written document properly reflects the agreement actually made between the parties, and secondly to prevent a party unfairly holding the other to a written contract which he knows does not accurately reflect the agreement reached. Cases on the rectification of building contracts are rare but one is *A. Roberts & Co Ltd* v. *Leicestershire County Council* (1961).

FACTS

The contractor's revised tender contained a completion period of 18 months. The county council decided that the period for completion should be 30 months, that is, the same date but 1 year later than the date put forward by the contractor. The county council did not refer to any date in its letter of acceptance. Instead, the formal contract, when drawn up, contained a completion date 1 year later than that put forward by the contractor. The contractor did not notice the change of year and sealed and returned the contract. Before the county council itself sealed the formal contract, it held a meeting with the contractor during which the contractor referred to his plans to complete in 18 months. The county council's officers did not mention the later date inserted in the formal contract. The county council subsequently sealed the formal contract.

HELD

Rectification would be ordered. A contracting party is entitled to rectification of a contract if he can prove that he believed a particular term to be included, and the other party concluded the contract without that term being included, in the knowledge that the first party believed that it was included.

5-109 It is important to distinguish the contract sum from the contractor's tender sum. The sum put forward by the contractor in his tender does not become the contract sum until the tender is accepted. Depending on the tendering procedures adopted, the tender sum may be adjusted by agreement between the parties, e.g. on the discovery of an error in the make-up of the tender sum. Alternatively, the contractor may be asked to elect to maintain the tender sum notwithstanding the error or to withdraw his tender.

Retention

5-110 Most standard forms of building contract which provide for payment by instalments will enable the employer to retain a certain percentage of the total value included in the interim certificate. The MPF does not provide for retention, though a similar result may in practice be achieved in the value placed on any stage or progress payment. From the employer's point of view it is seen as a useful system as it represents some protection against the inclusion of defective work in a valuation, which is reflected in the amount of an interim payment advice. It also provides security for the performance by the contractor of his obligations. Its main purpose, however, is to provide the employer with a fund during any defects liability or rectification period following practical or substantial completion, should the contractor fail to return and make good any defects of which he is notified.

In eschewing any formal retention provision, the MPF has departed from all other significant JCT and other standard forms in the industry. (Even that most modern and liberal of contracts, the Engineering and Construction Contract produced by the Institution of Civil Engineers has an optional retention clause (Option P)). During the course of a project the employer generally obtains some financial protection and security by reason of the fact that at any given time the contractor will have carried out work and added value to the project for which he will not yet have been paid. The position after practical completion is more of a problem for the employer, as the contractor should have been fully paid for the work carried out and there may be little or nothing left by way of an incentive for the contractor to return to rectify defects or by way of security for the employer if the contractor fails to do so.

Deductions from payments: set-off and counterclaim

AUTHORISED DEDUCTIONS

5-111 If a building contract provides for payments against interim certificates it will also usually entitle the employer to make certain deductions, generally in relation to specific ascertained amounts. The most important such entitlement will be the employer's right to deduct liquidated damages. However, other deductions may also be authorised, e.g. recovery of insurance premiums paid by the employer where the contractor has failed to take out any necessary insurance required by the contract; and the cost of employing another contractor to carry out work which the contractor has failed to do despite having received a valid instruction. This system of allowing a deduction from a certified sum, rather than reflecting it in the calculation of the sum itself, is understandable where the valuation and certification are in the hands of an independent certifier named in the contract, as the employer may or may not wish to make the deduction, e.g. of liquidated damages. Where, however, both valuation and payment are directly in the hands of the employer, there is no reason for not bringing such matters into the calculation by the employer of the sum which is to become due to the contractor. This is what the MPF does in relation to the examples given above (see clauses 22.5.4 and 22.6.2, dealt with later – paragraphs 5-148 and 5-152). From an employer's perspective, this is preferable to the approach in WCD 98 which treats such matters as deductible from sums due rather than being taken into account in determining the sums due. Taking account of such matters in calculating the amount which is to become due to the contractor has the added advantage of not requiring the service of a withholding notice pursuant to section 111 of the Housing Grants, Construction and Regeneration Act 1996. This is dealt with in detail later (see paragraph 5-163 *et seq*.).

5-112 If there is to be a withholding from sums due, e.g. discovery of defective work after a payment advice has been issued, including its value on the basis that it had complied with the contract, prior notice will be required pursuant to section 111 of the Housing Grants, Construction and Regeneration Act 1996. The notice must be given not later than the 'prescribed period' before the final date for payment. The contract may provide for such a period (in the case of the MPF it is not later than 7 days before the final date for payment of the sum from which the withholding is to take place – clause 23.2). If the contract provides no period then by virtue of the Scheme for Construction Contracts (England and Wales) Regulations 1998 (Part II) paragraph 10, the prescribed period is also not later than 7 days before the final date for payment under the contract. The notice must also set out the amount proposed to be withheld and the grounds for withholding it. This is dealt with in detail later when considering clause 23 (see paragraph 5-163 *et seq.*).

SET-OFF AND COUNTERCLAIM

5-113 Some cross-claims may be both counterclaims and set-offs. All set-offs can be counterclaims but not all counterclaims can be set-offs. A set-off can therefore be the subject of an independent cross-claim but may also be available to reduce the claimant's claim in its own right. Such set-offs when used in this way are available 'as a shield, not as a sword' (Cockburn CJ in *Stooke* v. *Taylor* (1880). They are in the nature of a defence to a claim so that they cannot overtop the claim. Any excess would have to be the subject of a separate counterclaim. Any such set-off can properly be regarded as a defence against liability rather than just a means of reducing the quantum of a claim.

It is often difficult to determine whether a cross-claim is so closely connected to the issue of liability on the claim that it is tantamount to a defence or partial defence to it. An authoritative and fairly modern case dealing with this issue is that of *Hanak* v. *Green* (1958).

FACTS

The plaintiff Bozena Hanak was a widow who bought a house from Green who was a builder. Green agreed to carry out certain work to the house. Mrs Hanak was dissatisfied with the work and sued for damages for breach of contract. Her claim was put in the sum of £266. The builder counterclaimed in respect of extra work and damages caused by Mrs Hanak's refusal to allow a workman into the house. There was also a small claim alleging trespass against Mrs Hanak in respect of Green's tools.

The County Court judge gave judgment for Mrs Hanak in the sum of £75 and for Green under the counterclaim in the total sum of £85. The judge then came to consider the question of costs. He awarded Mrs Hanak her costs on the claim as she had won her claim to the extent of £75. He then awarded costs to Green on the counterclaim as he had won £85 on the counterclaim.

Green was dissatisfied. He said that in net terms he was the winner by £10 and Mrs Hanak had got nothing. He said that on the question of costs his counterclaim should be treated as a right of set-off and usable as a defence as well as a counterclaim so that he should receive costs but not have to pay any.

In order to determine the proper basis upon which costs should be ordered, it was necessary for the court to consider the question of whether the counterclaim could also be used as a set-off which could be used as a defence. So for the sake of a net judgment worth £10, the Court of Appeal delved into the history of set-off. As Lord Justice Morris (at 1 BLR 4) put it:

> 'So it has come about that we have heard a learned debate, rich in academic interest ... on the subject as to whether certain claims could be proudly marshalled as set-off or could only be modestly deployed as counterclaim.'

Lord Sellers in the same case said (at 1 BLR 13):

> 'Some counterclaims might be quite incompatible with the plaintiff's claim, in no way connected with it and wholly unsuitable to be used as a set-off, but the

present class of action involving building or repairs, extras and incidental work so often leads to cross-claims for bad or unfinished work, delay or other breaches of contract that a set-off would normally prove just and convenient.'

HELD

The court held that in truth the counterclaim was a proper set-off which could be raised as a defence. Accordingly, Green obtained first his costs of defending Mrs Hanak's claim and secondly, the costs of his successful counterclaim.

5-114 Originally there was no common law right at all to set off a counterclaim. However, statute provided such a right in respect of mutual debt provisions on bankruptcy and liquidation.

5-115 Further, by procedural innovations, a defendant was allowed to raise a counterclaim in the same proceedings. Previously it was necessary to mount a separate action. This right is a statutory right and was originally introduced by the Supreme Court of Judicature Act 1873, section 24(3).

In equity the court could, as a matter of discretion, either:

- Restrain the execution of a judgment on a claim pending the outcome of a counterclaim, or;
- If the counterclaim was particularly closely related to the claim so as to justify it being used as a defence, it could permit an equitable set-off to stop judgment being given at all if the set-off was valid.

For example, in the case of *Morgan & Sons Ltd* v. *S. Martin & Johnson Co Ltd* (1949) the plaintiff stored vehicles for the defendant and subsequently claimed the agreed rent for so doing. The defendant admitted that the rent was due but for the fact that one of the vehicles had been stolen. The defendant counterclaimed in respect of the value of the vehicle. It was held that this should be permitted as a set-off and as a defence.

5-116 In summary therefore, the position is as follows:

- There can be a set-off of mutual liquidated debts.
- There can be a set-off of mutual debts in bankruptcy or liquidation situations. The current statutory provisions enabling the set-off of mutual debts are to be found in the case of individuals in section 323 of the Insolvency Act 1986, and in the case of companies, rule 4.90 of the Insolvency Rules 1986.
- In certain cases a setting-up of matters of complaint, which if established reduce or even extinguish the claim, may be raised by way of a defence in that they affect the value of the plaintiff's claim. In many cases this will amount in law to an abatement which is a pure defence and does not need the assistance of the law of set-off (*Mondel* v. *Steel* (1841)).
- There can be equitable set-off which can be used as a matter of defence.

5-117 The mere establishment of the existence of some form of counterclaim is insufficient. It must be so closely connected with the plaintiff's demand that it would be manifestly unjust to allow the plaintiff to enforce payment without taking into account the counterclaim. In the Court of Appeal case of *Dole Dried Fruit & Nut Company* v. *Trustin Kerwood Ltd* (1990) Lord Justice Lloyd made the point that:

> 'It may even be insufficient that claim and cross-claim arise out of the same contract or transaction, unless they are so inseparably connected that the one ought not to be enforced without taking account of the other.'

The key therefore is whether the matter raised as an equitable set-off so that it can be utilised as a defence, is so inseparably connected with the claim being made that the claim ought not to be enforced without taking account of the matter sought to be set off. The words of Lord Sellers in *Hanak* v. *Green* (1958) (see paragraph 5-113) would seem to suggest that in construction cases it is not too difficult to establish cross-claims as being set-offs of this type.

5-118 In procedural terms at least, the manner in which a set-off can be utilised as a defence is very broadly stated in the Civil Procedure Rules. Rule 16.6 provides:

'Where a defendant:
(a) contends he is entitled to money from the claimant; and
(b) relies on this as a defence to the whole or part of his claim, then
the counterclaim may be included in the defence and set off against the claim.'

However, as this is a procedural mechanism, there is some question over whether all such counterclaims are in truth set-offs in the nature of a defence against liability.

In conclusion, a set-off can be an equitable set-off and used in defence of liability if it is sufficiently closely connected with the claim as to make it unjust to allow judgment on the claim without taking the cross-claim into account. It has been held by the Court of Appeal that a cross-claim alleging delay, made by a contractor against a sub-contractor claiming the value of work carried out, was assumed to be capable of amounting to an equitable set-off (see *Mellowes Archital Ltd* v. *Bell Projects Ltd* (1997).

5-119 Much will depend upon the particular facts of any claim and the terms of the contract under which it is made. The fact that in many construction contracts, financial matters are for the most part intended to be settled by the issue of the final certificate or final payment advice, makes it more open to argument that a claim by an employer alleging that work has been overvalued because it is defective and that accordingly the final account should take this into account, could be marshalled as a set-off and defence in an action by the contractor for unpaid sums due under the contract. Certainly such a set-off must be an equitable set-off capable of being utilised as a defence if the contractor is seeking payment for other work properly carried out. Going further, the contractor may seek to set off an outstanding claim for loss and expense, which is not of course directly connected with the issue of defective work, but is nevertheless very closely connected with the proper calculation of the sum to be included in the final account. As such it is arguably an equitable set-off. The position could be different if the employer's claim is met by the contractor claiming, for example, damages for breach of contract based on, say, the employer's failure to indemnify the contractor under express indemnity provisions in the contract which do not form part of the calculation of the contract sum. The two issues would then be less closely linked.

Interestingly enough, however, under the MPF clause 22.7, an attempt is made to ensure that claims for an indemnity and even for breach of statutory duty or a duty of care in negligence are settled by the final payment advice provisions (see a detailed discussion of this later – paragraph 5-158). So, as a cross-claim by the contractor against the employer for an indemnity under the MPF clause 26.2 is intended to be included in the final payment advice (see clause 22.6.2 and clause 22.5.4), any such cross-claim is likely to be more arguably a set-off (if raised in time to defeat the conclusive effect of the final payment advice), than had it not formed part of the final payment advice valuation process. In the MPF, far greater use is made in bringing all financial matters into the calculation of the contract sum compared with other JCT contracts. This could well extend the area of set-off considerably. This was probably unintended.

Summary of clause 22

5-120 The payment provisions of clause 22 can be summarised as follows:

- The appendix to the contract provides for the dates when payment is to be made. This will generally be monthly.
- Not later than 7 days before each appendix date or whenever else the contractor considers a payment advice should be issued (e.g. after practical completion), the contractor must submit an application for payment.
- Before practical completion and whether or not the contractor has made an application, the employer must issue an interim payment advice on the dates required by the appendix, setting out the amount due from either party to the other.
- After practical completion the employer must issue further interim payment advices at intervals of not less than one month, but the employer is not obliged to do so if the amount is less than the minimum amount stated in the appendix.
- Any payment advice must set out the payment proposed to be made.
- The amount identified in any payment advice in favour of the contractor becomes due on receipt by the employer of a VAT invoice from the contractor. If the payment advice shows an amount in favour of the employer, this becomes due upon its issue.
- The final date for payment of any sum due is 14 days after that amount became due for payment (see previous point).
- An interim payment advice must state:
  - The proportion of the contract sum payable;
  - The value of any changes executed by the contractor;
  - The amount of any reductions in respect of:
    - Work not carried out in accordance with approved design documents
    - Services provided by pre-appointed consultants or named specialists where the required agreement with them has not been entered into by the contractor
    - Services omitted from agreements with pre-appointed consultants or named specialists and carried out by the contractor himself or others on his behalf without the prior consent of the employer;
  - Any amounts that either party is liable to pay the other under the contract.
- Previous payments are deducted to determine the amount payable.
- After the statement of rectification of defects is issued (or, at the employer's option, on the contractor's failure to rectify the defects) the employer is to issue a final payment advice to include:
  - The contract sum;
  - The matters set out above in respect of interim payment advices;
  - Appropriate deductions in respect of the contractor's failure to rectify defects.

- Previous payments are deducted to determine the final payment due.
- If the contractor disputes the final payment statement he must refer the dispute to an adjudicator or to litigation within 28 days of its issue.
- To the extent that the final payment advice is not so disputed, it is final and binding on the parties regarding amounts due from the employer to the contractor under or in connection with the contract and is inclusive of sums due to the contractor in respect of any breach of contract, breach of statutory duty, negligence or otherwise.
- There is a general provision that if the employer has notified the contractor that he is a 'contractor' for the purposes of the CIS, and the contractor has not provided vouchers in accordance with the CIS requirements for any payments received, then the employer is not obliged to make any further payments until the failure is remedied.

The provisions of clause 22 will now be considered in more detail.

The contractor's application

22.1 No later than 7 days before any date when the Contractor considers an interim payment advice should be issued by the Employer it shall submit a detailed application for payment to the Employer setting out the amounts it considers should be included within a payment advice and the amount that it considers due.

Requirement for an application

5-121 The contractor is obliged to make an application if he considers that an interim payment advice should be issued by the employer, However, the employer's obligation under clause 22.2 to issue an interim payment advice is not expressed to be dependent upon the contractor's application at all (c.f. clause 30.3.3 of WCD 98). The contractor's failure to make an application would be a breach of contract but it is difficult to see that any real loss would be caused to the employer unless he can argue that his own professionals, for example the quantity surveyor, have more work to do in producing the calculations etc. for an interim payment advice in the absence of the contractor's application, with the possibility that the employer will incur increased professional fees. The contractor would no doubt also suffer the risk of receiving less by way of the interim payment advice in the absence of an application than had an application been furnished with the appropriate details. In practice, contractors will invariably submit an application in order to influence the contents of the payment advice.

Timing of the application

5-122 The contractor is required to make an application for payment no later than 7 days before either the interim payment date referred to in or derived from the appendix, being monthly, or whenever else the contractor considers a payment advice should be issued. This would include, for example, the period following practical completion.

Details of the application

5-123 The application must be detailed and must set out the amounts which the contractor considers should be included in the payment advice as well as the amount it considers to be due.

The amount of detail required will depend to an extent on the pricing document and the work involved in determining the proportion of the contract sum which is payable having regard to the method of interim payments, i.e. stage, valuation or progress payments (see paragraphs P3, P4 and P5 of the pricing document – paragraph 5-179).

The application will generally need to have regard to all those matters set out in clause 22.5 which the employer must state in the payment advice. This will include in addition to the appropriate proportion of the contract sum, the value of any changes executed by the contractor; the amount of any reductions made as a consequence of the provisions of clauses 6.7, 18.5 or 18.6 (see commentary to clause 22.5 – paragraph 5-134 *et seq.*); and the amounts that either party is liable to pay the other in accordance with the provisions of the contract (see paragraph 5-148 for a list of these). The details in the application should therefore include amounts which the contractor considers due from the employer to him, and from him to the employer. It should then state the amount, i.e. the net figure, which the contractor considers due. The amount considered due will almost always, of course, be a sum due from the employer to the contractor. Nevertheless, it is possible in certain situations that the sum could be due in the other direction, e.g. a revision of the appropriate proportion of the contract sum following discovery of a significant amount of defective work. Strictly speaking, even if the contractor's detailed application would result in an amount due from the contractor to the employer, the contractor is still obliged to provide one, as a payment advice should still be issued in that situation.

Employer's interim payment advice

22.2 Prior to Practical Completion of the Project the Employer shall issue to the Contractor interim payment advices setting out the amount due from one party to the other on the dates stated in the Appendix. After Practical Completion of the Project the Employer shall issue further interim payment advices at intervals of not less than one month but shall not be obliged to issue a payment advice where the amount identified as due to either party would be less than the amount stated in the Appendix.

5-124 As already stated above, it should be noted that the employer's obligation to issue an interim payment advice to the contractor is not made dependent upon the contractor having made an application for payment under clause 22.1. If the payment provisions following practical completion are to be workable, it may be that in this period the contractor's application triggers an obligation to issue a further interim payment advice (see paragraph 5-127).

Status of interim payment advice

5-125 Of crucial importance to the parties are two dates: firstly, the date on which a payment becomes due for the purposes of section 110(1)(a) of the Housing Grants,

Construction and Regeneration Act 1996, by virtue of which all construction contracts must provide an 'adequate mechanism' for determining what payments become due under the contract and when they become due; and secondly, the final date for payment for the purposes of section 110(1)(b) which requires that every construction contract shall provide for a final date for payment in relation to any sum which becomes due. Where the amount identified as due is due from the employer to the contractor, the sum only becomes due for payment upon receipt of a VAT invoice for that amount sent from the contractor to the employer (see clause 22.4). Where, however, exceptionally the payment advice identifies amounts due from the contractor to the employer, the due date is the issue of the payment advice (see clause 22.4). In both cases the final date for payment is 14 days after the amount to be paid becomes due.

It is important to note therefore that it is not the issue or receipt of the contractor's application for payment which creates the due date. In this respect the MPF is different to WCD 98 which under clauses 30.3.1 to 30.3.3 inclusive provides that the receipt of the contractor's application for payment creates the due date for the purposes of section 110(1) of the Act (e.g. *Watkin Jones & Son Ltd* v. *Lidl UK GmbH* (2002)).

More is said in relation to the due date and final payment date when considering clause 22.4 (see paragraph 5-133).

Even if the obligation on the employer to issue an interim payment advice is not subject to the contractor having submitted an application, nevertheless it might have been expected that there would be a link between the two, and in particular that the employer's interim payment advice, where it differed from the contractor's application, should provide details to enable the contractor to appreciate any differences between the two.

Timing of employer's interim payment advice

PRIOR TO PRACTICAL COMPLETION OF THE PROJECT

5-126 Up until practical completion of the project, the employer is to issue an interim payment advice to the contractor on the dates stated in the appendix. The appendix (under its reference to clause 22) provides that the employer is to issue an interim payment advice each month. There is a space in the appendix for the insertion of a date which will then apply in respect of each month until practical completion. If no date is stated, the appendix entry provides that the interim payment advice is to be issued on the 28th of each month. If a date is chosen and inserted, care needs to be taken not to, for example, insert 31st as some months do not have 31 days. Whether in such a situation the fallback provision of 28th of each month would apply or alternatively, if not, that the payment mechanism would be held to be inadequate with the result that the Scheme for Construction Contracts would apply, is perhaps debatable, although it is more likely that the latter result would follow. If it is intended, therefore, for the date of issue of interim payment advices to be the last day of each month, then insertion of the words 'last day' should suffice.

POSITION AFTER PRACTICAL COMPLETION OF THE PROJECT

5-127 After practical completion, the employer is required to issue further interim payment advices at intervals of not less than one month provided that the amount identified as due is equal to or above the minimum figure stated in the appendix (see the second paragraph under the appendix reference to clause 22). If no amount is inserted in the appendix, the appendix provides that the amount is to be £10 000. The reference to the employer having an obligation to issue further interim payment advices 'at intervals of not less than one month' is lacking in precision. One construction of this requirement would be that the further interim payment advices can be issued at any interval which the employer wishes, of one month or more. In other words, on this construction the employer could take six months to issue the next interim payment advice following practical completion. The situation is only made worse by the fact that the employer's payment advice is not contractually linked to the contractor's application. This would be a most unfortunate construction, particularly as, unlike for instance, WCD 98 (see clause 30.3.1.2), the MPF does not provide for the monthly payments to include the one following practical completion. There is every likelihood that a considerable amount of money will have been 'earned' at the date of practical completion, which the contractor should expect to be included in a payment advice shortly after.

However, it is possible to interpret the employer's requirement in a more sensible, if slightly strained, manner. To begin with, the employer clearly has an obligation to issue further interim payment advices after practical completion. It would not make sense to impose an obligation to do this if there was no time at which the performance of the obligation was required. To make the obligation effective a time is required. In addition, even if the payment advice is not directly linked to the contractor's application, it is clear from clause 22.1 that the contractor can make an application to the employer whenever it considers an interim payment advice should be issued. If therefore a sum has been 'earned' since the last payment advice was issued, and provided at least one month has gone by and the amount identified as due is not less than the amount stated in the appendix, the employer will be under an obligation to issue the interim payment advice.

Contents of interim payment advice

5-128 The payment advice must set out the amount due from one party to the other. Although this is only expressly stated to be the requirement in connection with interim payment advices prior to practical completion, it is relatively clear that the same requirement applies after practical completion.

The requirement to give notice of the proposed payment

> 22.3 A payment advice shall set out, or be accompanied by a statement setting out, the amount of the payment proposed to be made, and the basis on which that amount was calculated.

5-129 Section 110(2) of the Housing Grants, Construction and Regeneration Act 1996, so far as relevant provides:

> 'Every construction contract shall provide for the giving of a notice by a party not later than 5 days after the date on which a payment becomes due from him under the contract, ... specifying the amount (if any) of the payment made or proposed to be made, and the basis on which that payment was calculated.'

Clause 22.3 is clearly intended to fulfil this requirement. However, this clause discloses a potentially serious problem in achieving this intention. Section 110(2) requires the notice to be given not later than 5 days after the date on which a payment becomes due. It seems implicit therefore that the notice of proposed payment must be given either at the same time as, or within 5 days after, the sum first becomes due. It is straining the construction of section 110(2) to suggest that the notice can be given before the sum becomes due. Where the payment advice discloses an amount as being payable to the contractor, that amount becomes due for payment only upon receipt by the employer of a VAT invoice for that amount (see clause 22.4). Clearly that invoice will be received by the employer some time after the payment advice and notice of proposed payment has been sent to the contractor. The notice of proposed payment will therefore pre-date the amount becoming due and there is a chance that this will render it invalid.

On this construction of section 110(2), the employer, to comply with section 110(2), would be required to serve a notice of proposed payment not later than 5 days after receipt of the appropriate VAT invoice. Having said this, the purpose of the section 110(2) notice is to inform the payee as early as possible if there is to be any difference between the amount due and the amount which will be paid by the paying party. In the particular circumstances of the MPF, where the VAT invoice is required to be for the same amount as the payment advice, the contractor will know that the notice of proposed payment will be the same as the payment advice even though he will know this before the amount disclosed in it becomes due. There is just a chance therefore that it will be effective. However to be safe, some employers may amend clause 22.3 to provide that a further notice of proposed payment will be given not later than 5 days after receipt of the VAT invoice.

This problem does not exist where the payment advice discloses an amount due from the contractor to the employer as in this case the amount becomes due on the issue of the payment advice (see clause 22.4).

Contents of proposed notice

5-130 The payment advice or an accompanying statement must set out the amount of the proposed payment and also the basis on which that amount was calculated.

Effect of failure to serve a valid notice of proposed payment

5-131 If the payment advice does not set out or is not accompanied by a statement setting out the amount of the payment proposed to be made and the basis upon which it was calculated, what is the effect? It is certainly a breach of contract on the part of the employer, but what is the loss to the contractor? The MPF does not provide that in the absence of a proposed payment notice the employer is obliged to pay the amount due. This can be compared with WCD 98 clauses 30.3.3 to 30.3.5 inclusive. Clause 30.3.5 expressly states that in the absence of either a proposed payment notice or a withholding notice, the employer must pay the amount stated in the contractor's application. This has been held to mean that the sum contained in the contractor's application becomes due and must be paid even if in terms of general law it could be claimed not to be due at all, e.g. because the application contained a claim in respect of defective work or work not yet carried out.

In *Watkin Jones & Son* v. *Lidl UK GmbH* (2002) Judge Humphrey Lloyd approved the Scottish case of *SL Timber Systems Ltd* v. *Carillion Construction Ltd* (2000) in which Lord Macfadyen held that a failure to give a section 110 or section 111 notice under the 1996 Act did not dispense with the need for the party seeking money to establish at least that the sum was due under the contract, and that this was to be done on the basis of the underlying contractual obligation to establish that the request for payment is justified in terms, e.g. of the work having been carried out, and carried out properly under the terms of the contract.

It appears that the judge only decided that this approach was not appropriate under the WCD 98 form because of the meaning he ascribed to clause 30.3.5. That clause provides:

> 'Where the Employer does not give any written notice pursuant to clause 30.3.3 and/or to clause 30.3.4 the Employer shall pay the Contractor the amount stated in the Application for Interim Payment.'

The judge said:

> 'The contract is thus precise. If a notice is not given under 30.3.3 or 30.3.4 then the amount applied for must be paid.
>
> ... Mr Neill's submission was otherwise right since the judgment of Lord Macfadyen in *SL Timber* and other authorities does establish what is in any event plain that, in the absence of provisions such as clauses 30.3.3 and 30.3.5, an adjudicator (and of course an arbitrator) will have to decide what sum is truly due as an interim payment if that is the dispute or an integral part of the dispute referred. For example, for the purposes of the HGCRA or contractual provisions giving it effect, one cannot withhold what is not due. Unless the amount due is agreed by the paying party it first has to be established, if necessary by the adjudicator...
>
> ... If its contract had not been subject to clauses 30.3.3 and 30.3.5 Watkin Jones would have had to justify its application number 11 in the first adjudication... In such circumstances the adjudicator would have to decide the sum to which the contractor was entitled which ought to be the amount for which the contractor applies.'

In the case of the MPF the whole structure of the payment mechanism is different as it is the employer's interim payment advice which will determine the amount to become due, and not the contractor's application. Even so, as under clause 22.4 the receipt of the contractor's VAT invoice renders the amount stated in the employer's payment advice due under the contract, the point made in the above still has some force in relation to the MPF. For instance, if the employer overvalued work and reflected this in the payment advice, what would have been an abatement argument claiming that the overvalued part was never due, ceases to be an option once the VAT invoice has been received. It would therefore require a notice of withholding. If, however, the overvaluation was not due to any default on the contractor's part, clause 23.1, which provides three grounds for withholding, appears not to apply. The employer could still serve a section 111 notice under the 1996 Act, but would need to make sure that the grounds were clearly stated and should not cite clause 23 or base the grounds for withholding on any of the three categories set out in clause 23.

5-132 If the assumption is made that compliance with clause 22.3 does in fact satisfy the requirements of section 110(2) of the Act so that it would amount to a valid notice of proposed payment, any failure to provide that information would amount to a breach of contract. It is difficult in such circumstances to determine what sort of loss this might possibly cause to the contractor, except perhaps that if the employer has not indicated the basis upon which the amount was calculated, the contractor may be engaged in wasted time and effort in trying to establish why there is a difference between the payment advice and his application. Even this could be problematical as there is little guidance as to what is meant by 'the basis on which that amount was calculated' in clause 22.3. Is it a detailed breakdown or is it adequate simply to refer to the fact that the proposed payment has been calculated in accordance with, for example, clause 22 or the pricing document?

If the employer does provide the required information in the interim payment advice or an accompanying statement, then even if the notice were to be invalid (see the point made earlier - paragraph 5-129), while this would be a failure to comply with section 110(2) of the Act, it would not be a breach of the contract itself. The Act itself does not provide for consequences as a result of such a failure.

The due date and the final date for payment

> 22.4 Any amount stated by a payment advice as being payable to the Contractor shall become due for payment upon the receipt of a VAT invoice for that amount by the Employer. Any amount stated by a payment advice as being payable to the Employer shall be due upon the issue of the payment advice. Subject to clause 32.4.4 the final date for payment shall be 14 days after the amount to be paid became due.

5-133 Section 110(1) of the Housing Grants, Construction and Regeneration Act 1996 provides as follows:

> 'Every construction contract shall –
> (a) provide an adequate mechanism for determining what payments become due under the contract, and when, and
> (b) provide for a final date for payment in relation to any sum which becomes due
>
> the parties are free to agree how long the period is to be between the date on which a sum becomes due and the final date for payment.'

Clause 22.4 is clearly intended to meet the 1996 Act requirement for a due date and a final date in respect of each payment to be made. Where money is stated in a payment advice to be payable to the contractor, it becomes due for payment when the employer receives a VAT invoice for that amount. Where on the other hand, exceptionally, the payment is in the other direction, the due date is the issue of the payment advice. In both cases the final date for payment is 14 days after the amount became due. If the employer wishes to withhold any amount from the sum due, then an appropriate withholding notice will be required to satisfy the requirements of section 111 of the 1996 Act. This is dealt with later under clause 23 (see paragraph 5-163 *et seq.*).

In requiring a VAT invoice as a condition of the payment becoming due, the MPF is reflecting a common practice and is consistent with accounting procedures. Contractors will need to issue a VAT invoice for accounting and VAT purposes and the employer will need to have one before payment is made.

The effective payment period is not therefore 14 days from the issue of the employer's payment advice, but 14 days from the receipt of the contractor's VAT invoice. In practice, this will mean a payment period of between about 16 and 21 days. Some employers may seek to extend this by amendment.

The issue by the contractor of his VAT invoice renders the amount included in the payment advice due under the contract. If the employer has made a mistake, e.g. having included any sum as a result of an overvaluation not attributable to any breach or default on the contractor's part, the employer appears to be unable to withhold it from the payment. This is discussed when considering clause 23 dealing with withholding notices (see paragraph 5-173).

Contents of the interim payment advice

22.5 Each interim payment advice shall state:

.1 the proportion of the Contract Sum to which the Contractor is entitled, calculated in the manner set out in the Pricing Document;

.2 the value of any Changes executed by the Contractor;

.3 the amount of any reductions made as a consequence of the provisions of clauses 6.7, 18.5 or 18.6; and

.4 any amounts that either party is liable to pay the other in accordance with the provisions of the Contract.

Payments previously made shall be deducted in order to determine the amount due from one party to the other.

5-134 This sets out the various entitlements to payment which the contract interim payment advices must state. These are:

- The appropriate proportion of the contract sum;
- The value of changes;
- The amount of reductions as a result of the operation of certain clauses;
- Amounts for which either party is liable to the other under the provisions of the contract.

These entitlements will be considered in turn.

Proportion of the contract sum

5-135 The interim payment advice is to state the proportion of the contract sum (being the figure referred to in the appendix) to which the contractor is entitled. This is calculated in the manner set out in the pricing document. The pricing document is considered separately (see paragraph 5-179), but note particularly paragraphs 3.1, 4.1, 5.1 and 6.1 of the pricing document.

It should be noted that the contract sum is not subject to any fluctuations provision covering increases or decreases in the cost of labour or materials during the course of the project.

Off-site and on-site goods and materials

5-136 The MPF conditions give no express right to payment for unfixed materials whether they are off-site or on-site. However, it may be that the way in which payment is made under the pricing document could in effect permit this, for example, where there are stage payments and a stage includes unfixed materials either on-site or off-site. If the effect of the pricing document rules for determining the manner in which the contractor is to receive payments for the contract sum do include the value of goods or materials which have not yet been incorporated, this raises issues in relation to ownership in such goods and materials.

5-137 The employer will wish to be sure that if he pays for materials or goods before they are incorporated into the works, the ownership will vest in him. There are often conflicting interests, and difficulties can arise.

5-138 Once materials or goods are physically incorporated into a building there will be no problem as the property will pass to the landowner. This legal rule carries the Latin tag *quicquid plantatur solo, solo cedit* (the property in all materials and fittings, once incorporated in or affixed to a building, will pass to the freeholder). Unless perhaps they are readily removable without causing damage, the employer can then safely pay for them.

5-139 Before such incorporation there is a risk. The MPF is lacking in this situation. It does not provide for the passing of title to goods and materials which may have been paid for but not incorporated. In any event, simply to state in a contract that property is to pass from contractor to employer is not a full protection for the employer, for instance where at the time that the contract says the property is to pass, e.g. the time of payment for it, the contractor does not own it. There is a general legal rule that no-one can transfer a better title than he himself possesses. The Latin tag is *nemo dat quod non habet*. There are exceptions to this rule, most of which relate to contracts for the sale of goods rather than a building contract. A building contract is a contract for work and materials which is not governed by the law relating to the sale of goods. However, these exceptions will be relevant to the position between the contractor and his suppliers and this in turn can determine the ability of the contractor to pass title to the employer even though he himself may not own the materials or goods. It is not possible in a book of this kind to deal with this rule and its exceptions in detail. Reference should be made to the appropriate textbooks, e.g. *Chitty on Contracts*, (28th edition, vol. 2 Specific Contracts, section 43, paragraphs 193–231).

Nevertheless, some contractual provision may be better than nothing. It could be located in the document which, for example, describes the stages, making it clear that title is to pass, so far as the contractor is able to pass it, upon a stage which includes unincorporated goods or materials being reflected in a payment advice.

5-140 If a contractor accepts payment from the employer under a contract which expressly provides that the property in materials or goods is to pass to the employer upon payment, when the contractor does not own such materials or goods at that time, clearly he is in breach of contract. Even if there is no express term in the contract relating to title, there will nevertheless be an implied term as to title by virtue of section 2 of the Supply of Goods and Services Act 1982 which applies, *inter alia*, to building contracts. Section 2 provides that:

> 'In a contract for the transfer of goods . . . there is an implied condition on the part of the transferor that in the case of a transfer of the property in the goods he has the right to transfer the property and in the case of an agreement to transfer the property in the goods he will have such right at the time when the property is to be transferred.'

5-141 There is also an implied warranty that:

(1) The goods are free and will remain free from any charge or incumbrance not disclosed or known to the transferee before the contract is made; and
(2) The transferee will enjoy quiet possession of the goods except so far as it may be disturbed by the owner or other person entitled to the benefit of any charge or incumbrance disclosed or known.

There are statutory restrictions on the ability of a contracting party to exclude or limit liability for breach of this implied condition or these implied warranties (see the Unfair Contract Terms Act 1977, section 7(3A)) and in relation to contracts with consumers (see the Unfair Terms in Consumer Contracts Regulations 1994 paragraph 4).

While it is clear therefore that an employer will have a right of redress against a contractor who is in breach of an express or implied term concerning the transfer of title, such a remedy is of little value if the contractor becomes bankrupt or goes into liquidation.

5-142 The most common reason why a contractor does not own the materials or goods at the time of their delivery to site is that the contract between the contractor and his supplier provides that title is not to pass to the contractor until payment in full has been received by the supplier. Such clauses are of various types and degrees of elaboration, e.g. purporting to deal with the situation where the materials or goods are mixed with others or become part of other materials or goods; making ownership dependent on the discharge of all debts from that contractor to the supplier; and so on. These retention of title clauses are sometimes called Romalpa clauses after the leading case of that name, *Aluminium Industrie Vaassen BV* v. *Romalpa Aluminium Ltd* (1976).

A successful retention of title clause will mean that if the contractor becomes bankrupt or goes into liquidation or a receiver is appointed, without the supplier having been paid, the supplier's title will hold good against the trustee in bankruptcy, liquidator or receiver and also as against the employer even though he may have paid the contractor for the goods.

However, such clauses, to be effective, require very careful drafting.

5-143 Where the sub-contract is not for the mere supply of materials or goods under a sale of goods contract but is a sub-contract for works and materials, the position can be even more precarious from the employer's point of view, as the sub-contract may well not expressly state when title in the materials or goods is to pass to the contractor, if at all. It could well be that on a true construction of the sub-contract, the title may never be intended to pass at all to the contractor. For instance, if the sub-contract assumes that the materials or goods will be incorporated into a building before payment is made to the sub-contractor by the contractor, then the title will transfer directly from the sub-contractor to the landowner. The sub-contract may, however, provide for payment on delivery to the site, and for the title to pass when delivery or payment is made. If there is no express contractual term dealing with the passing of title, then it will be a question of determining from the circumstances when the parties intended the title in the property to pass: see also the case of *Dawber Williamson Roofing Ltd* v. *Humberside County Council* (1979).

The value of changes

5-144 The interim payment advice is to state the value of any changes executed by the contractor. The wording of clause 22.5.2 is perhaps too economic to be perfectly clear. Literally, it could be interpreted as meaning that only when an accepted quotation for a change or change instruction had been completed ('executed'), could its value be included in the payment advice. In the overall context of the MPF and its provision for interim payments, it would be surprising if the contractor was to be required to await completion of a change before being paid anything in respect of it. It is far more likely that the intention is to include within the interim payment advice the value of changes to the extent that they have been executed by the contractor. In appropriate situations this will include preliminary design work. It will also, of course, include a proper proportion of all those matters which make up a fair valuation (see clause 20.6.1 to clause 20.6.4 inclusive - see paragraph 5-46 *et seq.*).

STAGE PAYMENTS

5-145 Where the pricing document states that Rule B - Stage payment, operates to determine the manner in which the contractor is to receive payments in respect of the contract sum, paragraph P4.1 provides that 'The proportion of the Contract Sum to be included in the interim payment advice shall be the total amounts identified in the contract sum analysis in respect of all of the stages completed'. What is the position if a stage is subject to a change which extends the time before the stage is completed? Take for example, piled foundations, where a change requires additional piles to be driven. It appears that clause 22.5.1 together with paragraph P4.1 of the pricing document, means that the stage as originally described and priced in the contract sum analysis remains fixed, and that any additional sum as a result of a change would be included separately under clause 22.5.2. This helps the contractor's cash-flow to some extent if the completion of the stage is delayed, though it will not produce a completely satisfactory result, e.g. a small change involving a long delivery date for materials, which delays the completion of a stage. No doubt the cost to the contractor of being kept out of money will be reflected in the contractor's claim for loss and expense as part of the valuation process (see clause 20.6.4). Presumably, where the change is by way of a straight omission of work only, the contractor will be entitled to the stage payment on completion of the diminished stage.

PROGRESS PAYMENTS

5-146 Where the pricing document states that Rule C - Progress payments, operates to determine the manner in which the contractor is to receive payments in respect of the contract sum, a literal interpretation of that Rule (P5) and clauses 22.5 1 and 22.5.2 would suggest that the scheduled payments are covered by 22.5.1 and that changes are paid for in addition. This might not tie in with any funding mechanism which is in place, though the employer and any funder will be well aware of the possibility of changes and will no doubt have made the necessary financial arrangements to cater for this.

The amount of any reductions as a consequence of clauses 6.7, 18.5 or 18.6

5-147 The interim payment advice is to state the amount by which any payment which would otherwise be included is to be reduced as a consequence of:

- The employer not being liable to pay for any work executed which has not been carried out in accordance with design documents marked (or deemed to be marked) 'A Action' or 'B Action' (see clause 6.7 and paragraph 2-149 *et seq.*).

- The employer not being liable to pay the contractor in respect of services provided by pre-appointed consultants or named specialists where the contractor has failed to enter into the required agreement with them (see clause 18.5 and paragraph 4-92 *et seq.*).
- The employer not being liable to pay the contractor in respect of services carried out by the contractor or someone else on his behalf as a result of the contractor amending his agreement with any pre-appointed consultants or named specialists or by waiving their strict compliance with their obligations without having obtained the prior written consent of the employer (see clause 18.6 and paragraph 4-98 *et seq.*).

These all relate to situations where the employer has received value under the contract, but the contractor has failed to obtain necessary approvals or follow required procedures, with the result that the employer's liability to pay for such value is removed. Mention has been made when discussing clauses 18.5 and 18.6 (see paragraphs 4-94 and 4-98) of the possibility that such provisions are in the nature of a penalty and could therefore be invalid.

To the extent that clause 22.5.3 applies, it will result in the proportion of the contract sum under clause 22.5.1 or the value of any change under clause 22.5.2 being reduced to take account of the fact that the employer has no liability to pay for the work concerned.

Any amounts that either party is liable to pay the other in accordance with the provisions of the contract

5-148 The interim payment advice must state any amounts by way of addition, for which the employer is liable to pay the contractor, or by way of subtraction, for which the contractor is liable to pay the employer, under the provisions of the contract. It therefore includes the following:

- Employer's additional costs of engaging others where the contractor has failed to comply with an instruction (see clause 2.3 and paragraph 2-36 *et seq.*).
- Contractor's liability to pay to the employer liquidated damages (see clause 10.1 and paragraph 3-29 *et seq.*).
- Employer's liability to pay the contractor any bonus for early completion (see clause 14 and paragraph 3-114 *et seq.*).
- Employer instructing that non-complying work, materials or goods may be used on the project subject to the contractor becoming liable to pay the employer an appropriate amount (see clause 16.2.2 and paragraph 4-28 *et seq.*).
- Employer's liability to pay the contractor, upon practical completion, the agreed proportion of any agreed financial benefit as a result of cost savings and value improvements (see clause 19.4 and paragraph 4-119).
- Employer's liability to pay to the contractor any ascertained loss or expense (see clause 21.7 and paragraph 5-99).
- The liability of either party to the other in respect of costs incurred in insuring following a failure to insure by the party having a responsibility to insure (see clause 27.3 and paragraph 6-35).
- The liability of the contractor to indemnify the employer under clause 26.1, or the employer to indemnify the contractor under clause 26.2 in respect of claims

arising out of personal injury or damage to real and personal property (see paragraph 6-08 *et seq*. and paragraph 6-16 *et seq*).
- The liability of either party, depending upon whose obligation it was to insure, to pay to the other any amounts that would otherwise have been paid by insurers, except for the fact that an excess provision has operated under the policy (see clauses 27.5 and 27.6 and paragraph 6-38 *et seq*.).

All of these except the penultimate one will generally be capable of quantification during the project or within a reasonable time afterwards and certainly in time for the issue of the final payment advice following the end of the rectification period. However, the penultimate matter concerning indemnities is in a different category. If a claim is made under clause 26 it will involve insurers and quite possibly lawyers. If liability or the amount of the indemnity is disputed, it could take many months or even years to determine. This has a number of possible implications including the following:

- If the liability issue is not resolved can the employer reflect his own contentions in the interim payment advice? Presumably yes, though the contractor will be able to get the issue reviewed by an adjudicator. The position would be the same if it was just the amount of the liability which was in issue.
- Does a requirement for the amount of this liability to be stated in interim payment advices and in the final payment advice, mean that the requirements of section 110(1) of the Housing Grants, Construction and Regeneration Act 1996 are not met, i.e. that all construction contracts are to provide for an 'adequate mechanism' for determining when payment becomes due under the contract and what payments become due? This must be a risk. While it is a laudable aspiration that all matters relating to sums which might become payable under or in connection with the MPF should be included within its internal payment mechanism, to include such indemnity payments may be taking it a step too far. It would have been sensible to exclude this head of liability and leave it as a separate claim or a deduction from sums due after a notice of withholding.
- As this head is to be included in the final payment advice, any amount included has the potential to become final and binding under clause 22.7 if not disputed within 28 days. This could easily catch out the contractor and his insurers.
- As this head is required to be included, what is the position if the issue is still live and it is simply not possible to include any realistic amount at the time when the final payment advice is to be issued? From the employer's point of view, as the final payment advice produces finality as to sums due (see clause 22.7), the employer could be prejudiced (see paragraph 5-157 where a similar problem could arise). However, if an indemnity is being claimed by the contractor from the employer, the contractor must dispute the absence of any sum to avoid being prejudiced.

Clause 32.6 provides that following termination by the employer of the contractor's employment, he may issue an interim payment advice in respect of amounts due under clause 32.5, which covers such matters as the costs of getting the project completed and any loss or damage suffered by the employer for which the contractor is liable, whether as a consequence of the termination or not.

Deduction of previous payments

5-149 The net sum payable to the contractor will be the total value of those elements referred to above less payments previously made by the employer. It should be noted that it is not payments previously stated by a payment advice as being payable which are deducted, but only those payments which have been made. The MPF seeks to include everything which could become payable under or in connection with the contract, up to and including the final payment, within a payment advice so that it should be relatively rare that the employer decides not to pay a payment advice in full and instead pays a lesser sum, having issued a withholding notice under clause 23. Even if the employer does elect to deduct a sum from the amount of a payment advice for which the contract makes the contractor liable to pay the employer, rather than it being calculated as part of it, this will generally sort itself out in the next payment advice in which the amount of that liability should be stated.

The fact that on each interim payment advice the total value of the project since its inception is stated, with previous payments being deducted to establish the net amount, confirms that interim payments are made on account of a final sum. This revaluation on each issue of a payment advice ensures that any defective work previously valued as conforming work will be taken into account on the revaluation exercise.

The final payment advice

22.6 Upon the issue by the Employer of the statement referred to by either clause 17.2 or clause 17.3, the Employer shall issue a final payment advice in respect of the total amount to which the Contractor is entitled. This total amount shall include:

.1 the Contract Sum;

.2 the matters set out in clauses 22.5.2 to 22.5.4;

.3 the appropriate deduction recorded by any statement issued under the provisions of clause 17.3 in respect of Defects that the Employer does not intend to rectify.

Payments previously made shall be deducted in order to determine the amount due from one party to the other.

5-150 The MPF in providing for the issue of a final payment advice pursuant to clause 22.6 and its binding effect under clause 22.7 does so in a matter of a few lines. This is to be contrasted with other JCT forms, for example WCD 98 which, in clauses 30.5 to 30.9 deals with the same subject matter in three and a half pages. With the simpler, more basic approach of the MPF, there is some risk of areas of uncertainty, particularly in relation to the binding nature of the final payment advice. Even so, the MPF approach is much preferred.

5-151 Clause 22.6 provides for the issue of a final payment advice. It has significant implications and these will be dealt with when considering clause 22.7 below.

The final payment advice is to be issued at the same time as a statement is issued under either clause 17.2 or clause 17.3, whichever applies. The final payment advice is therefore issued after the expiry of the rectification period under either:

- Clause 17.2 – employer issues a statement that the defects have been remedied; or
- Clause 17.3 – under which, where defects have not been remedied within a reasonable period after the expiry of the defects rectification period, the employer can issue a statement identifying those defects which will be remedied by others,

and those which are not to be remedied at all with an appropriate deduction being made from sums otherwise due to the contractor.

This means that the earliest date on which the final payment advice can be issued is the end of the defects rectification period, although in practice it is likely to be at least a little time after this if there are outstanding defects to remedy at the end of the period. The important point to note is that the timing of the issue of the final payment advice is in the hands of the employer, even where the contractor is failing to remedy defects. The employer can wait for these defects to be remedied and hold back the issue of a final payment advice, or alternatively can take the initiative under clause 17.3 by engaging others to remedy the defects and/or making an appropriate deduction from sums otherwise due to the contractor instead.

5-152 The calculation of the final payment advice is identical to that for interim payment advices except that where the final payment advice is issued following a clause 17.3 statement, an appropriate deduction will be made where the employer intends not to rectify the outstanding defects. Bearing in mind the final and binding nature of the final payment advice, there may be difficulties associated with these time-scales in so far as the advice must contain any amounts that either party is liable to pay the other in accordance with the provisions of the contract. These may not easily be determined in the time available. These difficulties were mentioned when discussing clause 22.5.4 above (see paragraph 5-148).

As with interim payment advices, any payment previously made will be deducted in order to determine the final amount due from one party to the other.

The effect of the issue of the final payment advice

22.7 The final payment advice shall be final and binding upon the parties in relation to amounts due from the Employer to the Contractor under or in connection with the Contract, including any sums due to the Contractor as a consequence of claims for breach of contract, breach of statutory duty, negligence or otherwise, unless within 28 days of the final payment advice being issued the Contractor disputes any aspect of it by reference to adjudication or litigation as provided in clause 35 (*Resolution of disputes*).

Final payment advice final and binding

5-153 The final payment advice becomes final and binding upon the parties in relation to amounts due from the employer to the contractor under or in connection with the contract, unless within 28 days of its issue the contractor disputes any aspect of it by referring it to adjudication or litigation as provided in clause 35. The following points are noteworthy.

NO EXCEPTION FOR ARITHMETICAL OR OTHER ERRORS

5-154 There is no express exception for arithmetical or other errors which may form part of the amount due in the final payment advice. Short therefore of some fundamental common mistake or misunderstanding between the parties, the parties will be bound despite the error.

FINAL AND BINDING, NOT JUST CONCLUSIVE (EVIDENCE)

5-155 The payment advice is expressed to be final and binding rather than as being 'conclusive evidence' (see for example clause 30.8.1 of WCD 98). Accordingly, it is more than just an evidential bar. It provides a substantive defence to a claim by either party that the amounts stated as due are disputed.

As the final payment advice is final and binding on the employer from the date of its issue, the employer will not be able to issue any notice of withholding under clause 23 or under section 111 of the Housing Grants, Construction and Regeneration Act 1996 in respect of anything covered by this finality. Oddly, the advice becomes final and binding on the employer even before any amount stated in it becomes due to the contractor (on receipt by the employer of the contractor's VAT invoice (see clause 22.4).

UNDER OR IN CONNECTION WITH THE CONTRACT

5-156 The finality extends to amounts due from the employer to the contractor under 'or in connection with' the contract. It can therefore potentially extend not only to amounts which the contract clearly identifies as due under its terms, but also anything which is to be reflected in the final payment advice which relates to any matter in connection with the contract. However, it is difficult to see to what this can relate in the context of the amounts to be included within the final payment advice. By virtue of clause 22.6.2 (and therefore clauses 22.5.2 to 22.5.4 inclusive) the only amounts which can be included are those which the contract expressly states are to be included. To take an example, a claim by the contractor against the employer for damages for breach of the contractor's copyright (see clause 7.1) could not be affected by the final payment advice as the amount of any such damages could never form part of its calculation.

Some clue as to what the reference to 'in connection with the Contract' could be intended to cover may be gleaned from the words which follow it, 'including any sums due to the Contractor as a consequence of claims for breach of contract, breach of statutory duty, negligence or otherwise'. What this purports to do is to provide that if there are claims for breach of contract, breach of statutory duty, negligence or otherwise which arise in connection with the contract, they are within the finality provision. Perhaps the clause is attempting to provide that, for example, a claim for damages for breach of statutory duty made in connection with the contract, rather than expressly under it, will be subject to the finality point. However, this does not work because such a sum would never have been included in the calculation in the first place. The drafting is far from clear and it is impossible to be sure what the reference to 'or in connection with' means. If the clause was to be interpreted as so wide ranging (highly unlikely when it would amount to a severe exclusion clause and would be interpreted very strictly against the party relying on it – see paragraph 2-11 *et seq.*), it might just be possible that this exclusion of liability could be attacked as failing to meet the requirements of reasonableness under the Unfair Contract Terms Act 1977 (see sections 2(2) together with section 11 and Schedule 2 of the Act). It is possible that a standard form such as the MPF could also be subject to section 3 of the Act as being the employer's written standard terms of business. See, for example, a case concerning the British International Freight Association Standard Trading Conditions: *Overland Shoes Ltd* v. *Schenkers Ltd* (1998) in which it was held that the section 3 requirement of reasonableness could apply to any contractual term seeking to exclude or restrict the employer's liability in respect of any breach by the employer of a contract term.

Latent defects

5-157 The phrase 'under or in connection with' could have a further meaning. The final payment advice is final and binding on both parties. It therefore binds the employer. In one sense this is not surprising since it is the employer who makes the final calculation and who should therefore have no reason to challenge it. However, the final payment advice becomes binding on the employer 'in relation to amounts due from the employer to the contractor under or in connection with the contract'. So, just as the contractor, unless he mounts a challenge as provided in clause 22.7 within 28 days of its issue, is bound and cannot seek to have the amount increased, so the employer cannot, having issued the advice, seek to thereafter claim that a lesser sum is due. Put another way, in contractual terms the contractor has been paid the correct sum in connection with fulfilling his obligation under the contract. The employer cannot later contend that the contractor's performance of his contractual obligations was worth less. Does this rule out a subsequent claim by the employer in respect of latent defects in the project as a result of a breach by the contractor of provisions contained in the requirements or proposals? The contractor can make a respectable case to say that it does. The calculation of the final payment advice determines (and it is a final determination) the sum due from the employer to the contractor and reflects the following:

- The value of work properly executed by the contractor (see, for example, clause P3 – Rule A – Interim Valuation – of the Pricing Document; that the employer considers work to have been properly executed is implicit if not explicit also in relation to Rule B – Stage payments, and Rule C – Progress payments). This probably, in context, includes design work.
- The amounts due as between the parties 'under or in connection with the contract'. The amount is final and binding as to amounts due not only under the contract but also in connection with it, for instance any claim for damages for latent defects. The final payment advice is clearly intended to put an end to claims in respect of indemnities under clause 26, so why not for other claims which would reopen the amount due to the contractor not just under but also in connection with the contract.

It is likely that clause 22.7 was never intended to have this effect but the danger is there and some employers may insert a clarifying amendment.

BREACHES OF STATUTORY DUTY, NEGLIGENCE OR OTHERWISE

5-158 The finality extends to sums due to the contractor as a consequence of claims for breach of contract, breach of statutory duty, negligence or otherwise. Are such sums ever reflected in amounts due 'under or in connection with the Contract'. It is of course possible that the employer could, in breach of contract, disrupt the contractor's regular progress of the project. It is also just about conceivable that such disruption could be caused as a result of actionable negligence or breach of statutory duty by the employer. In such a case the contractor could, of course, seek damages if he wished to as an alternative to seeking reimbursement of loss and expense under clause 21. However, the amount of such a claim is not required to be included as part of the calculation of the final payment advice. The calculation is limited to the reimbursement of loss and expense which is being claimed, so it is difficult to see how such claims can be caught by the finality provision. A more likely possibility is in connection with the indemnity provisions of clause 26, where if either party claims under those provisions seeking an indemnity as a result of breach of contract, breach of statutory duty or negligence (or otherwise), any liability to indemnify in respect of any expense, liability, loss, claim or proceedings is to be reflected as part of the calculation of the final payment advice, so that the finality provisions would be able to bite. It is possible, though it is a matter of speculation, that what was in fact intended by the reference to claims for breach of contract, breach of statutory duty, negligence or otherwise, was to provide a means of ensuring that any sums forming part of the final payment advice in accordance with the terms of the contract, which could alternatively have been claimed as damages for breach of contract, breach of statutory duty or negligence etc., cannot have the binding effect of the final payment advice sidestepped by a separate action. In other words, where there is an overlap between what is included in the final payment advice and what might be included in a separate claim for breach of contract etc., the finality provisions will equally apply to that other claim to prevent the finality of the final payment advice being sidestepped. However, this is not what clause 22.7 says; c.f. clause 30.8.1.3 of WCD 98 which talks about reimbursement of loss and expense being in final settlement of all and any claims arising out of the occurrence of any of the loss and expense matters 'whether such claim be for breach of contract, duty of care, statutory duty or otherwise'.

ADJUSTMENTS TO THE COMPLETION DATE

5-159 The finality provisions in clause 22.7 probably prevent either party from seeking to reopen any adjustments which have been made to the completion date. Adjustments to the completion date directly affect the period for which liquidated damages are payable. Liquidated damages are expressly included in the calculation of payment advices, including the final payment advice (see clause 22.5.4 and clause 10.1). It would not therefore be possible for either party to seek to revise any adjustments to the completion date as that would have the effect of varying the amount due in the final payment advice.

CHALLENGING THE FINAL PAYMENT ADVICE

5-160 It goes without saying that if the contractor wishes to dispute any aspect of the final payment advice, he should take the greatest care in the preparation of any notice of adjudication or claim form in court proceedings. It is only those matters identified within the served notice of adjudication or issued claim form within the 28-day period which will be excluded from the finality provisions. While the notice or claim form ought to be precise, it should nevertheless be as wide as possible to protect the contractor.

The operation of the time limit and relevant pre-action protocol

5-161 If the contractor wishes to dispute any aspect of the final payment advice in relation to amounts due from the employer to him, he has 28 days from the date of its issue in which to refer the dispute to adjudication or litigation to prevent the finality provisions applying. Any failure to do this will mean that the amount stated as due from the employer to the contractor will be subject to the finality provision. In terms of litigation, the 28-day limit will make it impossible for the contractor to comply with pre-action protocols such as the Pre-Action Protocol for Construction and

Engineering Disputes (August 2000), though by analogy the exception in relation to limitation of action would seem appropriate.

Having commenced adjudication within the 28-day period, it would appear that in relation to that dispute the 28-day time limit will not apply to any litigation which follows it.

Payment and the Construction Industry Scheme

22.8 Notwithstanding anything to the contrary elsewhere in the Contract, where the Employer has notified the Contractor that it is a 'contractor' for the purposes of the CIS the Contractor shall provide vouchers in accordance with the requirements of the CIS in respect of all payments received. If the Contractor fails to do so the Employer shall not be obliged to make any further payment to the Contractor until such time as the failure is remedied.

5-162 Clause 22.8 provides that where the employer is a 'contractor' for the purposes of the CIS and the employer has notified the contractor of this fact, then the contractor is to provide the appropriate vouchers in accordance with the CIS requirements in respect of all payments received. CIS is defined in clause 39.2 as:

'The Construction Industry Scheme established by the Income Tax (Sub-Contractors in the Construction Industry) Regulations 1993, as amended by the Income Tax (Sub-Contractors in the Construction Industry) (Amendment) Regulations 1998.

The clause goes on to provide that if the contractor fails to do this, the employer is not obliged to make any further payment to the contractor until that failure has been remedied. This provision overrides anything elsewhere in the contract requiring the employer to make payments to the contractor.

The MPF does not deal in detail with the CIS and it is not appropriate in a book of this nature to deal with it further. A useful Inland Revenue publication is IR14/15(CIS).

Withholding – clause 23

Withholding – requirements

23.1 Provided that an effective notice of withholding has been given, the Employer may withhold from any payment that is due to the Contractor:

.1 any amount that the Contractor is liable to pay the Employer in accordance with the terms of the Contract;

.2 any sums owed to the Employer by the Contractor as a consequence of any breach of the Contract; and/or

.3 where a final payment advice has been issued, a proper estimate of the cost to the Employer of rectifying any Defects referred to in clause 17.3.1.

23.2 To be effective any notice of withholding must be given to the Contractor no later than 7 days before the final date for payment of the sum from which the withholding is to be made and must identify the ground or grounds upon which the withholding is proposed to be made and the amount of withholding attributable to each ground.

Section 111 Housing Grants, Construction and Regeneration Act 1996

5-163 Section 111 of the 1996 Act provides:

'(1) A party to a construction contract may not withhold payment after the final date for payment of a sum due under the contract unless he has given notice of intention to withhold payment.

...

(2) To be effective such a notice must specify –

(a) the amount proposed to be withheld and the ground for withholding payment, or

(b) if there is more than one ground, each ground and the amount attributable to it,

and must be given not later than the prescribed period before the final date for payment.'

Under the MPF, most of the matters which could amount in law to a set-off are taken into account in determining the amount due in respect of interim payment advices or in the final payment advice. This will generally restrict the need to serve withholding notices. Presumably, clause 23 is aimed at the situation where the matter which gives rise to the intended withholding arises only during the 7 days after the issue of the relevant interim payment advice.

It is worth noting that clause 23 does not deal with the situation where the contractor wishes to withhold sums which may be due from him to the employer. This could happen, for instance, in the case of a negative payment advice against which the contractor wishes to withhold sums otherwise due. An example of such a withholding might be where, following the issue of a negative payment advice, a further adjustment to the completion date is made, which entitles the contractor to repayment of the appropriate liquidated damages (see clause 10.2). Section 111 would still apply to this situation and the contractor would need to serve an appropriate withholding notice within the prescribed period under the Act, i.e. 7 days before the final date for payment which is itself 14 days after the issue of the negative payment advice.

Clause 23 is clearly drafted on the assumption that a notice of withholding is required in all of the situations indicated and in order to comply with section 111. It is therefore appropriate to consider clause 23 against the background of section 111.

Grounds upon which a withholding can be made

5-164 Clause 23.1.1 to clause 23.1.3 set out the grounds which will justify a withholding, provided an effective notice is given. These are as follows.

Any amount that the contractor is liable to pay the employer

Clause 23.1.1 provides that any amount can be withheld that the contractor is liable to pay the employer in accordance with the terms of the contract. These matters have been listed when considering clause 22.5.4. Although clause 22.5.4 refers to the 'provisions', whereas clause 23.1.1 refers to the 'terms' of the contract, this appears to be a drafting slip rather than an intention to have two different meanings. The contract does of course require all of these matters to be taken into account in the calculation of the sum due. It is only exceptionally therefore that they will be the subject of a withholding notice (see for example earlier paragraphs 2-40, 3-34 and 5-112).

5-165 If the liability to pay the amount is arguable, is the contractor 'liable' to pay at all? And if not, does this mean that there is no right to set-off? In *Dawnays Ltd* v. *F.G. Minter Ltd* (1971) the Court of Appeal held that words in a sub-contract that the main contractor could deduct or set off against money due to the sub-contractor 'any sum ... which the sub-contractor is *liable* to pay to the contractor under this sub-contract', meant 'liquidated and ascertained sums which are established or admitted as being due' (per Lord Denning). This decision has generally been regarded as being overruled by the House of Lords in *Gilbert-Ash (Northern) Ltd* v. *Modern Engineering (Bristol) Ltd* (1973), though this was not expressly done and the point just may be worth arguing, especially as a rapid adjudication process is now available to resolve the issue of liability and its amount.

Sums owed to the employer as a consequence of breach of contract

5-166 Any sums owed to the employer by the contractor as a consequence of any breach of the contract by the contractor are covered. There could well be some overlap between this and the previous point. For example, liquidated damages, if they were not in the calculation of the amount due, could be treated as falling under either head.

Clause 23.1.2 could also extend to breaches of contract in respect of which any damages are not part of the calculation of the sum due. Examples would be a breach by the contractor of the copyright provisions in clause 7.1; possibly a breach by the contractor in failing properly to fulfil the roles of planning supervisor, principal contractor or designer under the CDM Regulations (see clause 1.3); the contractor's failure to make applications, give notices required by and to comply with statutory requirements (see clause 3); a damages claim in relation to the contractor's breach of design, workmanship or materials obligations (see clause 5) which does more than simply affect the value of the contract sum; and a 'Material Breach' by the contractor in respect of any of the matters listed under its definition in clause 39.2.

5-167 There may be some difficulty in interpreting exactly what the words 'any sums owed' to the employer mean. Is a claim for breach of contract by the employer against the contractor which is disputed and which has not yet been resolved by adjudication or litigation, a sum owed at all? Clause 23.1.2 does not, for example, refer to a claim made in good faith by the employer. A sum is only owed where there is a debt and the debt will not exist until it is a defined sum which is admitted, or found due in adjudication or litigation proceedings. This interpretation of the reference in clause 23.1.2 to sums owed is reinforced by the specific inclusion in clause 23.1.3 of a right to withhold based upon an estimate only. The employer might in this situation argue that if the sum was merely allegedly 'owed', this would take it outside clause 21.3 and provided a notice complying with section 111(2) was served the sum claimed could be withheld. This would produce the absurd result that a sum determined to be owed by an adjudicator or a court would need a clause 23 notice, whereas one that had not been so determined would not. The most sensible interpretation of clause 23 would suggest that the reference to sums 'owed' is intended to refer to sums which the employer claims to be owed.

5-168 Clause 23.1.2 does not expressly refer to sums owed as a consequence of breaches of tortious or other non-contractual duties such as negligence or breach of statutory duty. However, almost invariably such matters would be a breach of an express or alternatively implied term of the MPF. Even so if the employer, for some reason, wished to make an ex-contractual claim rather than a claim for breach of contract, he would be faced with the difficulty of not being able to withhold the claimed amount unless he served a notice under section 111 of the Act.

5-169 Finally, it may be that interest payable under clause 24 would fall within either clause 23.1.1 and/or 23.1.2 and so would require a withholding notice to be served.

Cost of engaging others pursuant to clause 17.3.1

5-170 Where the final payment advice has been issued, the employer may withhold a proper estimate of the cost to the employer of engaging others to rectify defects pursuant to clause 17.3.1 (see paragraph 4-59). As at the time of the issue of the final payment advice the actual cost of engaging others to rectify defects which the contractor has failed to attend to, will be unknown, clause 23.1.3 permits the employer in these limited circumstances to estimate the cost.

Requirements of the notice

5-171 Section 111 provides (see paragraph 5-163) that if a party intends to withhold payment after the final payment date of a sum due under the contract, then an effective notice of intention to withhold payment must be given to the other party within the 'prescribed period' under the Act (7 days) unless the parties have agreed what the prescribed period is to be. Under the MPF, by clause 23.2, it is not later than 7 days before the final date for payment of the sum due. The final date for payment from the employer to the contractor under a payment advice, whether interim or final, is 14 days after receipt of the relevant VAT invoice (see clause 22.4).

To be an effective notice under section 111 and clause 23.2 it must specify the amount proposed to be withheld and the ground or grounds for withholding payment. The section uses the word 'specify', whereas unfortunately for some reason clause 23.2 refers to 'identify'. There is a slight risk that these two words may not mean the same thing in this context, e.g. it might be possible to identify the amount by reference to another document without specifying the amount. On this basis a notice complying with clause 23.2 could nevertheless fall foul of section 111 and be ineffective.

The notice is a communication and must therefore be in writing or in some other agreed manner (see clause 38.1).

Position if there is no valid notice

5-172 What is the consequence if the employer fails to give a valid written notice? The employer will have to pay the amount due without deduction whatever may be the merits of the cross-claim. Even if the withholding notice is only required contractually rather than as a requirement of section 111, clause 23 appears to make the giving of such a notice a condition precedent to the right to withhold sums within the categories covered by clauses 23.1.1 to 23.1.3 inclusive (see paragraph 5-173).

The employer will be in difficulty if, at the point 7 days before the final date for payment, he is unaware of any basis upon which he may be entitled to withhold, but subsequently, when it is no longer possible to fulfil the notice requirements, a matter comes to his attention which would have entitled him to withhold or deduct. Presumably, in such a case he is nevertheless bound to pay having lost his right to withhold. It can be seen therefore that while the statutory requirements in section 111 are intended to regulate the exercise of common law or express contractual rights to withhold or deduct, and are not intended to affect the substantive law in relation to set-off, nevertheless its effect is to do so in a situation such as this.

Situations where no notice required

5-173 Considering the position apart from what the particular terms of a contract may say, if an employer intends not to pay the amount claimed as 'due' in full on the basis that the part not being paid had never, in law, become 'due' at all, e.g. an abatement in respect of work wrongly valued, then there would be no requirement under the Act for a notice of intended withholding. Such a notice is required where it is intended to withhold from a sum which is, in law, due. See *SL Timber Systems Ltd* v. *Carillion Construction Ltd* (2000) and *Watkin Jones & Son Ltd* v. *Lidl UK GmbH* (2002) (paragraph 5-131). The employer would simply be abating rather than setting off. However, clause 22.4 expressly states that 'Any amount stated by a payment advice as being payable to the Contractor shall become due for payment upon receipt of a VAT invoice for that amount'. It would appear therefore that the contract wording makes due what might otherwise in law not be due. In this position, in order to withhold sums it would be necessary for a notice of withholding to be served and for the employer to establish the grounds for withholding. Clause 23.1 contains three such grounds. If, for example, the issue relates to non-complying work then this will, at least arguably, be a breach of contract. (Some would argue not a breach of contract but only a temporary disconformity – see Lord Diplock's Speech in *Hosier & Dickinson Ltd* v. *P & M Kaye Ltd* (1972)). Where the proposed withholding is in connection with the employer's overvaluation, not linked to any breach of contract on the part of the contractor, it is not easy to see any basis upon which the employer can serve a withholding notice pursuant to clause 23. Instead, the employer would need to serve a withholding notice under section 111 of the 1996 Act, making sure not to refer to clause 23 or any of the grounds listed in it. The true basis of withholding, the overvaluation, would need to be stated. Generally, though not always in such instances, the matter can be put right in the next following payment advice. Where it is the final payment advice which contains the overvaluation, the employer appears to be in difficulties in being able to redress the situation as the final payment advice will be binding on the employer on its issue and even before the sum contained in it becomes due (and see clause 22.7, paragraph 5-153).

Interest – clause 24

Interest on late payments

24.1 If either party fails to make payment in accordance with the Contract the other party shall be entitled to simple interest calculated at a rate of 5% in excess of the Base Rate for the period until payment is made.

24.2 It is agreed that the provisions of clause 24.1 constitute a substantial remedy for the purposes of section 9(1) of the Late Payment of Commercial Debts (Interest) Act 1998.

5-174 If either party fails to pay in accordance with the contract, i.e. to pay in full the amount due under any payment advice or otherwise by the final date for payment, the other party is entitled to simple interest thereon for the period of non-payment. The rate of interest is 5% over the base rate. This is defined in clause 39.2 as:

> The rate set from time to time by the Bank of England's Monetary Policy Committee, or any successor.

This definition is unusual in not referring to the statutory source for the base rate. For example, in the JCT Standard Form of Domestic Sub-Contract 2002 Edition, 'Base Rate' is defined (in clause 1.4) as 'the official dealing rate as defined in Statutory Instrument 1998 No. 2765 (The Late Payment of Commercial Debts (Rate of Interest) No. 2 Order 1998). The base rate at the time of writing is 3.5%, so making a contractual interest rate of 9%.

This entitlement to interest in no way acts as a waiver by either party of his right to the proper payment at the proper time; nor does it affect the right of either party under section 112 of the Housing Grants, Construction and Regeneration Act to suspend performance of his obligations in connection with the non-payment. If it is the employer who is owed money, section 112 would seem also to entitle the employer to suspend performance of his obligations, though as the most important obligation is that to pay the contractor, the right to withhold (see clause 23) makes the right to suspend of less importance. Even so, it could affect some other obligations of the employer, e.g. to respond to a design document within 14 days under clause 6.3. Further, neither party's right to terminate his employment pursuant to the termination provisions in clauses 32.2 or 33.2. is prejudiced. The MPF does not contain an express contractual right to suspend performance for non-payment. The statutory right to suspend performance of obligations is dealt with in more detail in paragraph 3-72.

While clause 24.1 refers to a situation where either party 'fails to make payment', this does not automatically mean a requirement to pay over money. For example, if the reason why there has been no payment of money is due to the fact that the obligation to pay is adequately discharged in some other way, e.g. by an effective notice of withholding under clause 23, then there will have been no failure to make payment.

Clause 24.2 states that the provision for interest in clause 24.1 constitutes a substantial remedy for the purposes of the Late Payments of Commercial Debts (Interest) Act 1998. A simple statement in a contract that the contractual provision for interest constitutes 'a substantial remedy' under the late payment legislation will certainly not be conclusive that it does in fact provide a substantial remedy. The matter of interest on late payments under the late payments legislation is summarised below.

Late Payment of Commercial Debts (Interest) Act 1998 (as amended and supplemented by the Late Payment of Commercial Debts Regulations 2002)

This Act has been introduced in distinct phases. On 1 November 1998 it established a right to interest on late payment of commercial debts for the benefit of small firms owed money by large firms or the public sector. Then from 1 November 2000, the second phase of the legislation allowed small businesses to charge other small businesses statutory interest for late payment of debts. Finally, as from 7 August

2002 all businesses irrespective of size and also public sector bodies can claim statutory interest for the late payment of commercial debts. Furthermore, since 7 August 2002 creditors are entitled to 'reasonable' debt recovery costs. The Act applies to England and Wales. There is similar legislation in Northern Ireland and Scotland. The comments below are directed at the legislation applying in England and Wales. In addition, as from 7 August 2002, the same or similar rights are available throughout the European Union.

THE REMEDY FOR LATE PAYMENT

5-175 It should be noted that, as in the MPF, clause 24, it is possible for the contracting parties to provide within their contract for interest on late payment of sums due under it. However, if they do make their own arrangements for contractual interest the contractual remedy must be 'substantial' otherwise it would be void and the debtor would be unable to rely on it. It would then be struck down by the courts and the terms of the legislation would apply to the contract. Any such contract term is void if it provides for a rate of interest that is not a substantial remedy. The courts will, in determining whether or not the remedy is substantial, consider all the circumstances, including of course the rate of interest applied in the contract by comparison with that provided under the legislation (currently 11.5%). If the interest rate provided in the contract is less than the statutory rate, then if it fails to recompense the creditor for being kept out of his money because it is below the creditor's actual or theoretical cost of borrowing, it can be struck down by the courts.

However, the rate of interest is not the only factor which the court will consider. Other relevant factors are:

- The length of any credit period. In other words, debtors should not seek to include in their contracts payment terms which so extend the payment period as to effectively emasculate the late payment interest provisions. If a credit period is considered to be excessive the courts can strike it down and replace it with the 30 days default period provided by the legislation.
- Whether there is equality of bargaining position.
- Whether there is a standard form of contract in use.
- What is usual for the particular sector of business.

It is for the creditor to show that the contractual remedy is not substantial.

Clause 24 of the MPF provides that late payments attract a rate of simple interest at 5% per annum in excess of the 'Base Rate' for the period from when the payment became overdue until when payment is made. Base rate is defined in clause 39.2 as:

> The rate set from time to time by the Bank of England's Monetary Policy Committee, or any successor.

This rate is the same as in other JCT standard forms. At the time of writing, with the base rate at 3.5%, the result is a rate of 8.5%. This is 3% below that currently provided for in the late payment legislation. Whether this is in itself open to challenge as not being a 'substantial' remedy is perhaps debatable. At the time of writing, 8.5% probably does amount to a substantial remedy when compared with the likely actual borrowing costs for the industry generally. Clause 24.2 does of course expressly state that the parties agree that the contractual rate of interest constitutes a substantial remedy for the purposes of section 9(1) of the 1998 Act. This in itself will not be conclusive evidence that this is the case and neither will it create any sort of estoppel so as to prevent a party challenging the contractual provision for interest on the basis that it does not provide a 'substantial' remedy. It will simply be one factor which the court will take into account in determining this issue.

A payment will generally be late when it has not been paid by the final date for payment, which will be 14 days after the particular payment became due. So far as the contractor is concerned, the due amount will be the amount stated by a payment advice as being payable to the contractor upon receipt by the employer of a VAT invoice for that amount. This was considered in more detail when dealing with payment under clause 22 (see paragraph 5-133).

CLAIMING THE INTEREST AND ALL REASONABLE DEBT RECOVERY COSTS

5-176 Where the claim for interest is made under the 1998 Act, in the absence of a valid contractual provision, the creditor can claim statutory interest and all reasonable debt recovery costs. In making such a claim, it would be sensible to include the calculation of the interest to be added to the capital sum together with a daily rate claimed up until payment is made. The amount of compensation for debt recovery costs is determined by a table which currently ranges from £40 for a debt of less than £1000; £70 in respect of a debt between £1000 and less than £10 000; and £100 in respect of a debt of £10 000 or more.

Although a creditor has 6 years in England, Wales and Northern Ireland (5 years in Scotland) in which to make the claim, it is sensible of course to make the claim as early as the circumstances allow.

An extremely useful User's Guide to the late payment legislation can be obtained from DTI Publications, Order Line Admail 528, London SW1 W8YT. It can also be downloaded from the internet by visiting www.payontime.co.uk which also provides other information.

VAT – clause 25

Treatment of VAT

> 25.1 All amounts within the Contract are exclusive of any VAT that may be due to the Contractor in respect of the Project.
>
> 25.2 Where required by applicable legislation VAT shall be added to any payment by either party to the other and the party receiving payment shall provide any documentation reasonably necessary in order to permit such a payment to be properly made.

5-177 It is not proposed in this book to deal in any detail with the question of value added tax. The tax was originally introduced under the Finance Act 1972. VAT legislation has since been significantly amended on a number of occasions.

5-178 For the most part new buildings as well as works of alteration will attract value added tax at the standard rate. There will, however, still be some occasions on which the appropriate value added tax is at the zero rate, e.g. domestic residences and buildings for charitable purposes. Accordingly there will also on occasions be situations where part of the work is standard rated and part zero rated.

The MPF does not provide detailed clauses dealing with this topic, unlike many other JCT contracts which have a supplemental VAT Agreement (e.g. WCD 98 clause 14.1 and the Supplemental Provisions (the VAT Agreement).

Clause 25 makes it clear that all amounts within the contract are exclusive of any VAT (defined in clause 39.2 simply as 'Value Added Tax') due to the contractor and that VAT is to be added to any payment by either party to the other in accordance with the legislation.

If no express mention was made of amounts being exclusive of VAT, it would generally be treated as exclusive in any event, at any rate between those contracting parties who have a knowledge of the practice of the construction industry. This is despite section 19(2) of the Value Added Tax Act 1994 which states that if the supply is for a consideration in money, its value shall be taken to be such amount as, with the addition of the value added tax chargeable, is equal to the consideration. In other words, the Act assumes that a quoted price is inclusive of value added tax. However in the case of *Tony Cox (Dismantlers) Ltd* v. *Jim 5 Ltd* (1996), it was held that in the construction industry there was a notorious, certain and reasonable custom that prices were quoted exclusive of VAT, and this prevailed. On the other hand, if the contract is with an individual outside the construction industry and a price is quoted without reference to VAT, then section 19(2) will apply and it will be treated as VAT inclusive: see *Franks & Collingwood* v. *Gates* (1983) and *Lancaster* v. *Bird* (1998).

Bearing in mind that payment advices will include any amounts that either party is liable to pay the other in accordance with the provisions of the contract, issues are likely to arise as to whether parts of the payment advice will attract VAT, e.g. a judgment of a court that payment be made under the indemnity provisions in clause 26.

Both parties have an obligation to provide any documentation necessary in order to permit payment of VAT to be made and properly accounted for.

The pricing document

Content

5-179 Compared with the existing forms of JCT contract (e.g. WCD 98 and JCT 98), the MPF provides for a greater choice of payment method.

The pricing document is to be found at the very end of the MPF Conditions. It is defined in clause 39.2 as:

> The document identified in the Appendix containing the contract sum analysis and particulars of the manner in which the Contract Sum is to be paid to the Contractor.

It contains the following:

- the rules for determining the manner in which the contractor is to receive payments;
- the contract sum analysis; and
- the pricing information.

These will be dealt with in turn.

Payment rules

BACKGROUND

Although Clause 22 of the MPF sets out the payment mechanism, it is the pricing document which contains the method for determining the manner in which the contractor is to receive payments.

As with the majority of standard forms of building contract, the MPF envisages periodic payments. By completion of the MPF appendix the parties choose one of the four rules that will constitute the agreed method for payment. The options are:

- Rule A - interim valuations;
- Rule B - stage payments;
- Rule C - progress payments;
- Rule D - some other method, as the parties agree and set out in the pricing document.

The appendix provides for a default position when the parties fail to complete the appendix properly and fail to choose one of the options. The default position is that Rule A shall apply.

The MPF, through the pricing document, also envisages that the parties may choose to use advance payments, in which case there is provision for the contractor to provide a supporting bond.

The MPF differs from most other currently used standard forms of building contract in that its payment provisions do not provide for the deduction of retention by the employer from each periodic payment.

The MPF payment provisions do not provide expressly for the payment of unfixed materials (whether stored on or off site). However, since the pricing document allows for the parties to provide their own details in respect of the chosen payment method, such provision can be inserted if required, e.g. by including unfixed materials, whether off or on site, as part of the value of a given stage.

As with the rest of the MPF, the payment provisions (including the pricing document) are much shorter and simpler than other standard forms of building contract, including those of other JCT contracts. In light of this, careful consideration needs to be given by both the employer and contractor in completing and supplementing the pricing document to ensure all relevant issues are taken into account, and all relevant documents are properly prepared and appended.

THE RULES FOR DETERMINING THE MANNER IN WHICH THE CONTRACTOR IS TO RECEIVE PAYMENT

Unlike many JCT forms, e.g. WCD 98, the MPF provides considerable flexibility for the parties in their choice of payment method.

Rule A - Interim valuations

P3.1 The proportion of the Contract Sum to be included in an interim payment advice shall be the value of work properly executed by the Contractor up to a date 7 days prior to the issue of the payment advice, determined by reference to the rates and prices in the contract sum analysis.

This is the traditional and most commonly used method of payment in building contracts.

The essence in this sort of payment mechanism is that the employer, or in some contracts a third party certifier, will carry out an assessment of the contractor's completed work within the relevant period for the purpose of valuing it. That assessment will usually involve a site visit and inspection of the actual physical work undertaken on the project, and may involve the measurement and valuation of various quantities of work, materials and the like. In the case of the MPF, that valuation is conducted and determined in accordance with the rates and prices contained in the contract sum analysis. The valuation is to take place monthly and includes the value of work properly carried out up to a date 7 days prior to the issue of the payment advice under clause 22.2.

The interim valuation method is commonly chosen where the project is straightforward and the employer is self funded, and/or any significant funder is not concerned with absolute and certain drawdowns each month as part of any funding agreement.

This method of valuation places the least risk on the contractor as he is paid each month a proportion of the contract sum for the work he has completed.

The disadvantage for the employer with this type of payment method is that there is little certainty as to what sum he will have to pay in any month. There is also the possibility that the contractor may have artificially manipulated the pricing of the project so that he is paid a disproportionately large value in the earlier payment months to improve cash flow.

Rule B – Stage payments

P4.1 The proportion of the Contract Sum to be included in an interim payment advice shall be the total of the amounts identified in the contract sum analysis in respect of all of the stages completed by a date 7 days prior to the issue of the payment advice.

P4.2 Where the contract sum analysis indicates that any stage payment is to be treated as an advance payment then the issue of a payment advice in respect of that payment is conditional upon the receipt by the Employer of a bond in the form of the attached draft.

This method of payment is quite common in building contracts and is chosen where it is important to the employer that the contractor completes certain discrete elements or stages of the work in a certain sequence. An example of this might be a hospital project where it is critical to the employer that certain works, access to the same and key mechanical and electrical services are provided early in the project, with other non-critical elements following later.

So long as there is a carefully developed contract sum analysis appended to the pricing document which identifies such stages, and with the contractor required to price each stage separately, the monthly payment task can be a relatively simple one. As each stage is likely to be a tangible and completed element of the project, the necessity for detailed measurement of sub-elements of the works is likely to be dispensed with, so reducing the employer's involvement and consequent costs.

This method can also provide certainty to the employer as to the level of likely payments, if he monitors the contractor's progress sufficiently carefully to allow an accurate prediction to be made as to whether the stage is likely to be completed in any particular payment period. In the MPF, interim payment advices are to include any stages completed by 7 days prior to its issue.

If any stage payment is treated by the contract sum analysis as an advance payment, the issue of the payment advice for that payment is conditional on the receipt by the employer of an advance payment bond from the contractor in whatever form is attached to the pricing document. Advance payment is considered further in paragraph 5-180.

Where there are any changes to the project which affect the completion date for a stage and therefore the trigger for payment, it seems that these are valued and paid for separately in monthly payment advices (clause 22.5.2). While this can assist the contractor's cash flow where the effect is to delay completion of the stage, it may still create problems, e.g. a minor change involving materials on long delivery which prevents completion of the stage. No doubt the contractor would seek the cost of being stood out of money as part of the loss and expense element of the valuation of the change (see clause 20.6.4). Alternatively, the pricing document may expressly cater for the effect of changes.

Rule C - Progress payments

P5.1 The proportion of the Contract Sum to be included in an interim payment advice shall be as set out in the attached schedule.

P5.2 Where the schedule indicates that any progress payment is to be treated as an advance payment then the issue of a payment advice in respect of that payment is conditional upon the receipt by the Employer of a bond in the form of the attached draft.

P5.3 Where the Employer considers that Practical Completion will not be achieved by the date stated in the Appendix, whether due to some matter entitling the Contractor to an adjustment to the Completion Date or otherwise, then the Employer shall notify the Contractor of the date when it considers Practical Completion is likely to be achieved and, in consultation with the Contractor, determine the reasonable amendments necessary to the schedule.

P5.4 Any amendments made shall have regard to the prolonged duration over which future payments are to be made and reflect the proportions and intervals indicated by the schedule in respect of those payments that have not yet been the subject of a payment advice. No amendment shall be made in respect of payments that should have been the subject of a payment advice by the date of the Employer's notification to the Contractor under the clause P5.3.

P5.5 The amendments to the schedule shall take effect in relation to any payment advice issued by the Employer more than 14 days after the notification under clause P5.3.

Progress payments are often considered to be the best and most appropriate method of periodic payments where certainty is a key factor for the employer. Such a factor will often be driven by the employer's funder.

A schedule of periodic fixed payments is drawn up and included in the contract (in the case of the MPF in the schedule attached to the pricing document), and irrespective of the work completed in any payment period, the relevant fixed progress payment contained in the schedule for that particular period is the sum that the employer pays to the contractor, no more, no less.

An example of a variant of this method of payment is where the fixed sums in the schedule are used as a maximum sum payable unless the contractor actually completes a lesser value of work in which case he is paid the lesser sum in accordance with an interim valuation method. Variants of this sort are probably best included in the Rule D payment option.

The list of attachments to the pricing document is to include the payment schedule for use with Rule C.

Clauses P5.3 to P5.5 provide for the reasonable, necessary amendment of the schedule of fixed payments by the employer, in the event of delays to the project. This provision is sensible in that it prevents the situation occurring where, due to delays, the contractor may not have carried out sufficient work (in any one month) but, without an adjustment to the fixed payment, would nevertheless be due the full fixed payment. Though the contractor is to be consulted, his consent is not required to any adjustment. If the contractor is aggrieved with the adjustments made to the payment schedule, he could adjudicate or litigate the issue of whether they were 'reasonable amendments necessary to the schedule' (clause P5.3).

It should be noted by employers that it is their obligation under Rule C to give notice to the contractor stating when they consider practical completion will be achieved before they have the right to make amendments to the schedule.

In light of the risk, mentioned above, that the contractor (due to delays) may qualify for his full fixed periodic payment under Rule C even though he has not carried out a similar value of work, and the fact that in most situations it will be the contractor who has the most complete and up to date information on events which have delayed his progress on site, it might have been sensible for the clause to also require the contractor to notify the employer if he considered it unlikely that practical completion would be achieved by the contract completion date.

In the event of changes to the project, a literal interpretation of Rule P5 and clauses 22.5 1 and 22.5.2 of the MPF would suggest that the scheduled repayments are covered by 22.5.1 and that changes are paid for in addition. This might not tie in with any funding mechanism which is in place, though the employer and any funder will be well aware of the possibility of changes and will have no doubt made the necessary financial arrangements to cater for this. The pricing document may well cover the point specifically.

Rule D – Some other method

P6.1 The Contractor shall receive payment of the Contract Sum in the manner set out in:

____

____

Rule D provides the parties with the opportunity to devise and include their own hybrid method of payment, if Rules A to C do not provide the desired solution.

Rule D could be used for any variant of any of the payment methods referred to above.

When utilising Rule D, the method chosen is to be listed in the attachments to the pricing document and of course attached to it.

The contract sum analysis and pricing information

P7.1 The contract sum analysis and pricing information is attached.

P7.2 The contract sum analysis sets out the manner in which the Contractor has calculated the Contract Sum in the detail and contains such additional information as is specified by the Requirements.

P7.3 Where there is more than one Section, the contract sum analysis identifies the value of each Section.

P7.4 Where Rule B applies, the contract sum analysis also identifies the stages into which the Project is divided for payment purposes and the amounts applicable to each stage.

P7.5 The pricing information comprises such information as is specified by the Requirements or provided by the Contractor for use in the valuation of Changes.

Attachments

The form of advance payment bond to be provided by the Contractor:*

The Schedule for the operation of Rule C, comprising:*

The method referred to by Rule D:*

The contract sum analysis, comprising:

The pricing information, comprising:

---

*Delete as applicable

Form of contract sum analysis

No pre-determined form of contract sum analysis is provided within the pricing document, or elsewhere in the MPF.

There is guidance on the recommended form of the contract sum analysis (for use with JCT 81 WCD), including the preferred breakdown of the contract sum, in Practice Note CD/IB (August 1995), which the parties may find useful.

The form of contract sum analysis is important as it is the document with which the employer will assess and determine what payment is due to the contractor at each payment interval.

If the form of contract sum analysis that the contractor is asked to price and complete does not provide sufficiently certain, defined and detailed information and breakdowns (into relevant stages under Rule B, stage payments, and into relevant unit rates and prices for interim valuations under Rule A), the periodic payment task could be fraught with difficulties for the employer and could even lead to disputes as to the payment due.

Where there is a dispute between the employer and contractor over the value of a periodic payment (either due to the inadequate nature of the contract sum analysis, how it has been interpreted or otherwise), the parties will be free to refer their dispute under the contract to adjudication in accordance with Clause 37.

In light of the above, much care should be taken by the employer (and his professional advisors where employed) in the early formulation of the contract sum analysis. Similarly, contractors should give due attention to both the adequacy of the proposed contract sum analysis (as contained in the requirements) and to the pricing and completion of it.

Clause P7.3 provides that the contract sum analysis is to identify the value of any sections (identified in the appendix) into which the project is divided. This is relevant to the operation of the liquidated damages provision (clause 10) and any bonus payments (clause 14).

Pricing information

The pricing information is defined by clause P7.5 of the pricing document as being:

> . . . such information as is specified by the Requirements or provided by the Contractor for use in the valuation of Changes.

Again, as with the contract sum analysis, there is no form for the pricing information included within the MPF.

As it will form the basis of the fair valuation of changes under clause 20, the content of the pricing information is very important.

On some large and complex projects it may be that parties will consider that a bill of quantities (providing measured quantities and priced unit rates for all items of work) is appropriate as the pricing information for inclusion in the pricing document. Otherwise, a schedule of rates for typical work may be more appropriate. Either way, careful consideration needs to be given to both the form of pricing information required by the requirements and how it is to be priced by the contractor in his proposals.

Employers should consider how the pricing information will relate to the contract sum analysis. If the two are not sufficiently related, a situation could arise where the contractor can complete the pricing information without fear of making his tendered price uncompetitive. To avoid artificially high rates in the pricing information, and where no bill of quantities has been prepared which can total the full contract sum, the employer could consider preparing and including approximate quantities against the items in the schedule of rates which tendering contractors are required to price, the total of which is transferred to the tendered contract sum.

Other points for consideration

Errors in the contract sum analysis and pricing information

5-180 As considered earlier (see paragraph 5-57), the contractor and employer will generally be bound by errors and omissions in both the contract sum and contract sum analysis.

This will no doubt also be the case for errors or omissions that may be found in the 'pricing information', attached to the pricing document for the purpose of valuing changes. This is so whether the effect of such an error or omission has the result of providing the contractor with a windfall or loss (in the case of the valuation of a change) or a cash flow bonus or penalty (in the case of a periodic payment).

Advance payment bond

Payment Rules B and C (at clauses P4.2 and P5.2 respectively) include provision for the contractor to receive an advance payment, where and to the extent that any contract sum analysis (Rule A) or the payment schedule (Rule B) so provides. If this is so, the issue of a payment advice is conditional on the contractor providing the employer with a bond in a form attached to the pricing document.

There are various things that the employer should consider in light of these provisions. Firstly, there is no form of wording for the proposed bond included within the MPF form. It is, therefore, for the parties to agree to the form of wording of the bond and append it to the pricing document in advance of entering into the contract. The MPF differs in this respect from WCD 98, which includes a form of bond wording agreed between The British Bankers Association and the JCT. It may be that the parties to any MPF contract will adopt the same form. As with all agreements and contracts, the employer would be well advised to seek specialist advice concerning any form of bond he chooses to include in the pricing document. Similarly, and before entering into a contract, the contractor should seek advice from his usual surety and/or legal advisor concerning the suitability of the form of bond being proposed by the employer.

The provision for advance payment is a real benefit to contractors in that it assists cash flow, therefore reducing the cost of borrowing money to finance a project. In light of that commercial consideration, the sooner tendering contractors are made aware that advance payments will be permitted by the employer the better, as they can take this into account in preparing their tenders.

Contractors will often have no choice but to commit themselves to significant expenditure on a project well in advance of receipt of the first periodic payment from the employer. Examples of such early commitment on expenditure would be the payment to consultants employed by the contractor for early design work, and payments to suppliers of the contractor for long order material such as structural steel work.

Of course, there is always a risk to the employer in making an early or advance payment under a building contract. This is the case even when the advance payment is supported by a bond. The process will inevitably involve the employer in additional expenditure and management time in dealing with both the advance payment itself and agreeing the form of bond and ensuring it is procured properly by the contractor. The employer should therefore consider both the risk and any real benefit he might obtain before including the provision for an advance payment bond in the contract.

It is curious to note that, although the provision for an advance payment bond is provided in the pricing document for both Rules B and C, there is no such provision for Rule A. There seems to be no reason why an advance payment should not apply to Rule A and it may be that some users will amend the MPF form in this respect.

Where the pricing document does include for an advance payment, and where the parties intend to utilise the same, two further important issues need consideration. The first is that there appears to be no confirmation that the contractor will be responsible for meeting the cost of procuring the bond, and while this cost is likely to fall on the contractor, it may be that clauses P4.2 and P5.2 should be amended by way of clarification. Secondly, clause 22.5 (last sentence) provides that a payment advice will take into account any previous payments in determining the amount due. This means that any advance payment will be taken into account at the first opportunity in order to reduce the employer's risk. If this is not the intention, e.g. the employer's recovery is to be spread over a number of payments, this will have to be reflected in the contract sum analysis or by amending clause 22.5. Indeed, left as it is, clause 22.5 (e.g. where the advance payment is greater than the first valuation) will have the effect of requiring the contractor to repay money directly to the contractor rather than setting it against the value of work carried out.

'... work properly executed...'

It should be noted that the words 'properly executed' only feature in Rule A and not Rule B (Stage Payments). There appears no reason why Clause P4.1 of Rule B did not also include, in the third line, the word 'properly' before 'completed'.

Discrepancies between the completed appendix and pricing document

The Appendix to the MPF (see under the reference to clause 22 – Payments) assumes monthly payments. So far as Rule B is concerned, this can only work on the basis that monthly payment advices will include the value of completed stages together with the separate valuation of changes which will therefore not be treated as being within any stage unless the contract sum analysis provides otherwise.

The appendix provision in the MPF for monthly payments could give rise to a conflict where the parties choose Rule C and the frequency of periodic payments is other than monthly. If this appendix item is not completed so that the default provision for monthly payments on the 28th of each month applies, while it is likely that the particular provisions of the payment schedule under Rule C would prevail, care should nevertheless be taken to amend this appendix item. (Clause 4 of the MPF which addresses conflicts and discrepancies does not appear to deal with conflicts within the contract documents other than within and between the requirements and proposals.)

The effect of the pricing document in relation to taking over any part of the project and the applicable rate of liquidated damages

Clause 11.3 provides that on the employer taking over part of the project or part of any section of it, the pricing document will be used to calculate the relevant reduction in the rate of liquidated damages applicable. It is likely to be the contract sum analysis (and/or the values within any applicable milestone or stage payment in accordance with Rule B) that will be used for such calculation. Where Rule C (Progress Payments) is used, there may be little information with which to value the part taken over in relation to the contract sum. If there is a possibility of clause 11 applying where Rule C is to be used, the employer should consider requesting some form of contract sum analysis from the contractor (as part of his tender) for this purpose.

The MPF does not provide any required or suggested form of contract sum analysis. It is to the employer's benefit to consider what that form should be to enable such a calculation to be made during the project's execution stage. Some thought should be given therefore to those parts of the project that could lend themselves to early takeover by the employer, and to ensure that the contractor is presented with, and requested to price, a contract sum analysis which is in an appropriate format to enable such calculation.

On receipt of tenders from bidding contractors, the employer should take care to consider the pricing of elements within the contract sum analysis to ensure there has been no manipulation of the allocation of prices that might have the effect (whether intentional on the contractor's part or not) of placing the employer at a disadvantage, e.g. the value of the first part to be taken over being of a disproportionately high value to the remaining parts, giving the contractor a larger than justified payment and reducing liquidated and ascertained damages for the remaining parts disproportionately.

Other clauses and mechanisms in the MPF to which the pricing document is relevant

In addition to clauses already mentioned, the pricing document is referred to in the following further clauses within the MPF:

- Clause 16.2.2 – concerning the price that the contractor must pay to the employer for any work executed, or materials or goods supplied for the project which are not in accordance with the contract but which the employer allows the contractor to use.
- Clause 20.6.3 – concerning the fair valuation of any change using the prices and principles set out in the pricing document.
- Clause 22.5.1 – payments to the contractor from the employer are made in accordance with, and the proportions are calculated in, the manner set out in the pricing document.
- The first page of the appendix provides a space, to be completed by the parties, to define and identify the pricing document. It is suggested that the parties do this by entering the document's date and reference, including any relevant revision number where applicable.

Chapter 6
Indemnities and insurance

Content

6-01 This chapter looks at indemnities and insurances covered by clauses 26 to 28 of the MPF. These will be considered in turn. They deal with the following matters:

- Clause 26 – claims arising out of personal injury and loss or injury or damage to property as a result of the carrying out of the project, together with indemnities related to these.
- Clause 27 – liability insurance and insurance for the project. This is dealt with differently in the MPF form when compared with other JCT forms.
- Clause 28 – provides, by completion of the appropriate item in the appendix to the contract, for the contractor to obtain professional indemnity cover.

Indemnities – clause 26

Third party claims and indemnities

Background

6-02 Most standard forms of building contract will contain express clauses dealing with third party claims, i.e. third party claims from the employer's point of view, arising out of the carrying out of the contract project; in other words, the situation in which a stranger to the contract claims against the employer in respect of a wrong done to the stranger in the form of personal injury or physical damage to his property arising out of the carrying out of the project. Typically express contractual provisions will require the contractor in certain circumstances to indemnify the employer against any loss, damage, claims, proceedings, costs, etc. for which the employer becomes responsible to the third party. The MPF provides for this situation but also contains a cross-indemnity where the contractor is claimed against because of something for which the employer is responsible.

Such an indemnity will generally be coupled with an obligation on the contractor to insure himself against liability in respect of such claims. That liability – and accordingly the insurance to cover it – may well extend to sub-contractors employed by a main contractor. As will be seen when looking at insurances later (paragraph 6-17 *et seq.*), the MPF does not expressly provide that this or any other type of insurance will be provided; instead, it provides that any insurances required are to be specified in the appendix whether for public liability, employer's liability or damage to the project or existing structures. Clause 27 merely sets out general provisions in relation to any of these chosen insurances.

STRICT INTERPRETATION OF INDEMNITY CLAUSES

6-03 Indemnity provisions have tended to be construed by the courts very strictly against the party seeking to rely on them, especially where the provision would have the effect of indemnifying one party to the contract in respect of his own negligence. This is so even where the contract contains equivalent cross-indemnities so that either party can benefit or suffer: see *E.E. Caledonia Ltd* v. *Orbit Valve Co* (1993). It is worth stating that the Unfair Contract Terms Act 1977 does not purport to control indemnity provisions in non-consumer transactions so that the case law on this topic prior to that Act is still very relevant. However, exclusion clauses in respect of liability for negligence are subject to the Act, see sections 1 to 3 inclusive and section 11.

In *Walters* v. *Whessoe Ltd and Shell Refining Co Ltd* (1960) the Court of Appeal had to consider an indemnity clause in a contract between the first and second defendants following a finding that the second defendants had been negligent. The second defendants sought to recover the damages payable by them to the plaintiffs from the first defendants under an indemnity clause, despite the fact that the second defendants had themselves been at fault when an industrial accident caused the death of an employee of the first defendants. The clause required that the first defendants:

> 'shall indemnify and hold Shell [the second defendants] their servants and agents free and harmless against all claims arising out of the operations being undertaken by [the first defendants] in pursuance of this contract or order or incidental thereto.'

The court held that the reference to 'all claims' did not indemnify Shell against the results of their own negligence. Lord Justice Sellars said:

> 'It is well established that indemnity will not lie in respect of loss due to a person's own negligence or that of his servants unless adequate and clear words are used or unless the indemnity could have no reasonable meaning or application unless so applied.'

Lord Justice Devlin said:

> 'It is now well established that if a person obtains an indemnity against the consequences of certain acts, the indemnity is not to be construed so as to include the consequences of his own negligence unless those consequences are covered either expressly or by necessary implication. They are covered by necessary implication if there is no other subject matter upon which the indemnity could operate.'

In *Canada Steamship Lines Ltd* v. *The King* (1952), a Privy Council case, Lord Morton of Henryton set out the approach of the courts to indemnity (and exclusion) clauses as follows:

> '(1) If the clause contains language which expressly exempts the person in whose favour it is made (hereafter called "the proferens") from the consequence of the negligence of his own servants, effect must be given to that provision...
> (2) If there is no express reference to negligence, the court must consider whether the words used are wide enough, in their ordinary meaning, to cover negligence on the part of the servants of the proferens...

(3) If the words used are wide enough for the above person, the court must then consider whether "the head of damage may be based on some ground other than negligence", to quote again Lord Greene in the Alderslade case. The "other ground" must not be so fanciful or remote that the proferens cannot be supposed to have desired protection against it, but subject to this qualification, which is no doubt to be implied from Lord Greene's words, the existence of a possible head of damage other than that of negligence is fatal to the proferens even if the words are prima facie wide enough to cover negligence on the part of his servants.'

This test was subsequently applied in the case of *Smith* v. *South Wales Switchgear Co Ltd* (1978) and again by the Court of Appeal in *Dorset County Council* v. *Southern Felt Roofing Co Ltd* (1989). However, if the wording is such as to make it clear that one contracting party is taking the risk of loss or damage even if caused by the other contracting party's negligence, this will be given effect. To achieve this it is necessary to expressly refer to the 'negligence' or a suitable synonym (*Smith* v. *South Wales Switchgear Co Ltd* (1978)).

APPORTIONMENT

6-04 Where the indemnity provision does not expressly or by implication cover losses consequent on a person's own negligence, then if the loss is due in part to the fault of the contracting party against whom an indemnity is being sought, and in part due to the fault of the party seeking to enforce the indemnity, unless the indemnity provision expressly provides for an apportionment to be made, the indemnity provision will fail in its entirety: see *A.M.F. (International) Ltd* v. *Magnet Bowling Ltd and G.P. Trentham* (1968).

LIMITATION PERIODS

6-05 The application of the statutory limitation periods in connection with actions on indemnities should be noted. The limitation period under the Limitation Act 1980 will operate from the date when the cause of action on the indemnity accrues, i.e. most probably when the person claiming the indemnity actually incurred a liability, e.g. when the third party obtains judgment against him. In the case of *County & District Properties Ltd* v. *C. Jenner & Son and Others* (1976) Mr Justice Swanwick said:

'...an indemnity against a breach, or an act, or an omission, can only be an indemnity against the harmful consequences that may flow from it, and I take the law to be that the indemnity does not give rise to a cause of action until those consequences are ascertained.'

If this is a correct statement of the law, in a typical third party claim, while it is the date of the damage to the claimant third party which will start time running against the defendant, if that defendant in turn seeks to rely on an indemnity against another defendant, time will not start to run until the first defendant's liability has been ascertained. Only then will his cause of action under the indemnity provisions accrue. This can make such clauses extremely onerous. However, a contrary view – that the cause of action accrues, and therefore time begins to run, as soon as the original loss, damage, claim, etc. has been incurred, suffered or made – has received some support as a result of the House of Lords case of *Scott Lithgow Ltd* v. *Secretary of State for Defence* (1989).

This point arose in *City of London* v. *Reeve & Co Ltd and Others* (2000) before Judge

Hicks QC sitting in the Technology and Construction Court. The judge held that the answer depended upon the particular circumstances and the indemnity clause being considered. In paragraph 27 of his judgment he said:

> 'In dealing with the question when the cause of action on a contract to indemnify against liability to a third party accrues *Chitty on Contracts* comments with some justification that the modern authorities are conflicting... The conflict, put shortly, is whether the cause of action accrues as soon as liability is incurred or only at some later date, such as when it is quantified, or established by judgment, or realised by payment. If the generic term is to be used in each case the contrast is perhaps best conveyed by distinguishing between the ''inception'' and the ''realisation'' of liability.'

In this particular case the cause of action accrued upon the realisation of liability to the third party. The judge remarked in passing that generally it was more likely that an indemnity clause operated from the realisation of the liability to the third party rather than from the inception of it.

As a matter of convenience and expediency, a defendant can ask the court to join into the proceedings between the plaintiff and the defendant the person from whom an indemnity is being sought so that if any liability does attach to the defendant, the question of the indemnity can be considered immediately afterwards - see the Civil Procedure Rules Part 19.

OTHER POINTS ON INDEMNITIES

6-06 Contractual indemnities have a number of other advantages or disadvantages depending on whether the party concerned is the beneficiary under an indemnity or the giver of it. For example, if the act in respect of which the indemnity is given is a negligent act, the indemnity may be capable of extending to the recovery of purely economic loss, even though this may not otherwise have been recoverable against the negligent party. Additionally, the ordinary contractual rules as to remoteness of damage may not apply to the same extent in relation to indemnities which are worded so as to expressly cover all loss, expense, etc. resulting from the act complained of, whether or not the nature of that loss or expense is too remote in terms of ordinary contractual principles as to the measure of damages.

The provisions of clause 26

6-07 Clause 26 of the MPF provides firstly in clause 26.1 for the employer to be indemnified by the contractor, and then in clause 26.2 for the contractor to be indemnified by the employer in stated circumstances which will now be considered. Both sub-clauses are drafted in very similar fashion and are in the author's view a model of clarity and precision.

Contractor to indemnify employer

> 26.1 The Contractor shall be liable for and shall indemnify the Employer against any expense, liability, loss, claim or proceedings arising under statute or at common law in respect of:

.1 the personal injury to or the death of any person; and
.2 the loss, injury or damage to any property real or personal,

to the extent that such expense, liability, loss, claim or proceedings arise out of or in the course of carrying out of the Project and not as a consequence of some act or neglect on the part of the Employer or any person for whom the Employer is responsible (excluding the Contractor but including Others on the Site) but excluding any amount recoverable (or which but for any default by the Employer, excess or insurer's insolvency would have been recoverable) by the Employer under any policy required by clause 27.

6-08 Clause 26.1 deals with the indemnity of the employer by the contractor where the employer faces any expense, liability, loss, claim or proceedings arising under any statute or at common law in respect of personal injury or death and for loss injury or damage to any property, real or personal, arising out of or in the course of carrying out the project.

Basis of indemnity

6-09 It is interesting to note with regard to clause 26.1 (contractor indemnifying employer) that the contractor indemnifies the employer except to the extent that the expense, liability, loss, claim or proceedings is a consequence of any act or neglect of the employer or those for whom he is responsible; whereas, in clause 26.2 (employer indemnifying contractor), the position is reversed so that the employer only indemnifies the contractor to the extent that the contractor can prove that the expense, liability, loss, claim or proceedings arise as a consequence of any act or neglect of the employer, or those for whom he is responsible. The onus of proving that the employer should accept the liability or any part of it seems therefore to rest with the contractor whichever party is seeking the indemnity.

Responsibility

6-10 If the employer suffers any loss or expense or is held liable to a third party either under statute or at common law, the contractor must indemnify the employer in respect of such loss, expense or claim but not to the extent that the expense, liability, loss, claim or proceedings are due to the act or neglect of the employer or any person for whom the employer is responsible. This excludes the contractor but includes 'Others' on site. 'Others' is defined in clause 39.2 as:

> Persons whose presence on the Site has been authorised by the Employer, other than the Contractor, its sub-contractors and suppliers and any other persons under the control and direction of the Contractor.

Having regard to this definition, the prior reference in the clause to 'excluding the Contractor' appears superfluous and indeed even confusing as there is no express reference to sub-contractors and suppliers etc. This inclusion of authorised persons alters what would otherwise be the general law applicable in such circumstances. Generally, though with exceptions, the employer would not be responsible for the negligence of independent contractors appointed by him provided reasonable care was taken in their selection. However, clause 26 expressly provides that the employer is responsible for such independent contractors if their presence on site is authorised. Some employers may seek to amend clause 26 in relation to this.

'... but excluding any amount recoverable ... by the employer under any policy required by clause 27'

6-11 The contractor's obligation to indemnify the employer excludes any amount which the employer recovers (or but for his default, any excess or the insolvency of the insurer, he would have recovered) under any policy required by clause 27. It is important to appreciate that, unlike many other JCT contracts, e.g. WCD 98 (clauses 20.2 and 20.3), the contractor's indemnity includes loss or damage to the project itself. What the above words do is to limit the amount recoverable by the employer from the contractor to the extent that such amount is recoverable from the insurers. The effect is to prevent the insurers from paying out to the employer and then seeking to recover this sum from the contractor under rights of subrogation. This removal of subrogation rights will apply whether or not the contractor is a joint insured. The benefit of this limitation on recovery is not extended to sub-contractors. If any sub-contractor is not covered by the policy the insurers would therefore be able to exercise their subrogation rights against him.

IS LIABILITY REMOVED?

6-12 It is certainly arguable that contractor's liability, e.g. for negligence, is not removed, but only the ability of the employer (and his insurers) to recover certain amounts paid out under an insurance policy. This is a crucial point. If this is the position, then in circumstances such as those occurring in *Co-operative Retail Services Limited and Others* v. *Taylor Young Partnership and Others* (2002) (House of Lords) (see paragraph 6-29), any professionals or sub-contractors not covered by the project insurance policy would be able to claim a contribution from the contractor under section 1(1) of the Civil Liability (Contribution) Act 1978, as the contractor would be liable in respect of the same damage.

To remove the contractor's (under clause 26.1) and employer's (under clause 26.2) liability for negligence (and therefore any duty of care) would require clear and specific wording putting that issue beyond doubt. Clause 26 gets nowhere near this. The exclusion in relation to amounts recoverable under an insurance policy can be attributed to a specific purpose, namely to remove any subrogation rights which insurers might otherwise have. Further, even if the exclusion could be treated as an attempt to remove liability for negligence, it does not differentiate between loss or damage to property and personal injury. It seems clear from the wording of clauses 26.1 and 26.2 as a whole that liability is retained.

Apportionment

6-13 It should be noted that the words 'to the extent' ensure that an apportionment can be made. Clearly, if the expense etc. sustained by the employer is entirely due to his own 'act or neglect' or the act or neglect of those for whom he is responsible, then no liability rests with the contractor. If, however, the employer is, or those for whom he is responsible are, only partly at fault, liability will be apportioned between the parties, the degree of apportionment reflecting the contractor's share of that liability.

The question of apportionment was considered in a case involving similar wording as that in clause 26, including reference to the employer's 'act or neglect', namely *Barclays Bank Plc* v. *Fairclough Building Ltd and Others* (1993). At first instance (not affected on appeal), Judge Havery QC said:

> 'The clause offers no guidance how one should assess the extent to which a death or personal injury is due to an act or omission of the employer. It seems that culpability is irrelevant. In my judgment, the wording of the clause does not justify the court in apportioning liability on the basis provided for in section 1 of the (Law Reform (Contributory Negligence) Act 1945).'

The apportionment provision appears therefore to include situations where the employer is not at fault in a culpable sense. Accordingly, section 1 of the 1945 Act, which effectively deals with apportionment on the basis of fault, is not appropriate.

'... arising out of...'

6-14 Similar words were considered in the case of *Richardson* v. *Buckinghamshire County Council and Others* (1971) in connection with the ICE Conditions of Contract 4th Edition, clause 22(1), which is an indemnity clause. The plaintiff had been injured when he fell from his motorcycle at the point of some road works which were the subject of the contract. The defendant local authority successfully resisted a claim from the motorcyclist but the costs which were incurred in so doing were irrecoverable from the plaintiff as he was legally aided. The local authority therefore sought to recover these costs from the contractor under the wording quoted above, contending that the costs were incurred as a result of a claim arising out of the construction of the contract project. It was held that the local authority was not entitled to an indemnity from the contractor as the motorcyclist's claim was not one which arose out of or, as a consequence of, the carrying out of the construction of the works. This clearly makes good sense, for otherwise a contractor could find himself having to indemnify an employer in respect of the employer's costs in defending all manner of specious or far-fetched claims. As the motorcyclist could not establish that his injuries arose out of the construction of the road works, it could not be said that the costs arose out of the construction of the works.

Loss, injury or damage

6-15 Clause 26.1.2 deals specifically with the indemnity of the employer by the contractor against any expense, liability, loss, claim or proceedings in respect of any 'loss, injury or damage'.

While the words 'loss' and 'damage' have a reasonably precise meaning, the use of the term 'injury' is not so precise and there is consequently a question as to whether it could be held to apply to, say, interference with the reasonable access by third parties to their property or to a similar situation where the use of a third party's building is seriously affected. There is a possibility that it could also extend to such matters as the infringement of a copyright or patent.

Employer to indemnify contractor

> 26.2 The Employer shall be liable for and shall indemnify the Contractor against any expense, liability, loss, claim or proceedings arising under statute or at common law in respect of:

> 26.2.1 the personal injury to or the death of any person; and
>
> 26.2.2 the loss, injury or damage to any property real or personal,
>
> to the extent that such expense, liability, loss, claim or proceedings arise out of or in the course of carrying out of the Project as a consequence of some act or neglect on the part of the Employer or any person for whom the Employer is responsible (excluding the Contractor but including Others on the Site) but excluding any amount recoverable (or which but for any default by the Contractor, excess or insurer's insolvency would have been recoverable) by the Contractor under any policy required by clause 27.

6-16 Clause 26.2 is the reverse of clause 26.1 as here it is the employer who has a liability and an obligation to indemnify the contractor. Much of what is said above in relation to clause 26.1 is therefore relevant here also. The only difference is, as already mentioned (see paragraph 6-09), that the contractor has to prove the act or neglect on the employer's part so that whether it is the employer under clause 26.1, or the contractor under clause 26.2, seeking an indemnity, it appears to be the contractor who has to establish the act or neglect of the employer if the contractor is to reduce his own liability or establish that of the employer.

Again, as under clause 26.1, if the project is insured by the contractor against loss or damage, whether such insurance extends to the employer or not, the insurer's rights of subrogation are removed but other contribution claims are possible (see paragraph 6-12).

In the nature of things, as the contractor is the party responsible for carrying out the project, claims by the contractor for an indemnity from the employer are likely to be far less frequent than claims by the employer against the contractor.

Insurances – clause 27

Types and extent of cover

6-17 The MPF marks a major departure in respect of JCT contracts aimed at larger projects. All of the other relevant JCT contracts deal in some detail in the conditions with both liability insurance and works and existing structures insurance. For example, in WCD 98, clause 21 deals with insurance against injury to people and property (other than the works). Clause 22 deals extensively with insurance of the works. In both cases, the nature and extent of the insurance cover required is stated. For example, in relation to works insurance, there is a requirement for it be 'All Risks', and insurance for existing structures must be at least for 'Specified Perils'. The MPF is different. The type and extent of cover is set out in documents referred to in the appendix to the contract and attached to the contract itself. It does not in the conditions specify the type and nature of the cover required.

The rationale behind this approach is that in very many instances, large and complex projects have specific insurance arrangements tailored specifically for the needs of the project and the parties, particularly the employer. This can include a dedicated project policy offering a wide range of cover in respect of liability and property risks. However, it is important to make the point that where appropriate, the parties can readily make use of existing general policies which they may have in place, e.g. the contractor's all risks policy may be quite suitable in many situations. Experienced clients and contractors will already have their own practice and procedures in relation to their insurance arrangements, and these do not need to change because of the MPF. The MPF is intended to reflect current practice, not require the industry to change what is already familiar to it.

What the MPF does do in clause 27 is deal with the administrative aspects of any insurances. For example, the provision of documentary evidence of insurance; the remedies for a failure to insure; notification of claims; the treatment of excesses; and the implications of any cessation of insurance cover for acts of terrorism.

6-18 In clause 27.1 the MPF also deals with any failure of a party to comply with the Joint Fire Code. This is considered below (see paragraph 6-32).

6-19 Again, in a departure from any other JCT main contract forms, clause 28 requires the contractor, if the appropriate entry is made in the appendix to the contract, to maintain professional indemnity insurance in respect of any design and other associated professional obligations undertaken pursuant to the contract.

6-20 The existence of insurance cover against loss or damage to the project is of potential benefit to both parties. For the contractor, it guarantees a fund out of which he can meet his primary obligation to complete the project notwithstanding its damage or destruction. For the employer, it ensures that his right to have the project completed at no additional cost to him, despite its damage or destruction, is not merely a legal remedy with no substance, as it could be if the contractor was both uninsured and impecunious.

As stated earlier, the types of risk in respect of which insurance is required and the extent of cover required are not stated within the conditions. Instead, the insurances to be provided and maintained are to be indicated by virtue of the appendix item for clause 27.

Insurances required by clause 27

6-21 The insurances required by clause 27 are to be provided and maintained in the manner indicated in the appendix to the MPF. The appendix (against the reference to clause 27) provides as follows:

> The policies of insurance to be provided and maintained in accordance with the Contract are those defined by the documents listed in the following table, copies of which documents are attached to the Contract. The party responsible for providing and maintaining each policy of insurance is identified below.
>
> Type of insurance: ______________________________
>
> As detailed in attached documents reference: ______________________________
>
> Insurance to be provided and maintained by: ______________________________
>
> Amount of excess (Clause 27.6)[c] ______________________________
>
> [c] See Guidance Note
>
> [This entry is then repeated a further 4 times]

It can be seen therefore that the completed entries will identify the type of policies required (detailed in attached documents), who is to provide and maintain the insurance and the amount of any excess. In relation to the amount of excess, clause 27.6 provides that the party making a claim under the policy shall pay the excess. The Guidance Note (page 17) makes the point that if a policy provides for a substantial excess, either party may wish to consider taking out some form of separate cover in respect of that element of risk.

6-22 This approach to insurance requirements recognises the reality that many major projects give rise to their own bespoke insurance arrangements. The contract therefore allows for this and for the relevant documentation to be put in place. The contract does not in its conditions attempt to define the scope or terms of the insurances to be provided.

It is vital therefore that those using this contract ensure that appropriate insurance arrangements for the project are put in place. This will almost inevitably mean that the insurance arrangements should be reviewed at an early stage in the preparation of the contracts by insurance representatives of both employer and contractor.

6-23 Certain insurances such as public liability insurance and insuring the project itself will certainly require to be dealt with. Very often other insurances will need to be considered such as insurance of existing structures; insurance in respect of the contractor's plant, materials and equipment; insurances in respect of claims as a result of gradual pollution or environmental impact; and from the employer's point of view insurance in respect of consequential losses as a result of the delay to the project being caused by an insured event for which the contractor would receive an adjustment to the completion date, so removing any liability to liquidated damages which would have compensated the employer for such delay. In a good number of cases latent defect cover for the project may be an option. Additionally, professional indemnity insurance cover in respect of design and other professional liabilities covering consultants and possibly the contractor may also be part of these insurance arrangements. Alternatively, so far as professional indemnity insurance for the contractor is concerned, clause 28 provides expressly for this if selected through the Appendix to the contract.

6-24 Some form of overall project insurance cover will no doubt be considered in appropriate situations.

It is not intended in this book to deal in any detail with the various types of insurable risks and the nature and scope of any insurance which may be advisable to cover them. Parties entering into the MPF should ensure that they take specific advice from insurance brokers or other relevant specialists. However, one or two general points can be made.

Relationship between insurance against loss or damage to the project or existing structures and legal liability under the contract

Loss or damage to the project

6-25 As a general principle, subject to relatively few exceptions, a contractor's obligation to complete the project means that he will be obliged, at his own cost, to repair or reinstate the project should it be damaged or destroyed before completion. The main exception to this is in relation to damage or destruction which is sufficiently fundamental to cause the contract to become frustrated in law, in which case the contractor will be excused further performance. In practice, a building contract will often specify expressly who is to bear the risks in the event of loss or damage to the project before completion. In the case of the MPF there are no express provisions in the conditions themselves, except to a limited extent in relation to adjustments to the completion in respect of '*force majeure*' and the occurrence of 'Specified Perils' (see clauses 12.1.1 and 12.1.2).

If the words used in a contract are sufficiently clear, the contract can enable a party whose negligence has caused the loss or damage to nevertheless impose the responsibility for such loss or damage on the other contracting party. A case in point is *A.E. Farr* v. *The Admiralty* (1953).

FACTS

The plaintiff agreed to build a destroyer wharf on behalf of the defendants. The contract provided that the plaintiff should be responsible for, and should make good, any loss or damage arising from any cause whatsoever. A vessel belonging to the defendants collided with and damaged the wharf as a result of negligent navigation.

HELD

The words 'any cause whatsoever' included negligent navigation of a ship by the defendant's employee and the plaintiffs were accordingly liable under the contract to make good the damage at their own cost.

JOINT NAMES AND ALL RISKS

6-26 Insurance arrangements for a project can take a number of forms. Any such insurance, whoever takes it out, is likely to provide joint names cover, probably on an 'all risks' basis. Typically, all risks insurance provides cover against any physical loss or damage to work executed and site materials. It will also extend to cover the reasonable cost of removal and disposal of debris and any shoring or propping of the project which results from the physical loss or damage. It will be for the full reinstatement value of the project plus a percentage for professional fees. Such a policy will not generally include consequential losses. Express exclusions are likely to include damage to any part of the project resulting from a defect in that part's design, plan, specification, material or workmanship. This exclusion extends to other parts of the project which are lost or damaged in consequence, where such other part relied for its support or stability on the defective work. Resulting damage, for example by fire, to other parts of the project not so dependent will still therefore be covered. Other exclusions can include the following:

- Property which is defective due to wear and tear, obsolescence, deterioration, rust or mildew;
- Consequences of war, invasion, act of foreign enemy, etc;
- Disappearance or shortage if only revealed when an inventory is made or is not traceable to an identifiable event and if the contract is carried out in Northern Ireland then also:
  - Civil commotion;
  - Any unlawful wanton or malicious act committed maliciously in connection with any unlawful association.

There will also be some excepted risks such as ionising radiations or contamination by radioactivity from any nuclear fuel or from any nuclear waste from the combustion of nuclear fuel, radioactive toxic explosive or other hazardous properties of any explosive nuclear assembly or nuclear component thereof, pressure waves caused by aircraft or other aerial devices travelling at sonic or supersonic speeds.

Finally, there may be an exception in respect of cover for terrorist activities.

6-27 If joint names cover is obtained, even if the risk which materialised was due to negligence on the part of either contractor or employer, as they would both be covered by the policy of insurance, the insurers, having paid out under an insurance claim, could not seek recompense from any negligent party. In other words, any right of subrogation which the insurers have is removed.

6-28 Some JCT contracts may also expressly provide that sub-contractors are to be covered by a works policy or alternatively that any subrogation rights against them in favour of the insurers are waived.

EFFECT OF CONTRACT CLAUSES ON LEGAL LIABILITY UNDER THE CONTRACT

6-29 These arrangements can, where they are contained in the conditions of contract (e.g. WCD 98, clause 22) affect the liability of a party (and possibly others such as sub-contractors or if the policy is specific to the project, professionals) in respect of loss or damage caused by their negligence. The leading case which demonstrates this is *Co-operative Retail Services Limited and Others* v. *Taylor Young Partnership and Others* (2002) (House of Lords).

FACTS

This was a case concerning the effect of the joint names insurance requirement in a JCT contract, in this instance clause 22A of the JCT Standard Form of Building Contract 1980 Edition Private With Quantities incorporating amendments 1-2 and 4-11. The relevant sub-contract was based on DOM/1 1980 Edition with amendments 1-3 and 5-9.

A fire occurred at a site in Rochdale where a new headquarters building was being constructed for CRS. For the purposes of a preliminary issue, it was assumed that the fire was caused by the negligence of the main contractor, a sub-contractor and the architects and engineering consultants. The JCT contract provided that the contractor should have in force an all risks policy in the joint names of employer and contractor (which also protected sub-contractors) in respect of loss or damage to the works caused by any of the risks to be covered which included fire as one of the specified perils (defined in clause 1.3 of the main contract).

As the employer, main contractor and sub-contractor were all covered by the policy and therefore co-insured, there could be no subrogation claim on the part of the insurers. The employer therefore, through his insurers who paid out on the claim, sought to claim against the architects and consulting engineers, who were not covered by the joint names policy. The architects and consulting engineers in turn claimed a contribution from the main contractor and sub-contractor under section 1(1) of the Civil Liability (Contribution) Act 1978.

The main issue before the House of Lords was whether the contribution claim was between parties '*liable* in respect of the same damage' as required by section 1(1) of the Act.

HELD

The House of Lords agreed with the Court of Appeal that the effect of the relevant clauses in the JCT contract was such that neither the main contractor nor the sub-contractor were in fact *liable* for negligently caused fire damage. On that basis there could be no contribution claim, even if it produced an inequitable result so far as the consultants were concerned.

Lord Hope of Craighead (at paragraph 39 of the judgment) said:

> 'The issue which lies at the heart of this question is whether the effect of the contractual arrangements as between the parties is to be taken to be that [the main

contractor and the sub-contractor] were never under any obligation to pay compensation to CRS for fire damage caused by their negligence . . . or whether they were under an obligation to pay compensation for that damage to CRS until it was made good in the event that the insurance cover failed or proved to be inadequate . . .'

Lord Craighead in this respect agreed with the judgment of Lord Justice Brooke in the Court of Appeal in holding that under the provisions of the contract, the cost of reinstatement work and professional fees attendant on that work was completely provided for under the contractual scheme. In these circumstances there could be no question of the main contractor being liable to CRS for anything once the contractual scheme had worked itself out, even if otherwise allegations of negligence might have been sustained against them.

This case clearly demonstrates that the insurance arrangements are capable of fundamentally affecting a liability of a contracting party for negligent acts. The insurance funds were to be paid to the employer and the employer was to pay to the contractor (and effectively the sub-contractor) those funds for the purpose of reinstatement. If the funds were insufficient, then the main contractor and sub-contractor still had an obligation to reinstate and complete the works but that was a separate contractual obligation in substitution for any liability for negligence.

In the case of the MPF, the insurance arrangements put in place for each project would need to be considered to assess their effect on any underlying legal or contractual liabilities. However, as has already been discussed in relation to clause 26 (see paragraph 6-12), liability for loss or damage to the project appears arguably to be retained under the MPF and if the intention is that the insurance arrangements should remove legal liability for the contractor's negligence, it would be sensible to amend the MPF to make this perfectly clear. If it is thought that the MPF is ambiguous on this point, the actual insurance arrangements put in place may have the effect of resolving the issue.

Insurance of the project or existing structures and the position of sub-contractors

6-30 The insurance cover for the project may or may not extend to both domestic sub-contractors and named specialists. It may only extend protection in respect of specified perils and not to the extent of cover provided by all risks insurance. This protection is likely to be achieved by the policy either providing for recognition of sub-contractors as an insured party or by the inclusion in the policy of a waiver of any rights of subrogation which the insurers may have against any sub-contractor.

The same benefit might also be extended to sub-contractors in respect of any insurance of existing buildings and contents taken out by the employer. If it is not, it might be thought that if the loss or damage was caused by a sub-contractor's negligence, the employer could sue the sub-contractor in tort. However, there appear to be conflicting decisions on this: see for example *Norwich City Council* v. *Paul Clarke Harvey* (1989); *Ossory Road (Skelmersdale) Ltd* v. *Balfour Beatty Building Ltd and Others* (1993), all suggesting that the domestic sub-contractor does not owe a duty of care to the employer; and *National Trust* v. *Haden Young Ltd* (1994); *London Borough of Barking & Dagenham* v. *Stamford Asphalt Co Ltd and Others* (1997); *British Telecommunications Plc* v. *James Thomson & Sons (Engineers) Ltd* (1998) all suggesting that the sub-contractor does owe such a duty. This last mentioned case, being a

House of Lords decision, carries considerable weight and should now resolve this particular issue.

Obligation to insure and comply with Fire Code

> 27.1 Policies of insurance shall be provided and maintained in the manner indicated by the Appendix and each party shall comply with the terms and conditions of those policies to which it is a party including, where applicable, compliance with the Joint Fire Code. Where either party is notified of any remedial measures considered necessary by an insurer as a consequence of non-compliance with the Joint Fire Code, the other party shall be notified and the Contractor shall implement the remedial measures without delay and this shall not be treated as giving rise to a Change.

6-31 As previously mentioned, whatever insurance is put in place in respect of the carrying out of the project, it is to be maintained in the manner indicated by the appendix to the contract. The use of the word 'indicated' is hardly very specific. Some such word as 'specified' might have been more suitable. Where either the employer or the contractor or both are a party to any particular insurance policy, they have an obligation to comply with its terms and conditions. Failure to do so will therefore be a breach of the contract. Where the breach of the policy terms and conditions has the effect of enabling the insurer to avoid liability under the policy in the event of a claim being made, the party suffering the loss may well find a claim for damages for breach of contract a poor remedy if the other party has insufficient funds to cover the loss which should have been the subject of a valid insurance claim. The policy wording needs to be considered. It is not every breach of the policy conditions which will avoid the insurer's liability under it. Further, many policies still protect the innocent insured party where it is another insured party who has breached policy conditions.

Joint Fire Code

6-32 Where insurance is required in respect of construction work, many insurers require those taking out the insurance cover (often of course both employer and contractor jointly) to comply with The Joint Code of Practice on the Protection from Fire of Construction Sites and Buildings Undergoing Renovation.

Any failure to abide by the code could lead to insurance being withdrawn and this of course would mean one or other of the parties being in breach of their insurance obligations.

By clause 27.1 the parties must also comply with the Joint Fire Code. The wording in respect of this obligation is distinctly inelegant. A short separate sentence requiring compliance with the Fire Code would have been preferable. Nevertheless it is probably adequate to impose a contractual obligation on both parties to comply with the Code.

'Joint Fire Code' is defined in clause 39.2 as:

> The edition of the 'Joint Code of Practice on the Protection from Fire of Construction Sites and Building Undergoing Renovation', published by the Construction Confederation and the Fire Protection Association that is current at any particular time.

The reference at the end of the definition to 'current at any particular time' means that the parties have an obligation to comply with the Code as amended from time to time, and at all times at least until practical completion of the project, and possibly even beyond the rectification period under the contract (see clause 17.3).

If either employer or contractor or both have failed in some respects to comply with the Code, this will of course be a breach of contract. It may also result in the insurer, as a consequence of the non-compliance, stipulating that remedial measures are considered necessary. If so, whichever party is notified of this requirement must notify the other. Further, the contractor is required to implement the remedial measures without delay without this giving rise to a change. This appears to impose a significant risk on the contractor who seems, under this clause, to have to meet the costs of any remedial measures as well as not being able to obtain an adjustment to the completion date where compliance involves delay to the completion of the project. This is in addition to the fact that changes to the Fire Code during the course of the contract, which are more likely to affect the contractor than the employer, are also to be complied with on the assumption that any cost and time implications have been included in the original contract price and period. On the wording of this clause, it could be a breach of the Fire Code for which the employer is responsible, that leads the insurer to require remedial measures and the result is that the contractor is required to implement the remedial measures without this being treated as a change. However, is it the intention of the MPF that the contractor is to have no recompense for this breach? Any such meaning would make this clause, in such a situation, a powerful exclusion clause. It would have to be clear that this was its intended effect, otherwise the contractor would still be able to claim the cost of implementing the remedial measures. The clause is not worded sufficiently explicitly to prevent the contractor claiming damages for breach of contract to cover the cost of carrying out the remedial measures. This may also be the case where the non-compliance with the Code is due to acts carried out by 'Others' (see paragraph 4-12). In addition, there is no reason why, even if the contractor cannot obtain an adjustment to the completion date due to the express wording of the proviso to clause 12.1, he should not be able to obtain reimbursement of loss and expense under clauses 21.2.1 or 21.2.2 as appropriate, despite the provisions of clause 21.8.

Documentary evidence of insurance and failure to insure

27.2 Where a party is required by the Contract to provide and maintain a policy of insurance, the other party may request the production of documentary evidence that the policy has been taken out and remains in force and, apart from any policy required by clause 28 (*Professional Indemnity*), may also request a copy of the policy document.

27.3 Where a party fails to provide the documentary evidence referred to by clause 27.2 within 7 days of a request being made, the other party may assume that there has been a failure to insure. Where there has been a failure to insure by one party the other party may insure against any risk to which it is exposed as a consequence and the party that has failed to insure will be liable to pay the other any costs incurred in taking out and maintaining that insurance.

Documentary evidence

6-33 Clause 27.2 entitles the party who is not required to provide and maintain a particular policy of insurance, to request production of documentary evidence that it has been taken out and remains in force. That documentary evidence may be premium receipts in respect of maintaining a policy in force; a broker's certificate in relation to the scope of cover; or, if there is reasonable doubt as to whether the cover obtained and being maintained meets the agreed requirements, a sight of the policy itself. To a large extent the documentation attached to the contract as a result of the appendix entry for clause 27 will prove to be adequate so far as policy coverage is concerned and a broker's certificate confirming that such cover remains unchanged will be adequate.

6-34 It should be noted that the obligation to provide, on request, a copy of the policy document does not extend to any professional indemnity policy taken out by the contractor pursuant to clause 28. It is common practice for professional indemnity insurers to stipulate to their insured that they must not disclose the terms and conditions of the policy to any third party. Despite this, it is not unusual in practice to find clauses, in building contracts or consultant's appointments, requiring production of the originals of the policy and any other insurance documents as and when reasonably required. Some employers may therefore be tempted to amend this clause to remove the reference to clause 28.

Clause 27.2 is expressed in terms of a request rather than a requirement. In other words, it is probably not a breach of contract to fail to provide the documentary evidence or policy document as the case may be. However, clause 27.3 provides the strongest possible incentive for a party to comply with any such request.

Failure to insure

6-35 By clause 27.3, if a party fails to provide the documentary evidence referred to in clause 27.2 within 7 days of being requested to do so, the other party is entitled to assume that there has been a failure and can then insure against the uninsured risk at the cost of the other party.

In connection with clause 27.3, the first point to note is that the power of a party to insure may be based on either an assumption that there has been a failure to insure where documentary evidence has not been provided, or where there is some other actual evidence that there has been a failure to insure whether or not documentary evidence has been requested.

The express right given to a party to take out insurance in default of the other party insuring is a useful one. There is a need to be sure that there is an adequate fund to meet any claim should the risk materialise. The defaulting party may not be able to financially withstand any claim against him without insurance cover in place. A failure by a party to insure while being a breach of contract and entitling the other party to sue, will be a worthless right against an impecunious defendant. Without such an express right to insure, the party not in default would be placed in difficulty as, although it might be a breach of contract, there would be no damages flowing unless the risk to be insured against materialised. By then of course it would be too late to insure against it.

However, this right to take out the insurance may be easier to state than to implement. Insuring where someone else is the insured can prove very difficult, particularly in terms of information which an insurer may require and which would ordinarily be contained in a proposal form to be completed by the proposed insured.

Where the party not in default wishes to insure on the basis of an assumption that there has been a failure to insure, it is important to note that this assumption is based upon the failure to provide documentary evidence only. It does not extend to a failure to provide a copy of the policy document.

Despite what the first sentence of clause 27.3 provides, if the documentary evidence is in fact provided outside the 7 days, or maybe not provided at all but before the party not in default takes out insurance there is evidence that the insurance is in fact already in place, the assumption of a failure to insure cannot be relied on. In such a case it would be absurd for the party not in default to nevertheless carry on as if there had been a failure to insure when it was evident that there had not been such a failure.

Where the party not in default has incurred costs in taking out and maintaining the required insurance, the other party is liable to meet that cost. It will therefore be taken into account in calculating any interim payment advice (see clause 22.5.4) or, though unlikely, in calculating the amount due under the final payment advice (see clause 22.6.2).

The occurrence of an event

27.4 Upon the occurrence of an event giving rise to a claim under any policy of insurance required to be provided by the Contract the party intending to make the claim shall notify the other party.

27.5 The occurrence of an event giving rise to a claim shall be disregarded in the computation of the amount due to the Contractor in accordance with the Contract and, subject to clauses 26 (*Indemnities*) and 27.6, neither the Employer nor the Contractor shall be entitled to receive any payment from the other in respect of the event giving rise to the claim.

Notification

6-36 If an event occurs which gives rise to a claim under any of the policies required to be provided by the contract, then the party who intends making the claim must notify the other party. The use of the phrase 'occurrence of an event' is familiar language in the insurance world. The relationship between the occurrence of an event and it giving rise to an intention to make a claim in terms of when notification is required is far from clear in the drafting of clause 27.4. It might be that the event must actually give rise to an intention to make a claim before notification is required. If so, it is not therefore a requirement for notice to be given if a party merely has reason to believe that there may be a claim or if that party decides not to make a claim. However, this would not be a very purposive interpretation as notification of the occurrence of an event which could possibly give rise to a claim ought to be the trigger for notification. On its literal wording, however, the clause seems only to require notification once there is a positive intention to make a claim. If a party who could claim has decided not to, presumably no notice to the other party would then be required. Having said this, the fact that the notification is required 'Upon the occurrence of an event' must mean that a party who is able to claim must make up his mind quickly whether or not he intends to do so.

It could of course be that an event occurs which gives rise to a claim under the policy which, depending upon the nature of the policy, will involve a claim by someone who is protected under the policy but who is neither employer nor contractor. For example, in certain circumstances a sub-contractor may be able to claim under a project policy, in which case no notification is required as the person making the claim is not a party to the MPF (where appropriate any sub-contract, terms of appointment, etc. ought sensibly to require the claiming party to notify all other insured persons under the policy).

As notification is required upon the occurrence of an event, clause 27.4 makes no allowance for the fact that the event giving rise to a claim may have occurred and yet the party who is entitled to claim under the policy is at that point unaware of it. The clause does not deal expressly with the situation where the discovery of the event is later than its occurrence. This is very likely to happen in relation to claims under professional indemnity policies. Presumably any late notification in these circumstances, though it may be a technical breach of clause 27.4, is most unlikely to result in any loss of the benefit of the policy due to late notification.

Often the 'event' will be physical loss or damage occurring. However, it could of course be related to economic loss if the particular policy protects against this risk.

Financial implications

6-37 Clause 27.5 provides that the occurrence of an event which gives rise to the claim is to be disregarded in computing the amount due to the contractor in accordance with the contract. It goes on to provide that apart from preserving any rights to an indemnity under clause 26 and the obligation on the claiming party to make up any excess which may exist under the relevant insurance policy (clause 27.6), neither party is to be entitled to receive any payment from the other.

Clause 27.5 raises interesting and important issues:

- There is no express requirement that the proceeds of the insurance claim are to be spent on reinstating the project where it has suffered loss or damage or, as appropriate, a design defect. All other JCT standard forms expressly provide that the contractor will reinstate following loss or damage to the works and provide in one way or another for the insurance monies to be used for that purpose. It has to be remembered that the general obligation of the contractor will be to build the project in accordance with the contract for the contract price, unless the contract specifically provides for additional payment. Under the MPF, clause 27.5 expressly provides that the event, e.g. the causing of loss or damage from an insurable event is to be expressly disregarded in computing the amount due to the contractor. It is therefore the contractor's responsibility under this contract to reinstate without compensation under the contract. The contractor's only recourse therefore will be to the insurance proceeds. The MPF does not expressly provide that the contractor will be entitled to these proceeds. If the insurance were to be in the employer's name only, it would pose grave difficulties for the contractor as there is no requirement for the employer to expend the insurance monies on the reinstatement or to pay the contractor in respect of it. Even if a relevant policy is in joint names so that the contractor is an insured, the contract does not deal with who should receive the payment under the policy. Presumably it will be the party making the claim. If this is the contractor then the employer may well prefer the sum to be paid to him and for him to release the money to the contractor as expense is incurred on the reinstatement of the project. On the other hand, the contractor will not be happy for the employer to receive the money unless this is coupled with an express obligation to pay it to the contractor under some separate arrangement. This serves as a reminder to those using the form that such matters as this need to be clearly resolved before the contract is entered into and after any appropriate specialist insurance advice has been taken.

- The provision that neither employer nor contractor is to be entitled to receive payment from the other in respect of the event giving rise to the claim (with the exception of any rights to an indemnity under clause 26 and the claiming party making up the excess under clause 27.6), may be an attempt to ensure that rights of subrogation are removed. If the relevant policy is in the joint names of employer and contractor, this will be the effect anyway; however, if the policy, unusually, is in the name of either party alone, any rights of subrogation which the insurers may have will be taken away by this clause. It means that even if the insured loss or damage has been brought about by the breach of contract or negligence or other breach of duty on the part of one of the parties, the other party (through its subrogated insurers) cannot seek a claim against the party in default to recover the insurance monies. This is probably the effect of clause 26 in any event (see paragraph 6-11). It also makes it clear that the contractor's obligation to complete means that he will have to meet the costs of any shortfall in insurance funds where the reinstatement costs exceed the policy proceeds, subject however to any indemnity which might be available under clause 26.2.
- Clause 27.5 appears generally to relate to damage to the project during the period up to practical completion. This is reinforced by the reference to disregarding the event giving rise to the claim in computing the amount due to the contractor in accordance with the contract. However, there could be a claim against the contractor many years later under the contractor's professional indemnity policy taken out under clause 28, where so stated in the appendix to the MPF. Assuming it is the employer making the claim, rather than a funder, purchaser or tenant taking the benefit of the insurance under their third party rights (see paragraphs 7-65 and 7-77), clause 27.5 would literally prevent the employer from making a claim, certainly for losses other than those relating to reinstatement etc., and possibly from making any claim at all once the computation of amount due to the contractor has been the subject of the final payment advice (see paragraph 5-153 *et seq.*). The situation is far from clear and looks like an oversight in the drafting. Employers will no doubt amend this clause to exclude professional indemnity claims from the ambit of this payment restriction.

Excesses

> 27.6 Where any policy of insurance required to be provided by the Contract contains an excess, the party making a claim under the policy shall pay or bear the excess stated in the Appendix.[c]
>
> [c] See Guidance Note

6-38 The party making the claim pays or bears the excess under the relevant policy. As the contractor is responsible for completing the project after it has sustained loss or damage without there being any adjustment to the contract sum, it will often be the contractor making the claim. The Guidance Note (page 17) makes the point that if the excess is substantial, the possibility of obtaining an additional layer of cover to protect against this might be considered.

It is likely in practice that cover for loss or damage to the project will be in the joint names of employer and contractor and that there will be an excess under the policy. By clause 27.6, where there is an excess, the party making the claim under the policy is to pay it. The excess is required to be stated in the appendix to the contract. As it is the contractor who has the general obligation to complete the project (subject in extreme cases to a right to terminate his own employment under clause 34.1 where the event is a *force majeure* or a defined specified peril which has resulted in a prolonged suspension to the project – 13 weeks or such other period stated in the appendix), it would pay the employer never to claim but to leave the contractor to do so where this is at all feasible. In this way the contractor would need to claim to obtain the funds to support the cost of reinstatement in order to complete the project, and would be left short of funds as a result of the excess. He would not be able to claim this excess from the employer even if the employer had caused the event to occur as a result of breach of contract, negligence or breach of duty. This all seems very unsatisfactory and potentially problematical. Matters such as this need to be carefully addressed by the parties or their insurance advisors before the contract is entered into.

Terrorism cover

27.7 Where any part of the Terrorism Cover ceases to be available the party responsible for providing and maintaining the relevant policy shall immediately notify the other.

27.8 From the later of the date of the cessation of such Terrorism Cover or the date of any required notification to the Employer by the Contractor under clause 27.7 the risk of any loss that would otherwise have been covered by a policy of insurance required by the Contract shall rest with the Employer. Any additional works necessary to complete the Project as a consequence of a loss due to terrorism that would otherwise have been covered by a policy of insurance required by the Contract shall be treated as a Change.

Background

6-39 In 1992 reinsurers indicated to the insurance industry that they would not reinsure in respect of the risk of loss or damage due to fire and explosion caused by terrorism. Insurers informed the Government that they could not therefore cover damage to commercial and industrial buildings resulting from terrorism. The Government agreed to act as insurers of last resort and a new method of providing terrorism cover was introduced. Standard policies accordingly now exclude terrorism cover and then bring it back upon payment of a standard premium fixed for all policies according to graded risk zones in the UK. These premiums are then paid into a reinsurance pool administered by a company formed by the Government, Pool Reinsurance Co Ltd.

However, the Government retained a right to terminate its agreement with Pool Reinsurance Co Ltd to act as an insurer of last resort. As a result, there remains a possibility that during the course of the project insurers could remove terrorism cover. Clauses 27.7 and 27.8 deal with this eventuality.

Following the bringing down of the twin towers at the World Trade Centre in New York on 11 September 2001, terrorism reinsurance and insurance cover was withdrawn across significant areas of the insurance industry. The problem for commercial property insurance was particularly acute, as Pool Reinsurance Ltd provided reinsurance cover for commercial property damage and business interruption costs only from acts of terrorism which caused fire or explosion and not against other forms of terrorist attack. This gap in cover resulted in the establishment of a working group under Treasury chairmanship involving representatives of the insurance market and buyers of commercial property together with Pool Reinsurance Ltd itself.

The result of these discussions was that Pool Reinsurance Ltd agreed to extend the scope of terrorism cover beyond fire and explosion, e.g. to cover a major flood or contamination as a result of terrorism. However, the extension of terrorism cover to 'All Risks' is reflected in the rating charged for cover, and the Pool Reinsurance Ltd cover is subject to a retention under which insurers bear the first amount of claims, currently £100 000 per head of cover.

Any big increase in retention levels would be difficult for insurers to bear. An alternative model for retention was therefore put forward: to move to a per event retention, combined with an annual aggregate limit for each insurer based on the overall terrorism market share of each insurer. The intention is that each insurer would have its retention set annually as a proportion of an industry-wide figure. This would provide insurers with certainty, leaving Pool Reinsurance Ltd, and if necessary the government, to bear the cost of any major incident or terrorist campaign involving a sustained series of incidents.

From 1 January 2003 the maximum retention was set at £30 million per event (£60 million per annum), with individual insurers' retention being based on market share, with the intention over 4 years to increase the retention steadily bringing in commercial reinsurance to cover insurers' retentions or permitting insurers to retain this element of risk themselves. The retention is to increase between 1 January 2003 and 1 January 2006 from £30 million to £100 million per event and from £60 million to £200 million per annum.

None of this removes altogether the possibility that terrorism cover will cease to be available some time during the period of construction of the project.

Definition of terrorism cover

'Terrorism Cover' is defined in clause 39.2 as:

> Cover under any policy required to be provided by the Contract against the physical loss or damage to work executed or site materials caused by an act of terrorism as defined by the Terrorism Act 2000.

6-40 The cover required in respect of an act of terrorism extends to any physical loss or damage to work executed or site materials. This, of course, goes beyond just fire or explosion. An act of terrorism is defined by the Terrorism Act 2000 (see paragraph 3-65). However, it is important to point out that cover through Pool Reinsurance Ltd (see paragraph 6-39) is limited to an Act of Terrorism as laid down by Pool Reinsurance:

> 'an act of any person acting on behalf of or in connection with any organisation with activities directed towards the overthrowing or influencing of any government *de jure* or *de facto* by force or violence.'

Notification

6-41 Under clause 27.7, if any part of the terrorism cover ceases to be available the party responsible for providing and maintaining the relevant policy must immediately notify the other party. This requires immediate notification where any part of the cover ceases. Hopefully the party responsible for notification will be informed in advance that the cover is to cease and so can give adequate warning to the other party.

Consequences of cessation of all or part of the terrorism cover

6-42 Clause 27.8 provides that either from the cessation of the cover or, where it is the contractor who is responsible for taking out and maintaining the cover, the date of any required notification to the employer if that is later, the risk of loss which would otherwise have been covered rests with the employer.

The result of the employer taking the risk of the cessation of cover means that any additional work required to complete the project due to any loss as a result of uninsured terrorism is to be treated as a change. The contractor is therefore entitled to the cost of carrying out any rectification works, to an adjustment of the completion date and recovery of loss and expense as appropriate.

In many other JCT contracts, e.g. WCD 98, the possibility of cessation of insurance cover for acts of terrorism is covered in an optional amendment issued in 1994 which the parties can elect to include or not as they wish. If the amendment is incorporated, the employer is given the additional option, not available in the MPF, of determining the employment of the contractor. Such an option may on occasions be useful, e.g. where the project is being constructed in an area which is at high risk of acts of terrorism.

Professional indemnity – clause 28

Background

6-43 Contractors who carry out design and associated professional service functions will obviously have contractual responsibility and liability in connection with the carrying out of those functions. If the contract which includes these obligations also requires the contractor to construct the project, then, to the extent that such a contract does not deal expressly with the nature of the contractor's responsibility in connection with such design etc., there will be an implied term incorporated into the contract that the contractor will provide a project reasonably fit for its intended purpose. This is a greater responsibility than that which ordinarily falls upon an independent consultant, e.g. architect or engineer, who carries out these services independently of any obligation to provide the physical structure. In the latter case the obligation will generally be to use reasonable skill and care. This topic has been discussed in some detail earlier (see paragraph 2-67 *et seq*.).

6-44 Under the MPF, by virtue of clause 5.3 (see paragraph 2-90 *et seq*.), the contractor warrants that he will exercise the skill and care to be expected of a professional designer. The clause goes on to provide specifically that the contractor does not warrant that the project will be suitable for any particular purpose. There is a footnote to clause 5.3 referring to a Guidance Note (page 7) which contains an alternative model clause where it is agreed between the parties that the contractor's responsibility will be to provide a project which is reasonably fit for its intended purpose.

6-45 Contractors generally wish to have, and employers generally insist that they have, some form of insurance cover in the event of any claim being made as a result of a breach of a contractual or some independent professional duty of care. Under the MPF, this may be obtained by virtue of some form of total project cover policy. Alternatively, it may be the subject of a separate requirement. Clause 28 of the MPF, if selected through an entry in the appendix to the contract, provides for this situation. This requirement, even though only optional in the sense that an appendix entry is required, is the first occasion on which the JCT has provided in a main contract for a contractor to obtain professional indemnity insurance cover. Almost all contractors of experience and standing in the design and build market will in any event have such cover. However, because the insurance market can become 'hard' it was thought appropriate to keep the requirement as an option rather than a standard provision.

Such policies are likely to cover the design and build contractor for breach of a duty to exercise reasonable skill and care in design and any other normal professional services associated with the design function. Sometimes contractors can obtain insurance cover which extends further than this and covers them for a failure to provide a project which is reasonably suitable for a known intended purpose. Such cover is less readily obtained and more expensive. However, some contractors do carry such cover even if, for reasons fully understandable to them and to the insurance industry, they are not likely to broadcast this fact. Nevertheless, some employers do impose a contractual obligation on contractors to obtain cover of this type. Such provisions may or may not have to be modified for any given project.

6-46 Professional indemnity insurance policies are almost universally issued on a claims made basis. In other words, cover only exists if at the time when a claim is made (not when the actual breach of duty occurred) a policy is in force. The result is that to be effective the insurance cover needs to be maintained for a considerable period of time after the design etc. has actually been undertaken. This is why contract clauses typically require the insurance to be kept in place for a minimum period of years following practical completion of the project. Care needs to be taken where there is a change of insurer as, on occasions, subsequent insurers will not be prepared to pick up liability for prior breaches of duty.

We now turn to consider clause 28.

Requirements for professional indemnity cover

28.1 The provisions of this clause only apply when so stated in the Appendix.

28.2 The Contractor shall take out and maintain professional indemnity insurance for the amount stated in the Appendix. Provided that it remains generally available at commercially reasonable rates, such insurance shall be maintained until the expiry of 12 years from the date of Practical Completion of the Project.

28.3 Where the Contractor considers that any insurance required by clause 28.2 is no longer generally available at commercially reasonable rates it shall notify the Employer and cooperate with the Employer in seeking means by which the Contractor can be protected against professional liability claims arising out of the Project.

Optional provision

6-47 Clause 28.1 states that clause 28 will only apply where it is stated to apply in the appendix to the contract. The appendix provides as follows:

> Clause 28 does/does not* apply. Where no selection is made, clause 28 shall not apply. The limit of indemnity shall be not less than £. . . for any one claim or series of claims arising out of one event/in aggregate for any one year*.
>
> *Delete as applicable

6-48 Clause 28 therefore only applies if the appendix states that it is to apply. The appendix requires action in the following respects:

- The deletion of either 'does' or 'does not'. Where no selection is made, clause 28 does not apply. This is simple enough.
- The monetary limit of the indemnity has to be stated. No default position is provided for. In other words, if the parties leave this blank it is left as a matter of interpretation exactly what is intended by the parties. Employers may argue that as there is no limit, it is unlimited. However unlimited insurance in respect of these risks is never available. It is clearly therefore an unlikely interpretation to adopt when considering the overall commercial objective and purpose of the contract. Contractors would argue that the insertion of a £ sign with nothing next to it is the equivalent of £nil. In other words the limit of indemnity is to be not less than zero. The effect of this is that it would require a policy with zero cover. This again is a commercial nonsense. Yet a third alternative could be that it demonstrates an intention that, even if a deletion of 'does not apply' has been made, nevertheless the parties, having not completed something which clearly requires completion, have left the issue as so uncertain that it can be given no effect at all. This is the most likely interpretation. It might have been sensible to expressly provide that where no figure is inserted then clause 28 does not apply.
- A choice has to be made whether the financial limit inserted is to be in respect of either any one claim or series of claims arising out of one event or in an aggregate sum for any one year. Again, no default position is stipulated and this can clearly lead to uncertainty if no choice has been made. It may well render the provision sufficiently uncertain in meaning to make it void. It would have been sensible to expressly provide for what is to be the position if no choice has been made. The choice therefore is between cover in respect of either any one claim or a series of claims arising out of one event on the one hand, and an aggregate limit for any one year on the other. Two points can be noted:
  - Clearly cover on an each and every claim basis is likely to be considerably more expensive than where there is an aggregate limit. In the latter case there is a known maximum liability that the insurer is undertaking, which is not the case with the former. Employers clearly prefer the former cover to be obtained and kept in place;
  - Restricting the financial level of cover to 'any one claim or series of claims arising out of one event' is intended to limit the possibility of dividing losses from an event into a number of claims in order to try and overcome the financial cap.

Contractor's obligation

6-49 As this is a claims made policy (see paragraph 6-46), the requirement is for the contractor to both take out and maintain the cover. Clause 28.2 fails to specify in respect of what risk the cover is required, e.g. in respect of any defect or insufficiency in design. It would have at least been sensible to cross-refer to the contractor's design-related obligations in clause 5.3. Cover is to be maintained until expiry of 12 years from the date of practical completion of the project. This period ties in generally with the limitation period for actions for breach of contract where the contract is executed as a deed. Some contractors may seek to get this reduced to 6 years. It is a significant obligation on the contractor to maintain cover, possibly at considerable expense, for many years after the project is completed.

There is an exception to this requirement where the cover required is no longer generally available at commercial rates. This is dealt with in clause 28.3.

Cover no longer available

6-50 Clause 28.3 provides that if the contractor considers that the cover is no longer generally available at commercially reasonable rates, he must notify the employer and then co-operate in seeking a means by which the contractor can be protected against professional liability claims arising out of the project. Clauses such as this are notoriously uncertain in meaning and effect. Having said this, it is difficult to provide for certainty if a reasonable stance is to be taken in the event that maintaining the insurance becomes prohibitively expensive. There is inevitably considerable debate as to what would amount to a commercially reasonable rate and what evidence is required to demonstrate that cover is not available at such rates.

Additional matters

6-51 Clause 28 generally is very basic in nature. This is no doubt the result of a compromise between competing interests. In the market place many employers introduce far more detailed provisions covering these matters in a more substantial way and also covering many other matters. Examples of matters often covered are:

- The commencement date for the policy if not already in force;
- A requirement for cover to be on the customary and usual terms prevailing for the time being in the market;
- The cover to be with reputable insurers (possibly carrying on insurance business in the UK);
- A requirement that the policy must not include any provision that the contractor must discharge any liability before being entitled to recover under the policy or any term or condition which might adversely affect the rights of any person to recover from the insurers under the Third Parties (Rights Against Insurers) Act 1930;
- A provision that the contractor is not, without the prior approval in writing of the employer, to settle or compromise with insurers any claim which the contractor may have against the insurers which could prejudicially affect the employer;
- A provision that if any increase in premium is the result of the contractor's own claims record or acts or omissions then this will not form part of any consideration of what can be regarded as commercially reasonable rates;
- A provision that in discussing what is to happen in the event of insurance not being available at commercially reasonable rates, the position of the employer as well as the contractor is to be taken into account;
- A requirement for the contractor to produce documentary evidence (sometimes including the policy) that his professional indemnity insurance has been obtained

and is being maintained. It should be noted that many policies prohibit such disclosure by their insured;

- A provision that the obligations in respect of professional indemnity insurance continue even if the contract is terminated and even if the termination is the result of a breach of contract by the employer.

Chapter 7
Assignment and third party rights

Content

Clauses 29 and 30 cover the issues of assignment and third party rights.

Background

7-01 Assignment in relation to a contract means the transfer of contractual rights or benefits by one party to the contract to a person who was not originally a party to it, i.e. a stranger to the contract.

A contract is made up partly of benefits and partly of burdens. What amounts to a benefit to one party will very often be a burden to the other, e.g. under a building contract the employer obtains the benefit of a building and the contractor assumes the burden of building it. The law treats the transfer of benefits and burdens differently.

The benefits

7-02 Subject to any express contractual provisions, a contracting party may assign the benefits due to him under the contract, e.g. the contractor under a building contract may assign the benefit of the retention money or the employer may assign the partly constructed building together with the right to have it completed. If a party to a contract assigns any benefit under it, he should give notice of that fact to the other party. Once notified the other party must, if the benefit being assigned represents a burden to him, discharge that burden in favour of the new beneficiary, i.e. the assignee.

Certain benefits of a truly personal nature cannot be assigned, e.g. the right to litigate where the cause of action is purely personal.

The burdens

7-03 The law does not permit the unilateral assignment of contractual liabilities. The burdens under a contract can only be transferred with the consent of the other contracting party. Once that consent is forthcoming the original party can be released from the burdens. The person to whom the burdens are transferred becomes liable in the place of the person who transferred them. This in law is called a novation (see paragraph 4-71 where the difference is illustrated). It must be distinguished from vicarious performance of contractual liabilities, i.e. someone else actually performing the obligation while the contracting party remains fully liable under the contract (see paragraph 4-78).

Assignment and building contracts

7-04 The law in relation to assignment stems essentially from simple contracts such as money debts or simple contracts of sale where any discussion of benefits and burdens is straightforward. However, in relation to building contracts and other complicated contracts, these rather elementary concepts are not always suitable. In particular, a distinction can sometimes be drawn between an assignment of rights under the contract, such as a right to claim damages or to payment on the one hand, and the right to have the contract performed on the other. Depending on the wording of any particular assignment clause in the contract, the former may be assignable while the latter is not. However, when a court comes to consider a clause prohibiting assignment, it will require very clear words before it construes the clause as allowing assignment of the fruits of the contract such as damages or payment on the one hand, while prohibiting assignment of the right to performance of the contract on the other: see *Linden Gardens Trust Ltd* v. *Lenesta Sludge Disposals Ltd* (1993). As Lord Browne-Wilkinson said in this case:

> 'In my view they [the parties] cannot have contemplated a position in which the right to future performance and the right to benefits accrued under the contract should become vested in two separate people.'

The likely result therefore is that a prohibition against assignment will be total rather than partial.

Assignee's right to sue

7-05 A valid assignment will enable the assignee to sue on the contract in respect of the assigned rights. Whether or not he can sue in his own name or must use the name of the assignor depends upon whether the assignment is statutory or equitable. For the assignee to be able to sue in his own name the assignment must be absolute, i.e. a transfer of all the assignor's rights in the contract without qualification. For a statutory assignment, appropriate notice in writing must be given to the other party to the contract. If the assignment is not absolute, then if the assignee wishes to take action the assignor must be named as a co-claimant, or if he refuses then joined as a co-defendant.

For statutory assignments, section 136 of the Law of Property Act 1925 provides that an absolute assignment in writing of which express notice in writing has been given to the other contracting party, is effective to transfer the legal right as from the date of the notice, enabling the assignee to sue on the contract in his own name without joining the assignor in the proceedings.

Even if an assignment does not take effect as a statutory assignment, e.g. because it is not in writing or notice is not given or because it is conditional or is an assignment of part only, nevertheless it can be a valid equitable assignment. An equitable assignment need not be in writing and notice of it need not be given although obviously the assignee would be well advised to ensure that any assignment is in writing and that notice is given, particularly as, by giving notice, priority against any other assignee will be obtained. While an absolute assignee of an equitable right can sue in his own name, this is not the case if the right is not a legal one and equity will require the assignor to be joined. Only if section 136 is complied with in all respects can the assignee of a legal right sue his own name.

Any assignment is subject to equities. This means that the other contracting party can raise against the assignor all those defences and equitable set-offs that can operate by way of defence, which he would have had against the assignor. A good example is the case of *Young* v. *Kitchin* (1978) where a builder validly assigned his right to payment and the assignee sued upon the contract. The building owner was permitted to set off damages due to delay by the builder as an equitable set-off (or defence), to the extent of the assignee's claim, but not to recover any excess for which the builder–assignor remained liable.

Prohibition on assignment

7-06 It is common to find construction contracts providing for a restriction or even complete prohibition on assignment, particularly on the part of the contractor. It is now established that there is no public policy objection to provisions prohibiting assignment (*Linden Gardens Trust Ltd* v. *Lenesta Sludge Disposals Ltd* (1993)). A complete prohibition on the assignment of any benefits under a contract will prevent assignment of two basic kinds of rights: the right to future performance of the contract (i.e. to have the works carried out); and what are known as choses in action, namely, rights which have already accrued to a party under the contract, such as the right to sue for damages for breach of contract.

The 'no loss' arguments

Whether there is a valid assignment or not, there can be what are termed 'no loss' arguments.

FOLLOWING VALID ASSIGNMENT

7-07 For example, the assignee/purchaser of the employer's/developer's interest in the building may seek to claim against the contractor in respect of work which has, post assignment, been found to be defective. The contractor contends that the only loss for which he can be liable and for which the purchaser can claim is that suffered by the original developer. The developer sold the building to the purchaser at full market value unaware of any latent defects. The answer to this issue is probably based on something called remoteness of damage (touched upon in other contexts – see paragraphs 3-18, 5-71 and 8-24). Was the loss sufficiently foreseeable by the contractor at the time of entering into the building contract as likely to result from the breach being complained of? Was any assignment foreseeable? The answer will depend on the facts. What was known to the contractor at the time of entering into the contract is highly relevant. If it was known that the employer was a developer intending to sell on, for what purpose or purposes was the building intended to be used? Even if the possibility of assignment was foreseeable, the nature of any assignee's/purchaser's loss may not have been. The purchaser may, even if defeating the 'no loss' argument, be restricted to claiming for the loss which would have been suffered by a typical purchaser, rather than the actual purchaser. Often provisions are inserted into assignment clauses aimed at preventing the contractor from raising the 'no loss' argument. Also, if the definition of 'employer' in the building contract expressly includes assignees, this may have the effect of preventing the 'no loss' argument from getting off the ground.

The chances of being able to sue in the tort of negligence in such circumstances have now just about disappeared following *D & F Estates* v. *Church Commissioners* (1989) and *Murphy* v. *Brentwood District Council* (1990) (see paragraph 7-16).

WHERE THERE IS NO VALID ASSIGNMENT

Claims by subsequent purchasers who have not taken a valid assignment have in recent years given rise to problems in many areas, including causation and measure of damages. A particular area of concern and debate again has been in relation to the 'no loss' argument. This can happen where, for example, there is a prohibition on assignment under a building contract and a purchaser acquires the development from the original developer employer under the building contract having paid full value for it on the assumption that it is defect free. If the building is subsequently found to be defective, the purchaser cannot sue the builder because he does not have a contract with him. The chances of being able to sue in the tort of negligence in such circumstances are negligible following *D & F Estates* v. *Church Commissioners* (1989) and *Murphy* v. *Brentwood District Council* (1990) (see paragraph 7-16). If the developer sues, he may be faced with the argument that he has sold on at full value and therefore suffers no loss. This has been referred to as a 'legal black hole' into which the right to damages disappears. (Lord Keith of Kinkel in *G.U.S. Property Management Ltd* v. *Littlewoods Mail Order Stores Ltd* (1982) '... the claim to damages would disappear ... into some legal black hole, so that the wrongdoer escapes scot-free.')

In such situations the courts strive to find a remedy. This is exemplified in the House of Lords judgment in relation to two cases heard at the same time, *Linden Gardens Trust Ltd* v. *Lenesta Sludge Disposals Ltd* (1993) and *St Martins Property Corporation Ltd* v. *Sir Robert McAlpine & Sons Ltd* (1993).

One of a number of issues raised in these two cases concerned the ability of a party (such as the employer/developer) to claim damages for breach of contract when that party suffers no actual loss as a result of having sold on the subject matter of the contract at full market value or otherwise having disposed of the subject matter. This can apply to both chattels and buildings. The general principle of course is that if no loss is suffered as a result of a breach of contract, only nominal damages can be claimed. The question which arose in this case is whether nevertheless in certain circumstances substantial damages could be claimed. In this connection the major speech given by Lord Browne-Wilkinson stated that such substantial damages can be claimed by means of an exception to the general rule, namely, that referred to by Lord Diplock in *The Albazero* (1977) where Lord Diplock said (at AC 846B):

> 'Nevertheless, although it is exceptional at common law that a plaintiff in an action for a breach of contract, although he himself has not suffered any loss, should be entitled to recover damages on behalf of some third person who is not a party to the action for loss which that third person has sustained, the notion that there may be circumstances in which he is entitled to do so was not entirely unfamiliar to the common law and particularly to that part of it, which under the influence of Lord Mansfield and his successors, Lord Ellenborough and Lord Tenterden, had been appropriated from the Law Merchant.'

Lord Diplock then mentioned exceptions in relation to bailment and trespass and also consignment under a bill of lading.

Lord Browne-Wilkinson stated that in construction contracts and indeed in other contracts under which the property in goods passes, there could also be an exception to the general rule based on the *Albazero*.

The nature of this exception is difficult to analyse. It would appear that in this type of situation, where there is no express contractual obligation as between the original employer under the building contract and the subsequent purchaser to whom the building is transferred, to pursue such an action for the benefit of the purchaser, the original employer under the building contract cannot be forced to do so. If nevertheless he does so, presumably there will be some form of trust operating in favour of the purchaser once funds are in the hands of the original employer under the building contract. It is a very unsatisfactory situation leaving many questions unresolved, e.g. as to the exact nature of any obligation; whether the unrecovered costs of taking such proceedings can be taken into account; and whether the 'beneficiary' under any trust has any sort of right to control the way any such action is taken.

On the other hand, Lord Griffiths in his speech faced the matter head on and rather than classify the right to claim substantial damages where no loss is suffered as an exception to the general rule, treated it instead as really being part of a qualified general rule. He gave a typical example:

> 'To take a common example, the matrimonial home is owned by the wife and the couple's remaining assets are owned by the husband and he is the sole earner. The house requires a new roof and the husband places a contract with a builder to carry out the work. The husband is not acting as the agent for his wife. He makes the contract as a principle because only he can pay for it. The builder fails to replace the roof properly and the husband has to call in and pay another builder to complete the work. Is it to be said that the husband has suffered no damage because he does not own the property?'

Lord Griffiths then gave a further example of the husband owning the property when placing the contract but at the time when performance was due and was in fact defectively carried out, the husband had transferred the asset to his wife as part of his inheritance tax planning. This should be of no concern to the builder who should not be able to argue that as the husband had now transferred the asset he had suffered no loss.

Lord Griffiths referred to the speech of Lord Browne-Wilkinson and to his remarks that before finding a general principle allowing a claim for substantial damages in such circumstances, he would welcome academic debate. Of this Lord Griffiths said:

> 'Whilst I always welcome and find the view of academic writers most helpful, I am prepared even without the benefit of their views to adopt the direct route to the award of damages.'

The key may be to distinguish between cases where there is a pure claim for damages, and cases where it is a claim based upon someone's failure to perform their side of the bargain which has resulted in the subject matter of the performance being worth less than that agreed to be paid for it.

In the later House of Lords case of *Alfred McAlpine Construction Ltd* v. *Panatown* (2000), the contractors had, on the same day as entering the building contract with Panatown, entered into an agreement with the building owners by virtue of a duty of care deed. The House of Lords held that the existence of the duty of care deed was a crucial factor making it clear that the *Albarezo* exception did not apply. The parties had addressed the problem of the legal black hole. The original building owner suffered no loss and could therefore only recover nominal damage. However, it is worth bearing in mind that the majority was only 3 to 2.

We still appear to be left with two different approaches, the narrow and the wide. The narrow approach seeks to remedy what might otherwise be a legal black hole, by relying upon an exception (the *Albarezo* exception). The wide approach is based upon the general principle that a loss capable of monetary compensation has been suffered where someone has been deprived of the interest they had in the performance of a contract. In this case it remains unclear as to whether an independent warranty in favour of the purchaser/tenant etc., giving a right to claim directly against the contractor, affects the application of the wider approach. The law in this respect is yet to be clarified. For the time being, the safest course is to assume that the narrow approach applies.

All that has been said above in relation to a building contractor applies also to professionals involved in providing services such as design.

Much of what has been said above has to be considered in the light of the Contracts (Rights of Third Parties) Act 1999 which will be considered below in relation to rights of third parties (see paragraph 7-20 *et seq.*).

Assignment – clause 29

Rights of the parties in connection with assignment and novation

29.1 The Contractor may not assign either the benefit or the burden of the Contract without the consent of the Employer.

29.2 The Employer may assign the benefit of the Contract at any time without the consent of the Contractor.

29.3 The Contractor hereby consents to an assignment by the Employer of both the benefit and the burden of the Contract to the Funder at any time.

Prohibition of contractor assignment

7-08 Clause 29.1 provides that the contractor may not assign either the benefit or the burden of the contract without the consent of the employer. This clause accurately refers specifically to the benefit and burden of the contract rather than as for example in WCD 98, clause 18.1.1, to the assignment of the contract itself. Such imprecise wording probably dates back to leasehold conveyancing documentation out of which it is possible that the forerunners of the JCT forms of building contract grew.

The effect of this prohibition is that the contractor will not be able to assign absolutely, or by way of charge, the benefit of any payments due under the contract. It is not unusual for contractors to wish to assign by way of charge whatever may be the value to the contractor of the final account. Any such assignment will require the consent of the employer. This consent is not qualified by any reference to the employer not unreasonably withholding it. It is therefore an absolute discretion. As any such consent is in the nature of a communication, it should be in writing or under any electronic communications procedure agreed by the parties (see clause 38.1). However, it is possible that an oral consent, or even consent given by conduct might bind the employer if the contractor has, to the employer's knowledge, relied upon it to his detriment.

Employer can assign the benefit at any time without the consent of the contractor

7-09 The employer's right to assign the benefit of the contract at any time and without the consent of the contractor should be contrasted with other JCT contracts, e.g. WCD 98, clause 18.1.1 which provides that neither the employer nor the contractor shall, without the written consent of the other, assign the contract. Clause 18.1.2 of that contract then goes on to provide a limited exception enabling the employer, by means of the selection of the appropriate option in the appendix to that contract, to assign to someone to whom the employer is transferring his interest in the whole of the premises comprising the works, the right to bring proceedings in the employer's name to enforce terms of the contract made for the benefit of the employer. Such limited rights are almost invariably inadequate for the employer's needs, particularly if the employer is a developer. The employer may wish to offer the benefit of the contractor's duties and liabilities in respect of parts only of the project to many different parties, e.g. purchasers or tenants. Further, the employer may well wish to do this during construction of the project, i.e. before practical completion, and accordingly will wish to confer on purchasers, tenants or funders the right of future performance under the contract as well as merely a right to bring proceedings in the name of the employer.

7-10 In recent times, rather than assigning, whether by way of legal or equitable assignment, if a developer has wished to dispose of all or part of a development to one or more purchasers or tenants, or to charge the partly completed project to raise finance, it is highly likely that this has been achieved by requiring the contractor to provide a direct warranty or, though rarely before the publication of the MPF, by means of the creation of third party rights using the Contracts (Rights of Third Parties) Act 1999 in favour of purchasers, tenants or funders. However, clause 29.2 provides for the employer to assign the benefit of the contract without the consent of the contractor. Though not absolutely clear from the wording, it is probable that this, however, refers to the assignment of all of the benefit of the contract to a third party rather than different benefits to different third parties, e.g. to various purchasers or tenants of part only of the project.

7-11 If such an assignment takes place, the question could be raised as to whether the assignee, e.g. a purchaser of the project, can bring an action for damages in respect of his own loss rather than any loss suffered by the employer. This has already been discussed (see paragraph 7-07). To help overcome this argument, it is not unusual to find an additional provision in assignment clauses in construction contracts stating that the contractor will not be entitled to contend that any person to whom the benefit is assigned is to be precluded from recovering their loss resulting from any breach of the contract, by reason that he is an assignee and not the original named employer. Another option is to expand the definition of 'employer' under the building contract. Under the MPF, the 'Employer' is defined in clause 39.2 as:

> The party to the Contract named as such or any assignee permitted by clause 29 (*Assignment*).

This inclusion of an assignee within the meaning of the 'Employer' may make such an amendment unnecessary. Contractors should appreciate therefore the likelihood that a purchaser or tenant taking the benefit of the contract by way of assignment is likely to be able to overcome any argument that the original employer has suffered no loss. The corollary of this is that the purchaser or tenant will be able to recover his own losses rather than being limited to any losses which the original employer may have suffered. In other words, any rules of remoteness based on *Hadley* v. *Baxendale* principles (see paragraph 7-07), which might otherwise have limited any damages claim by the assignee to that loss which the contractor could have contemplated would have been suffered by the original employer, will not apply.

7-12 Clause 29.2 does not require any assignment by the employer to be in writing or for written notice of it to be served on the contractor. However, as failure to do either would take any assignment outside the provisions of section 136 of the Law of Property Act 1925, it is clearly sensible for this and other reasons that the requirements of section 136 are complied with (see paragraph 7-05).

Employer's right to assign both benefit and burden of a contract to the funder at any time

7-13 Many projects will be funded by means of the employer borrowing the money. Funders will look for various means of protection, and these are likely to include not only a charge on the land including the partially completed or wholly completed project, but also by means of being able to effectively step into the employer's shoes under the contract in order to protect its interest. The funding agreement will deal with when this is to happen as between the funder and employer. So far as the contractor is concerned, this clause provides his consent in advance to any such assignment of both the benefit and the burden of the contract at any time.

Contractors need to fully appreciate that this clause operates in the form of a novation under which the employer will step out of the contract and be replaced by the funder who will become the contractor's new paymaster. The risks for the contractor in this are obvious. His decision to tender at all may well have been affected by his knowledge or perception of the employer. The funder's attitude in crucial areas, e.g. willingness to compromise where disputes arise, could well be crucial to the profitability of the contract so far as the contractor is concerned. The 'Funder' is defined in clause 39.2 as:

> The person or syndicate providing funding for the purposes of the Project, as identified in the Appendix.

As a permitted assignee the funder will fall within the definition of 'Employer' and accordingly can effectively novate again, e.g. to another developer, still retaining his right to take over yet again under the terms of the new funding arrangements if he wishes to.

In the normal course of events any novation would be entered into by employer, contractor and funder, as all are parties to it. However, clause 29.3 contains the contractor's consent so it would be possible to effect the novation by an agreement between the employer and funder, possibly contained in the funding agreement and brought into operation by notice by the funder to the employer even without a formal three-party agreement being executed. However, this is uncertain in effect and it is usual to obtain the execution of a formal tripartite novation agreement. This novation in practice is more often achieved by means of a separate collateral warranty from the contractor in favour of the funder, under which the contractor agrees to the step-in arrangement and agrees also to execute a formal novation agreement if required.

Rights of third parties – clause 30

Background

The purpose of collateral warranties

7-14 Purchasers and tenants of the completed (or even sometimes the partially completed) project are clearly concerned to protect their interests in it. Similarly, funders who have an interest in the project have an interest in its successful completion. These interests need to be protected. From the early 1980s, such interests have been protected firstly by duty of care letters and subsequently more formally by the use of collateral warranties between those responsible for the design and construction of the project, including construction professionals, contractors and sub-contractors, and purchasers, tenants or funders as the case may be.

The need for collateral warranties

7-15 The need for such collateral warranties is to overcome the problems faced by such third parties as a result of the doctrine of English law known as 'privity of contract'. Essentially this means that any rights or obligations created by a contract exist solely for the benefit of the parties to that contract. So, the absence of any contractual link between the purchaser, tenant or funder and the construction professional, contractor or sub-contractor meant that if any of the latter were in breach of their own contracts, there was no method by which the purchaser, tenant or funder could sue for breach of that contract in respect of their own losses.

No remedy in tort

7-16 For some time, there existed the possibility that a duty of care in the tort of negligence might be owed by those responsible for design and construction in favour of those subsequently acquiring an interest in the property such as purchasers or tenants – see *Anns* v. *Merton London Borough Council* (1978). However, that case marked the zenith of such a duty of care being owed in terms of defective design or defective workmanship and materials. This possibility gradually diminished and was virtually removed by the House of Lords decisions in *D & F Estates Limited and Others* v. *The Church Commissioners of England and Others* (1989) and *Murphy* v. *Brentwood District Council* (1990), effectively precluding such a remedy for non-parties to professional appointments, building contracts or sub-contracts against professionals, contractors and sub-contractors respectively in the tort of negligence for 'economic loss'.

Economic loss

7-17 Damage to the structure of a building caused by negligent design or workmanship is for this purpose regarded as economic loss rather than as physical damage to property. The result is that a purchaser, or a tenant under a full repairing lease, could be left without a remedy. Under contracts for sale and agreements for lease the developer or landlord has an interest in seeking to improve its position by taking no risks in respect of present or future defects in the building. Such contracts for sale or lease therefore regularly place the risk of defects in the building on the shoulders of the purchaser or tenant. It is obvious in these circumstances that the purchaser or tenant will seek protection by a direct contractual link with those responsible for causing the defect.

Advantages of collateral warranties

7-18 It should be remembered that while the curtailment of an effective remedy through the breach of duty of care in negligence route added momentum to the use of collateral warranties, such warranties existed before this turn of events. The use of such collateral warranties could always significantly improve the position of a purchaser or tenant when compared to being able to bring a claim in negligence. Examples of such benefits are:

- The limitation period within which an action can be commenced could well be longer under a contractual warranty than in negligence;
- Under a contractual warranty, the contractor can be made responsible not only for his own defaults but also those of his sub-contractors. In negligence the contractor would not be vicariously responsible for the negligent performance of an independent sub-contractor;
- The obligation under contract can be framed as an absolute warranty irrespective of any want of care and skill on the part of the giver of the warranty.

The position of the funder

7-19 Funders of the project also seek protection by means of collateral warranties. They wish to protect the value of the security which they hold in the project and may also wish to be able to step in and take over as the employer under the building contract and as client under professional appointments where otherwise the satisfactory completion of the project is in jeopardy.

Contracts (Rights of Third Parties) Act 1999

7-20 The need for collateral warranties as a result of the effect of the doctrine of privity of contract has potentially been fundamentally affected by the coming into force of the Contracts (Rights of Third Parties) Act 1999.

There follows a brief summary of the Act, after which there is a detailed consideration of the third party rights contained in the MPF.

Introduction

7-21 This Act, which received the Royal Assent on 11 November 1999 and came into force six months later on 11 May 2000, fundamentally affects the rule of privity of contract. A very brief synopsis is set out below.

This Act reforms the rule of 'privity of contract' and sets out the circumstances in which a third party is to have a right to enforce a term of a contract. As such it can clearly be of benefit to funders, purchasers and tenants in relation to construction projects. The Act also deals with circumstances in which that term can be varied or rescinded by the contracting parties themselves; and also the defences available when a third party seeks to enforce the term.

Some significant features

THE RIGHT TO ENFORCE A TERM

7-22 In order for the third party to be able to enforce a term of a contract, the contract must either:

- Expressly provide that he may (section 1(1)(a)); or
- Contain a term that purports to confer a benefit on him (section 1(1)(b)). This does not apply if on a proper construction of the contract it appears that the parties did not intend the term to be enforceable by the third party (section 1(2)).

The third party must be expressly identified in the contract by:

- Name; or
- As a member of a class; or
- As answering a particular description.

The third party need not be in existence when the contract is entered into (section 1(3)).

Any right of a third party to enforce a term is subject to any provisions in the contract relating to such right (section 1(4)).

The third party has all the usual contractual remedies such as that for damages, an injunction or specific performance. The third party is also subject to the normal rules relating to such remedies, such as issues of causation, remoteness and the duty to mitigate (section 1(5)). It is worth pointing out that the third party has no right to treat a repudiatory breach of the term by the promisor as giving rise to a right for the third party to bring the contract to an end.

The third party can obtain the benefit of any limitation or exclusion clause in the contract which is expressed to benefit the third party, e.g. a contractual term which excludes or limits a contracting party's liability for negligence and expressly states that the exclusion or limitation is also for the benefit of that party's agents, servants or sub-contractors, would enable such third parties to enforce the exclusion clause if sued by the other contracting party (section 1(6)).

Section 2(2) of the Unfair Contract Terms Act 1977 (restrictions on excluding liability for negligence other than personal injury or death) is not to apply where the negligence consists of the breach of an obligation arising from a term of the contract and the person seeking to enforce it is the third party. Thus, presumably, where the negligence arises only due to a breach of a contractual term (and not independently of contract), the exclusion clause cannot be challenged by the third party on this ground (section 7(2)).

VARYING THE TERM

7-23 Any term which the third party is entitled to enforce may not be varied or rescinded by the contracting parties in such a way as to extinguish or alter the third party's entitlement without the third party's consent if:

- The third party has communicated his assent (which can be by words or conduct) to the term to the promisor (section 2(2));
- The promisor is aware that the third party has relied on the term; or
- The promisor can reasonably be expected to have foreseen that the third party would rely on the term and the third has in fact relied upon it (section 2(1)).

It is important to note that the contract itself can contain conditions by which the term can be varied by the contracting parties themselves without reference to the third party (section 2(3)).

DEFENCES ETC. AGAINST CLAIMS BY THE THIRD PARTY

7-24 If sued by the third party, the promisor can make use of any defences or rights of set-off which are relevant to the term or which the contract expressly provides are available in respect of claims by a third party and would have been available against the other party (the promisee) to the contract (sections 3(2) and 3(3)).

In addition, the promisor will also be able to make use of any defences or rights of set-off or indeed counterclaims which would have been available had the third party been the other party to the contract. In other words, defences set-offs or counterclaims are available against the third party as if he were the promisee (section 3(4)).

AVOIDING DOUBLE RECOVERY

7-25 If a third party sues the promisor in circumstances where the other party to the contract (the promisee) has received from the promisor a sum in respect of:

- The third party's loss; or
- The expense to the promisee of making good to the third party the default of the promisor,

then the court or arbitral tribunal are to reduce the award to the third party to such extent as it thinks appropriate (section 5).

EXCLUDED CATEGORIES

7-26 A number of matters are excluded from the operation of the Act such as:

- Bills of exchange;
- Contracts binding on a company and its members under section 14 of the Companies Act 1985;
- Employment and other related contracts;
- Carriage of goods by sea, road or air (except that exclusions of liability can still be enforced (section 6(5)).

ARBITRATION

7-27 Where the contract under which the third party has a right to enforce a term contains a valid arbitration agreement referring disputes etc. to arbitration, the third party is bound by that provision (section 8(1)).

Where the third party has a right to enforce a term to the effect that disputes between him and the promisor go to arbitration, then the third party is to be treated for such purposes as a party to the arbitration agreement (section 8(2)).

COMING INTO FORCE

7-28 The Act received Royal Assent on 11 November 1999 and came into force six months after it was passed on 11 May 2000. However, if a contract was entered into after 11 November 1999 but before 11 May 2000 and expressly provided for the application of the Act, then the Act applied (section 10(3)).

The Act extends to England, Wales and Northern Ireland (section 10(4)).

The Act in the context of construction projects

7-29 There is no doubt that the Act was intended to provide a means by which third party rights could be granted in relation to construction contracts. The prevalent use of collateral warranties in the construction industry was discussed in the Law Commission Report 242: Privity Of Contract: Contracts For The Benefit Of Third Parties (e.g. paragraph 2.18 of the Report). The Report makes clear how the Commission consider that the provisions of the Act enabling third party rights to be granted could replace the existing practice of using collateral warranties. For example, the Report says:

> '3.15 – A typical collateral warranty given by an architect, engineer or main contractor excludes consequential economic loss and limits the defendant's liability, having regard to other claims of the warrantee, to a just and equitable proportion of the third party's loss. A typical warranty in favour of a finance house will also contain provisions permitting the finance house to take over the benefit of the contractor's or architect's or engineer's appointment contracts on condition of payment of liabilities, if the main finance contract is determined for any breach on the part of the employer, so that the finance house could ensure the continuance of work on the development notwithstanding some breach of the loan agreement by which the original employer was financed. There will also be provisions permitting the finance house, purchaser or tenant a licence to copy and use for specified purposes any designs or documents that are the property of the contractor or architect or engineer, and a clause undertaking that the contractor, architect or engineer will maintain professional indemnity insurance in a specified sum for a specified period. Finally, the warranty will normally permit assignment by the finance house, purchaser or tenant without any consent of the warrantor being required. These collateral warranties are generally supported by separate nominal consideration or are made under deed and thus are not tied to consideration in the main contract.
>
> . . .
>
> 3.17 – Our proposed reforms would enable contracting parties to avoid the need for collateral warranties by simply laying down third party rights in the main contract. Moreover, our proposed reforms would enable the contracting parties to mirror the terms in existing collateral warranties . . . there is no reason why the architects' engineers' and contractors' liability to the third party could not be limited, as it presently is under collateral warranty agreements, so as to exclude consequential loss and so as to be limited to a specified share or a just and equitable share of the third party's loss. As regards defences, a claim by a third party under our proposed legislation, . . . will be subject to defences and set-offs arising from, or in connection with, the contract and relevant to the particular

contractual provision being enforced by the third party and which would have been available against the promisee. But this is a default rule only and the contracting parties can provide for a wider or a narrower sphere of operation for defences and set-offs, if they so wish. So the present position under collateral warranties, whereby the claim is subject to defences arising under the main contract, is, or can be, replicated.

3.18 – So, in our view, our proposals would enable the contracting parties to replicate the advantages of collateral warranties without the inconvenience of actually drafting and entering into separate contracts. Moreover, our proposals may carry a limited degree of extra flexibility as regards variation.

3.19 – A further advantage of our legislative reform, as against collateral warranty agreements, is that it would not be necessary to assign the benefit of a provision extending the contractor's or architect's duty of care to sub-financiers and other purchasers and tenants down the line, since these persons could simply be named as potential beneficiaries of the clause by class. Thus, the difficulties caused by quantification of damages in claims under assigned collateral warranty agreements would be entirely removed.'

The JCT approach to the Contracts (Rights of Third Parties) Act 1999

7-30 The JCT, along with other bodies responsible for producing standard forms of construction contract, e.g. the Institution of Civil Engineers, have excluded the operation of the Contracts (Rights of Third Parties) Act 1999 by providing that nothing in its contracts is intended to confer third party rights. For example, clause 1.9 of WCD 98 following amendment 2 of January 2000 provides:

> 'Notwithstanding any other provision of this Contract nothing in this Contract confers or purports to confer any right to enforce any of its terms on any person who is not a party to it.'

One reason for this may be that until there has been a thorough review of all JCT contracts, following the coming into force of the Act, terms possibly beneficial to a third party may unintentionally create enforceable third party rights.

The approach under the MPF

7-31 The MPF marks the first departure by the JCT from this policy of excluding the provisions of the Act. The MPF form takes advantage of the Act by providing a Third Party Rights Schedule. This Schedule provides for third party rights in favour of funders, purchasers and tenants. It is no doubt intended that this should reduce or eliminate any need for separate collateral warranties from the contractor in favour of such funders, purchasers and tenants.

As the rights in favour of funders, purchasers and tenants are by way of terms included in the MPF itself by taking advantage of the Act, it is clearly the intention to do away with the need for separate collateral warranties. Such a separate warranty requires all the usual elements of a binding contract such as offer, acceptance and consideration passing from each party to it. By using the MPF itself, these requirements are removed. However, it is important to note that by making use of the Act, the parties, including any third party, will be subject to its provisions. This is particularly significant in relation to clause F2 and clause PT4 dealt with below (equivalent rights in defence – see paragraphs 7-52 *et seq.* and 7-73).

From the employer's point of view, it is important to appreciate that as the contractor is not responsible for the contents of the requirements or the adequacy of the design contained in the requirements (clause 5.1 of the MPF), the employer will in any event want to be able to offer funders, purchasers and tenants adequate direct remedies, probably through the use of collateral warranties, against those who may have been engaged by the employer in connection with design or the preparation of the Requirements such as pre-appointed consultants or named specialists (see clause 18 of the MPF). Even if the contractor, by an amendment to the MPF, is made fully responsible for all design whether carried out by him or by others before his appointment, the existing practice in such circumstances of collateral warranties being obtained in any event from professionals and specialist sub-contractors will no doubt continue as before.

We now turn to consider the MPF enabling clause for the third party rights schedule, clause 30, followed by a consideration of the third party rights schedule itself.

Limited exclusion of the Act

30.1 Other than such rights of the Funder, Purchasers or Tenants as take effect pursuant to clause 30, nothing in the Contract confers or is intended to confer any right to enforce any of its terms on any person who is not a party to it.

7-32 The third party rights schedule is clearly intended to make use of the opportunity presented by the Act. What clause 30.1 does is to provide that apart from what is provided as a result of clause 30 and the schedule, nothing else in the contract either confers or is intended to confer any right to enforce any of its terms on third parties. This is sensible as otherwise there could be an unintended benefit given to someone who is not a party to the contract but who might argue that in some way a benefit was intended to be conferred upon them.

However, it is arguable that clause 30.1 operates to exclude a funder's assignee's rights pursuant to F14, and a purchaser's or tenant's assignees' rights under PT9. Clause 30.1 states that only funders, purchasers and tenants are to have the rights conferred by the schedule. The definitions of 'Funder', 'Purchasers' and 'Tenants' in clause 39 do not extend to cover their assignees. It is not unusual to find the definition of these beneficiaries in collateral warranties extended to refer to their 'successors in title and assignees'. This would remove this potential ambiguity and some employers may make such an amendment. The alternative view is that assignees are not in fact third parties to the contract at all, but claim by virtue of becoming a party to it, so that there is no inconsistency or ambiguity.

When compared with JCT clauses excluding the operation of the Act in relation to the contract as a whole, e.g. WCD 98, clause 1.9 (see paragraph 7-30 above), the only change is that previously reference has been made to there being nothing in the contract conferring or 'purporting to confer' rights. This has now been replaced in the MPF with 'intended to confer'. Though the words 'purports to confer' reflect the wording in section 1(1)(b) of the Act, stating that a third party will have a right to enforce a term of the contract if it purports to confer a benefit on him, in truth the contract itself either will or will not purport to confer a benefit even if the contract has a clause which states that it does not purport to confer any benefit. Accordingly it is more sensible to state directly that it is not intended to confer any right to enforce any terms over and above those which are expressly included by virtue of clause 30 and the schedule.

Position of the Funder

7-33 30.2 The rights set out in clauses F1 to F18 of the Third Party Rights Schedule are hereby vested in the Funder.

30.3 Where rights have vested in the Funder pursuant to clause 30.2:
.1 no amendment or variation shall be made to the express terms of clauses 30.2 or 30.3 or of clauses F1 to F18 of the Third Party Rights Schedule without the prior written consent of the Funder; and
.2 neither the Employer nor the Contractor shall agree to rescind the Contract and the rights of the Contractor to terminate its employment thereunder or to treat the Contract as repudiated shall in all respects be subject to the provisions of clauses F7 to F9 of the Third Party Rights Schedule

but, subject thereto, unless and until the Funder gives notice pursuant to clause F6 or clause F10 of the Third Party Rights Schedule, the Contractor shall remain free without the consent of the Funder to agree with the Employer to amend or otherwise vary or to waive any term of the Contract and to settle any dispute or other matter arising out of or in connection with the Contract in each case in such terms as they think fit without any requirement on the part of the Contractor to obtain any consent from the Funder.

Vesting of rights

7-34 The funder will be in place when the contract is entered into. 'Funder' is defined in clause 39.2 as:

> The person or syndicate providing funding for the purposes of the Project, as identified in the Appendix.

The appendix provides a space for the funder (if any) to be identified. Accordingly, it could be a person or a syndicate. Section 1(3) of the Act requires that any third party must be expressly identified in the contract by name, as a member of a class or as answering a particular description. Clearly a named person would satisfy this requirement. If the funder is a syndicate, care should be taken to properly and clearly identify that syndicate to avoid confusion. While the Act (see again section 1(3)) provides that the person identified by name or as a member of a class or as answering a particular description need not be in existence when the contract is entered into, so that a simple reference to a funder providing finance in connection with the carrying of the project may suffice, the MPF requires the identity of the funder to be stated. This is a practical problem where the identity of the funder is unknown at the time of entering into the contract.

Section 1 of the Act confers third party rights. Section 1(4) provides:

> 'This section does not confer a right on a third party to enforce a term of a contract otherwise than subject to and in accordance with any other relevant terms of the contract.'

As the vesting is automatic, it is important that prior to the contract being entered into, all relevant matters such as the level of professional indemnity cover required, the period of limitation of liability and so on, are included in the Schedule either by standard provisions, as is the case with funders, or in part in that manner and in part by providing items within the MPF appendix to deal with such issues, e.g. the extent of recoverable losses from the contractor by a purchaser or tenant (see the schedule clauses PT2 and PT3 and the MPF appendix entry relating to the schedule). Some of the practical implications of this are dealt with below (see paragraph 7-83 *et seq.*).

If the schedule is used without also seeking to obtain separate collateral warranties, the level of administration will significantly diminish as will the risk of project-specific variables not having been inserted at the right time, or at all.

No amendment or variation without consent after vesting

7-35 Clause 30.3.1 provides that neither clause 30.2 nor clause 30.3 themselves, nor the provisions of the schedule relating to funders, namely F1 to F18, may be amended or varied without the prior written consent of the funder. This mirrors the effect of sections 2(1) and (3) of the Act which limits the ability of the original contracting parties to vary the contract in such a way as to extinguish or alter the third party's entitlement unless the consent of the third party is obtained. While section 1(4) makes provision for the court in certain circumstances to dispense with a party's consent, this is highly unlikely to be relevant in the case of a funder as it applies to the situation where the third party's whereabouts cannot be ascertained or where he is mentally incapable of giving his consent.

The consent of the funder is in no way fettered or conditioned. Does this mean that the employer and contractor are not free to agree to vary or change the design, quality or quantity of anything required in accordance with the contract? Clause F1 of the schedule provides that the contractor warrants that it will comply with the contract. The 'Contract' is defined as being the conditions, appendix, Third Party Rights Schedule, requirements, proposals and the pricing document. The requirements and proposals are identified as specific documents in the appendix to the MPF. It might be thought by some therefore that both the requirements and the proposals have to be complied with as part of the contractor's obligation to comply with the contract and that therefore any departure from this would require the consent of the funder. However, the obligation of the contractor under the contract is to execute and complete the project in accordance with the contract (clause 1.1) and this must mean in accordance with the conditions which themselves expressly permit changes of the kind referred to above under clause 20. Indeed, under clause 30.3.1, by implication, the employer and contractor are free to amend or vary without the consent of the funder, any of the clauses in the contract conditions other than clauses 30.2 or 30.3. Any control to be exercised by the funder in relation to this possibility will need to be governed by the finance agreement, defined by clause 39.2 as 'The agreement between the Funder and the Employer for the provision of finance for the Project'. The contractor should not be concerned with seeking the consent of the funder in relation to such matters.

The employer and contractor cannot agree to rescind

7-36 By clause 30.3.2 neither employer nor contractor can agree to rescind the contract. This reflects section 2(1) and section 2(3)(b) of the Act which provides that where a third party has rights to enforce a term of the contract, the parties to the contract cannot agree to rescind the contract without the consent of that third party.

The contractor's right to terminate his employment or treat the contract as repudiated is subject to the operation of step-in rights

7-37 Clause 30.3.2 also provides that the rights of the contractor to terminate his employment or to treat the contract as repudiated is to be subject to the provisions of clauses F7 to F9 of the schedule which requires the contractor to give advance notice of any such intention with the opportunity for the funder or his appointee to step into the employer's shoes under the contract.

Saving provisions prior to funder's notice to step in

7-38 The final paragraph of clause 30.3 provides that at any time before the funder has given notice pursuant to the schedule clause F6 or clause F10 that he is stepping into the contract in place of the employer, the contractor remains free, without the consent of the funder, to agree with the employer to amend or otherwise vary or waive any other terms of the contract, i.e. other than clause 30.2 and 30.3 and to settle disputes in such terms as they think fit without any requirement on the contractor's part to obtain the funder's consent.

The effect therefore is that unless the funder has stepped into the contract in place of the employer, the employer and contractor can agree to amend, vary or waive terms of the contract and to settle any disputes in connection with it as they see fit. This would allow the employer and contractor to vary the contract introducing new defences to liability and in effect limiting the contractor's liability to the employer and therefore the funder (see clause F2 of the schedule).

Situations not covered by the MPF

7-39 The MPF does not cater for the situation in which a subsequent funder becomes involved in the project in place of the named funder otherwise than by way of an assignment pursuant to clause F9 of the schedule. The same is true where an additional funder appears on the scene, or where the MPF is entered into without a funder but one is subsequently used during the course of the contract. No doubt amendments can be incorporated to cover this situation if it is thought to be a possibility. Some of the practical aspects of these issues are dealt with below (see paragraph 7-83 *et seq.*).

Position of the purchaser

7-40 30.4 The rights set out in clauses PT1 to PT13 of the Third Party Rights Schedule shall vest in a Purchaser or Tenant on the date on which the Employer serves on the Contractor a written notice identifying such person and the nature of its interest in the Project.

30.5 No right of the Employer and/or the Contractor

.1 to terminate the Contractor's employment under the Contract (whether pursuant to clauses 32, 33 and 34 (*Termination*) or otherwise) or to agree to rescind the Contract;

.2 to agree to amend or otherwise vary or to waive any terms of the Contract; or

.3 to agree to settle any dispute or other matter arising out of or in connection with the Contract, in each case in or on such terms as they in their absolute discretion shall think fit,

shall be subject to the consent of any Purchaser or Tenant.

30.6 Notwithstanding the provisions of clause 30.5, where rights have vested in any Purchaser or Tenant under clause 30.4 the Employer and the Contractor shall not be entitled without the consent of all such Purchasers or Tenants to amend or vary the express provisions of clauses 30.4 to 30.6 or of clauses PT1 to PT13 of the Third Party Rights Schedule.

Who are purchasers and tenants?

7-41 'Purchasers' is defined by clause 39.2 as:

Any and all first purchasers of all or any part of the Project.

'Tenants' is defined by clause 39.2 as:

Any and all first tenants of all or any part of the Project.

As discussed above (see paragraph 7-32), the definitions of 'Purchasers' and 'Tenants' will not cover their successors in title and assignees. This means that subsequent purchasers or tenants taking from first purchasers or tenants are not given third party rights under the schedule. However, by clause PT 9 of the schedule a limited assignment is possible (dealt with later – see paragraph 7-78). There is no restriction on the number of first purchasers or tenants entitled to the benefits contained in the schedule. Clearly in some developments, e.g. large-scale office or retail developments, the contractor could be faced with a considerable number of potential claimants if there has been a breach by the contractor of the provisions of the MPF which are widespread throughout the project rather than being localised.

The benefits of contractually granted third party rights

7-42 A major advantage of having provision for third party rights within the contract itself is that the administration involved in sorting out collateral warranties, sometimes involving different terms for different purchasers or tenants, can be avoided. Further, the fact that all of the terms containing these rights have to be in place by the time the contract is entered into can be a useful discipline.

Most of the matters which would, when using a collateral warranty, be required to be agreed and then included in each warranty given, are fixed by the MPF by being included in the schedule. However, the MPF does have an appendix item dealing with the choice to be made as to the extent of losses for which the contractor is liable to purchasers and tenants. Clauses PT1, 2 and 3 of the schedule deal with the nature and extent of the losses for which the contractor is to be liable to the purchaser or tenant in the event of not carrying out the project in accordance with the contract. Such losses may be restricted to liability for the reasonable costs of repair, renewal or reinstatement or alternatively can extend to other losses up to a stated maximum. These clauses will be considered later when dealing with the schedule itself (see paragraph 7-72 *et seq.*). However, the point is made here that the MPF itself makes provision in the appendix for this question to be resolved. If the appendix entry is not completed, the contractor's liability is limited to the reasonable costs of repair, renewal or reinstatement.

A potential disadvantage of making provision for third party rights for purchasers and tenants within the contract itself is that, as happens regularly in practice, the form of the collateral warranty which the contractor has agreed to enter into with future purchasers or tenants when entering into the contract, becomes the subject of negotiation and change. In such cases, once agreement has been reached between purchaser or tenant and contractor, there is little difficulty in producing the appropriate collateral warranty for execution. However, it is significantly more cumbersome to have to go through the process of a deed of variation to the MPF form itself. This can become even more of a problem if the clauses of the schedule have to be amended differently for different purchasers or tenants. Some of the practical implications of this are dealt with below (see paragraph 7-83 *et seq.*).

Vesting of rights

7-43 Clause 30.4 provides that the rights in the schedule relevant to purchasers and tenants, clauses PT 1 to PT 13, vest in the purchaser or tenant on the date on which the employer serves a written notice on the contractor identifying the purchaser or tenant and the nature of his interest in the project. The vesting therefore takes place once the identity of the purchaser or tenant has been notified in writing together with the nature of that person's interest in the project. The rather simple reference in clause 30.4 to the nature of the relevant person's interest in the project is rather vague. No doubt the notice must refer to whether it is a freehold or leasehold interest. To be on the safe side it should also refer to the specific part or section of the project, if it is other than the whole of the project in which the purchaser or tenant is acquiring an interest.

Clearly, in many cases a purchaser's or tenant's identity will not be known at the time the contract is entered into so that the vesting can only take place when that identity is known and when the relevant interest has been acquired.

Position prior to vesting of rights

7-44 Clause 30.5 provides in effect that the employer and contractor are free to exercise their contractual rights to terminate the contractor's employment or to agree to rescind the contract at any time whether before or after the purchaser's or tenant's rights have vested under clause 30.4. There are of course no step-in provisions in the schedule in favour of purchasers or tenants as they are in favour of funders. This may present a practical problem as often purchasers will require step-in rights. No doubt an amendment could be incorporated to introduce step-in rights for purchasers only. However, if these were introduced it would be necessary to deal with the priority of such step-in rights as between the funder and any purchaser. The alternative would be to allow the step-in rights to be exercised on a first come first served basis.

Similarly, the employer and contractor can agree to settle any disputes they may have on such terms as they think fit whether this takes place before or after the vesting of rights in any purchaser or tenant. However, once vesting has taken place, the employer and contractor are not entitled to agree to amend or vary clauses 30.4 to 30.6 of the MPF or clauses PT1 to PT 3 of the schedule without the consent of all purchasers and tenants whose rights have vested.

It can be seen, therefore, that even upon the vesting of third party rights in favour of a purchaser or tenant, the employer and contractor still have considerable freedom to vary the contract or indeed to agree to bring it to an end completely. Both also have the right unilaterally to terminate the contractor's employment in accordance with the provisions of the MPF or by accepting a repudiation of the contract by the other party. If a purchaser or tenant whose rights have vested could be affected by such matters, the place in which some influence can be brought to bear, at any rate to the extent that it is based upon the agreement of the employer and contractor, is in the agreement for sale or agreement for lease. In this instance, section 2(1) of the Act is not fully reflected. That section provides that once a third party has acquired a right to enforce a term of the contract, the parties may not agree to rescind or vary the contract in such a way as to extinguish or alter the third party's entitlement under that right without the third party's consent. Rather, clauses 30.5 and 30.6 reflect the provisions of section 2(3) of the Act which provides that the contract can expressly entitle the parties to vary, to any degree they choose, the effect of section 2(1).

Position after vesting of rights

7-45 While clause 30.6 seeks to prevent the contractor and employer from unilaterally amending clauses 30.4 to 30.6 or the third party rights included in the schedule once vesting has occurred, it does not prevent the employer and contractor from amending other terms which could nevertheless affect the position of the purchaser or tenant, e.g. an agreement that a defect be left unremedied in consideration of a reduction in the price. In addition, the employer and contractor can still agree to bring the contractor's employment to an end or to bring the contract itself to an end by rescinding it. The purchaser's or tenant's opportunity to influence such matters will be by means of the agreement for sale or agreement for lease. This may seem harsh for purchasers or tenants, but on the other hand, contractors would argue that they should not have to seek the consent of purchasers or tenants in such circumstances.

Where the employer and contractor wish to amend clauses 30.4 to 30.6, or Part PT of the schedule, the consent of all tenants and purchasers whose rights have vested must be obtained. Strangely, and possibly as a result of an oversight in the drafting, whereas the funder's consent under clause 30.3.1 must be both in writing and prior to any amendment or variation, under clause 30.6 there is no reference to the consent being in writing and no mention of it being obtained before any amendment or variation takes place. If the project is for multi-occupation and a significant number of purchasers' or tenants' rights have vested, it is likely to be an administrative quagmire to obtain all of the necessary consents.

If a purchaser or tenant seeks an amendment to the schedule by the inclusion of step-in rights, similar to that of a funder, then the enabling provisions should mirror those in clause 30.3 rather than those in clause 30.5.

Similarly to the position in relation to funders where the employer and contractor agree to alter design or quality (see earlier paragraph 7.35), the provisions of clause PT1 of the schedule under which the contractor warrants that he has carried out the project in accordance with the contract, do not prevent the employer and contractor agreeing a change to the requirements or proposals even if this is to the detriment of a purchaser or tenant.

The third party rights schedule

7-46 The third party rights schedule containing the rights in favour of the funder and purchasers or tenants is to be found near the end of the MPF between the appendix and pricing document. It is divided into two parts, the first being Part F (for funder), and the second Part PT (for purchaser/tenant). The various clauses are then prefixed with F or PT.

The third party rights granted by the schedule in favour of funders, purchasers and tenants are necessarily the product of compromise between the client and contracting interests represented on the JCT. They may not therefore reflect the bargaining position of the parties on any given project. It is to be anticipated that the schedule will be frequently amended in practice and usually in favour of the employer.

The schedule begins with a statement that it forms part of a contract and there is then space for inserting the date of the contract, the name of the employer and the name of the contractor.

Third party rights from the contractor in favour of the funder (Part F of the schedule)

The section of the schedule dealing with funder's rights is made up of clauses F1 to F18, which will now be considered in turn.

Contractor to comply with the contract

F1 The Contractor warrants that it will comply with the Contract.

Scope of obligation

7-47 This clause states in simple terms that the contractor warrants that he will comply with the contract. Clause F1 operates from day one of the contract as the funder will already be in place and the third party rights in favour of the funder vest immediately. The funder is clearly crucially interested in how the project progresses.

While clause F1 means that the contractor will be required to fulfil all of his obligations under the contract, not just those in relation to design workmanship and materials, e.g. to comply with the Construction (Design and Management) Regulations 1994, this is not unreasonable as at least there is no requirement for the contractor to enter into obligations which he does not already owe to the employer. What could be objectionable would be if obligations greater than that were placed upon the contractor. This is often done by what appear to be quite innocuous requirements such as a requirement to build in a good and workmanlike manner. In truth, while this may be an implied term where the contract is silent, it may be that the requirements place specific obligations on the contractor which differ from this implied term. In the normal way any such implied term would be subject to any express terms of the contract which differed from it, but including it as an express provision whether in a collateral warranty or within the contract itself would make it pre-eminent so far as the third party (funder) and the contractor are concerned.

Materials and workmanship

7-48 Clause 5.4 of the MPF provides that the contractor shall use materials and goods of the kinds and standards described in the contract. Only if they are not described are the materials and goods to be reasonably fit for their intended purpose. It also provides that materials and goods used for the project are to be of satisfactory quality (see commentary on clause 5.4 – paragraph 2-102 *et seq.*).

It should be noted that, unlike in many collateral warranties, there is no obligation on the part of the contractor not to select or use deleterious materials. Under clause F1, the contractor's obligations in this regard are the same as in the MPF itself. Clause 5.2.3 in effect achieves a similar result by providing a warranty from the contractor in favour of the employer that the project will include materials selected in accordance with 'Good practice in the selection of construction materials', prepared by Ove Arup & Partners and sponsored by the British Council for Offices and the British Property Federation, as current at the base date. This particular requirement has been discussed earlier (see paragraph 2-89).

By clause 5.5, the contractor warrants that all workmanship will meet the standards described in the contract. If no standards are described, work is required to be executed in a good and workmanlike manner. Clause 5.5 has been discussed earlier (see paragraph 2-118 *et seq.*).

Design

7-49 The contractor's warranty to comply with the MPF also conditions his design liability so far as the funder's rights are concerned. In this connection clauses 5.1 to 5.3 inclusive are relevant and have already been discussed (see paragraph 2-79 *et seq.*).

No limitation in respect of damages

7-50 Collateral warranties sometimes seek to restrict the extent of any claim under the warranty in various ways. The most common of these is to limit the liability of the contractor to meeting the reasonable costs of repair renewal or reinstatement of any part of the works affected by the contractor's breach of warranty. This does not appear in the JCT Main Contractor Warranty in Favour of a Funder (MCWa/F 2001), nor in the schedule so far as the funder is concerned, though it does feature in the schedule in respect of a purchaser or tenant (see clauses PT1 to PT3).

No restriction on liability or net contribution clauses

7-51 Some warranties in favour of funders, e.g. MCWa/F 2001, provide for a net contribution clause. Such a clause effectively limits the contractor's liability for defects in design where consultants described by discipline or sub-contractors described by trade or speciality are also at fault. The clause effectively shifts the risk of such specified consultant's or sub-contractor's insolvency from the contractor to the funder. The intention of such clauses is that the effect of 'several' liability at common law will be negated. At common law, where the same loss is caused by more than one wrongdoer, the claimant can seek the whole of his loss from any of the wrongdoers; it is then possible for any wrongdoers selected by the claimant to seek a contribution from wrongdoers liable to the claimant for the same loss. This is not too much of a problem for the selected wrongdoers if the other wrongdoers are all solvent. However, if there is an insolvency among them then, while the claimant can still recover all of his losses from whichever of the wrongdoers he chooses, a selected wrongdoer will not be able to obtain a just and equitable proportion of that loss from the insolvent wrongdoer. The net contribution clause therefore switches the risk of this insolvency from the giver of the warranty to the claimant or beneficiary under the warranty. Each specified consultant or sub-contractor is deemed to have paid to the beneficiary under the warranty his fair share of the loss having regard to his share of the blame. The MPF schedule does not contain such a provision, and therefore the funder's claim is not affected in this way.

Equivalent rights in defence of liability

F2 The Contractor shall be entitled in any action or proceedings by the Funder to rely on any term in the Contract and to raise the equivalent rights in defence of liability as it would have against the Employer under the Contract.

Introduction

7-52 Clause F2 (as does clause PT4 in relation to purchasers and tenants) enables the contractor, if subject to a claim by way of action or proceedings from the funder, to rely on any term in the contract and to raise the equivalent rights in defence of liability as he would have if it had been the employer making the claim.

While this may relate to any claim made by the funder which could have been made by the employer under the MPF, its most likely application is in relation to defects in design, workmanship or materials.

The ability for the contractor to rely upon any term in the contract would clearly include such matters as the possible effects of the issue of the final payment advice pursuant to clause 22.7; or the decision of the employer under clause 17.3.2 to leave a defect unrectified and instead to make an appropriate deduction from amounts otherwise due to the contractor.

The reference to 'equivalent rights in defence of liability' in clause F2 is interesting, in the light of the Contracts (Rights of Third Parties) Act 1999, particularly when compared with what is effectively identical wording in clause 1(b) of the JCT Main Contractor Warranty for Funders (MCWa/F 2001) (and see also clause 1(d) of MCWa/PT 2001). These warranties are applicable to a number of JCT contracts including WCD 98. The effect of these words in relation to the ability of the contractor to raise a set-off by way of defence to a funder's claim is different when used by reference to the Act and when they are contained in a separate collateral warranty. It is worth detailed consideration and this now follows.

The meaning of 'equivalent rights in defence of liability' and the set-off issue

THE ISSUE

7-53 Although clause F2 is effectively in identical terms to clause 1(b) of MCWa/F 2001 (see previous paragraph), the effect of the words 'equivalent rights in defence of liability' in clause F2 is different to their effect when used under a separate collateral warranty. This is because of the effect of the Act on clause F2. First of all we need to consider what these words mean. It is certainly arguable that these words mean that while the contractor can use any defence which would be available in relation to a claim by the employer in order to reduce or remove *liability*, these words would not enable the contractor to raise issues which would have the effect of limiting the *quantum* of the claim rather than liability itself. The position is far from clear, however, and it is equally arguable that *liability* in this context means liability for the *amount* being claimed by the funder in any action or proceedings; in other words that it does relate to quantum. If the former meaning is the correct one then under a separate collateral warranty such as MCWa/F 2001, the contractor would be unable to rely upon what would amount to a set-off which would have been available against the employer; on the other hand, if the latter meaning is correct, then the contractor would certainly be able to rely upon a set-off, which is in the nature of a defence.

A collateral warranty is a contract between the contractor and the funder, and is separate from the underlying building contract. The right of the contractor to raise, in answer to any claim made under the warranty by the funder, defences, set-offs or counterclaims which would have been available if the action had been taken by the employer under the building contract, would not be available unless the warranty specifically gave such rights. Therefore a provision entitling the contractor to equivalent rights in defence of liability can only be provided by way of an additional provision introduced into these collateral warranties as a specific benefit to the contractor. Accordingly, unless such a separate collateral warranty expressly provides that the contractor can raise equivalent rights of defence, set-off or counterclaim, he will not be entitled to raise them.

In this situation, where third party rights are created by means of a separate collateral warranty, it becomes important to determine whether 'in defence of liability' can include at least certain types of set-off, e.g. equitable set-off. If these cannot be brought within this phrase, then the contractor has no right to raise them. Assuming for the moment that these words do not include all set-offs, but only those which can be used as a defence against liability, the position will be very different under the MPF which uses section I of the Act to grant third party rights.

THE EFFECT OF THE ACT

7-54 Turning then to the MPF and its reliance upon section 1 of the Act, the use of the same words, 'in defence of liability', will have a different effect from that under a separate collateral warranty. This is because the Act, by sections 3(2) and 3(5), permits defences and set-offs to be raised against third party claims (and in limited circumstances allows counterclaims also) unless they are expressly removed or limited by the contract.

Before considering in more detail the provisions of clause F2 (and PT4) of the schedule in the light of the Act, the issue of using a set-off by way of defence is worth recalling. A summary of the law is to be found in Chapter 5 (see paragraph 5-111 *et seq.*).

It is against this background that we need to consider the provisions of clause F2 (and PT4) of the schedule.

Clause F2 (and clause PT4) and section 3 of the Act

7-55 The rights given in the schedule are given pursuant to section 1 of the Act. While clause F2 (and clause PT4) contain effectively identical words to those found in separate JCT collateral warranties such as MCWa/F 2001 clause 1(b) (MCWa/PT 2001 clause 1(d)), the 1999 Act governs the MPF schedule of rights whereas it does not govern those in a separate collateral warranty. Section 3 of the Act is relevant here and it provides as follows:

> '(1) Subsections (2) to (5) apply where, in reliance on section 1, proceedings for the enforcement of a term of a contract are brought by a third party.
> (2) The promisor shall have available to him by way of defence or set-off any matter that
> (a) arises from or in connection with the contract and is relevant to the term, and
> (b) would have been available to him by way of defence or set-off if the proceedings had been brought by the promisee.
> (3) The promisor shall also have available to him by way of defence or set off any matter if –
> (a) an express term of the contract provides for it to be available to him in proceedings brought by the third party, and
> (b) it would have been available to him by way of defence or set-off if the proceedings had been brought by the promisee.
> (4) The promisor shall also have available to him –
> (a) by way of defence or set off any matter, and
> (b) by way of counterclaim any matter not arising from the contract,
> that would have been available to him by way of defence or set-off or, as the case may be, by way of counterclaim against the third party if the third party had been a party to the contract
> (5) Subsections (2) and (4) are subject to any express term of the contract as to the matters that are not to be available to the promisor by way of defence, set-off or counterclaim.'

As any claim by the funder will need to rely upon section 1, sections 3(2) to (5) will apply.

It should be noted at once that section 3(2) does not refer to counterclaims which are not also set-offs. The contractor will not therefore be able to raise such counterclaims to resist a claim from the funder except in the limited circumstances permitted under section 3(4) (dealt with later (see paragraph 7-58) or if the contract expressly permits it. The MPF does not.

Defences and set-offs allowed unless the contract provides otherwise

7-56 Section 3(5) provides that subsection (2) is subject to any express term of the contract as to the matters that are not to be available to the promisor by way of defence or set-off. It would be possible therefore for the contract, and clearly the appropriate place would be in Part F of the schedule, to provide that no matters at all were to be available to the contractor by way of set-off. However, it does not do this. The schedule to the MPF, rather than state what matters are not to be available, instead takes the opposite approach of stating what is to be available, namely equivalent rights in defence of liability that the contractor would have had against the employer under the contract. This is the wrong way round. In relation to the JCT collateral warranties such as MCWa/F 2001 and MCWa/PT 2001, being a separate contract, no defence or set-off would be available unless expressly included, and the inclusion of an equivalent defence clause there means that defences other than those limited to liability (whatever they may be) and set-offs (to the extent that they also do not amount to a defence against liability) are not available. Precisely the same words used in the schedule to the MPF have the opposite effect in that unless these other matters are excluded they are, by section 3(2), included. The effect is to let in a range of defences and set-offs which prevents funders, purchasers or tenants from obtaining a 'clean' set of rights. In other words, they will be faced in answer to any claim they make for, e.g. defective design, workmanship or materials, with cross-claims relating to issues between the contractor and employer which, while they will need to be relevant to the term being enforced (section 3(2)(a)) (see paragraph 7-57), may have little or nothing at all to do with the issue which concerns them. This can be particularly startling where a project is for multi-purchasers or tenants and, for example, a purchaser claims against the contractor in relation to defects in his part of the project only to find that the contractor can raise matters between himself and the employer which have no bearing at all on that part, e.g. a dispute about the proper value of a change in relation to some other part of the project, or even a loss and expense claim by the contractor.

First of all, therefore, the contractor can clearly raise the equivalent rights of defence of liability as he would have had against the employer, and this will include pure defences and equitable set-offs which can be used by way of defence; but secondly and very importantly, the contractor can also raise set-offs generally as they are not expressed to be excluded by any terms of the MPF. The JCT is unlikely to be unaware of this issue, and if this is the case it looks to be almost disingenuous to use precisely the same words in its collateral warranties and its schedule when the difference in effect is so stark. This position will almost certainly in practice lead to employers amending the schedule to expressly exclude such claims.

There may be some argument that including rights of defence of liability in clause F2 (and clause PT4) is implicitly excluding other rights such as that of set-off, but the words do not appear clear enough to remove what would otherwise be the contractor's rights under section 3 of the 1999 Act.

Relevant to the term

7-57 Section 3(2)(a) provides that any defence or set-off raised must be 'relevant to the term', that is, the term being enforced by the third party (the funder, purchaser or tenant). Clause F2, however, expressly provides for equivalent rights as there would be against the employer under the contract 'in defence of liability'. This is therefore covered by section 3(3) of the Act which provides that the promisor shall have available by way of defence or set-off any matter if an express term of the contract provides for it to be available in proceedings brought by the third party. Clause F2 does not restrict defences as to liability by requiring them to be relevant to the term being enforced. In this latter case, therefore, so far as the contractor raising equivalent rights in defence of liability is concerned, including equitable set-offs, any possible restrictive effect that the words 'relevant to the term' may have in relation to any other defences or set-offs do not apply to clause F2 (or clause PT4).

The words 'relevant to the term' are however vitally important in relation to the scope of defences or set-offs not covered by clause F2.

The HMSO explanatory notes to the Act prepared by the Lord Chancellor's Department in referring to section 3 on the issue of defence or set-off arising out of the contract 'and relevant to the term being enforced' provide an illustration of where this would be applicable. In the illustration, the buyer who is buying from the seller has agreed to pay the price to a third party. The seller delivers goods that are not of the standard contracted for. The third party sues the buyer for the price of the goods. In such a case the buyer is entitled to reduce or extinguish the price by reason of the damages for breach of contract. What is clear from this is that the term to be enforced is the obligation of the buyer to pay the price over to the third party. The defence or set-off arising out of the contract and relevant to that payment obligation is the fact that the goods are defective. It is quite likely that the illustrated cross-claim raised by the buyer is in the nature of a pure defence, being an abatement of the price. Whether it is a pure defence or an equitable set-off, it is clearly related to the term (the right to payment for the goods) being enforced. It would also fall within a clause such as F2 in any event. If the factual situation had been somewhat different, for example the seller agreeing to and delivering the goods to the third party, if the third party subsequently sued the seller because the goods were defective, the seller could of course raise as relevant to that term the fact that he had not been paid or paid in full for the goods as a set-off against such a claim. Again, this is arguably a pure defence or at least an equitable set-off. Even if it is a set-off falling short of one which can be used as a defence, it would seem to be relevant to the term being enforced, e.g. that the goods were not compliant with the contract. If therefore a funder or purchaser or tenant took action or proceedings against the contractor in respect of defective work, would the contractor be able to raise as a set-off, as being a matter relevant to the term, the fact that he had not been paid or not been paid in full? If so, would it include such matters as a failure to pay loss and expense; failure to discharge any liability incurred by the employer under the indemnity provisions; failure to pay any bonus; or any other matter intended to be included in payment advices under the MPF?

Matters which are relevant to the term to be enforced might therefore be construed liberally in favour of the contractor.

So what does this all mean in practical terms? This can be considered by way of a hypothetical example in relation to a project using the MPF:

The funder, purchaser or tenant (the third party) claims in respect of defective work. More particularly, the sealant in connection with the glazing has rapidly degraded as a result of it being contaminated with a chemical that ought not to have been there. As a result the window system lets in a significant amount of water. The 'term' of the contract of which the third party has the benefit, and which he wishes to now enforce, is clause 5.4 of the MPF which so far as relevant provides:

> 'The Contractor shall use in the execution of the Project materials and goods of the kinds and standards described in the Contract or, if no such kinds or standards are described, materials and goods that are reasonably fit for their intended purpose. All materials and goods used for the Project shall be of satisfactory quality.'

In this situation the contractor wishes to raise four cross-claims:

(1) The sealant is part of the window system. The window system incorporated was as a result of a change instruction under clause 20. The contractor has not been paid for the value of the window system. If the contractor is required to meet the cost of the reinstatement works in connection with the sealant, the third party will have obtained the benefit of a compliant window system without the contractor being paid anything for it.

Clearly here the contractor's cross-claim must be relevant to the term, i.e. clause 5.4. While it is not a pure defence to liability, it is clearly an appropriate set-off by way of defence against it. It is relevant to the term. It may also in any event be an equitable set-off and covered by clause F2 (and clause PT4).

(2) The window system was the subject of a change instruction under clause 20. As a result of this instruction the contractor incurred loss and expense. The valuation of this loss and expense was included in the value of the change (see clauses 20.4.3 and 20.6.4). The contractor has still not been paid for this element of the value of the work.

Though less related to the claim than (1) above, under the MPF it is closely connected with the claim and may be a defence by way of equitable set-off and covered by clause F2 (and clause PT4). It is in any event a cross-claim in the nature of a set-off which is relevant to the term, being, under the MFP contract, an essential ingredient of the valuation of the window system.

(3) The contractor has a claim in respect of being unpaid for other work properly carried out under the contract. Adopting the Lord Sellers approach (see paragraph 5-113), in a claim for defective work, i.e. in effect the cost of remedying or reinstating it, or its diminution in value, a cross-claim on the part of the contractor that he was still unpaid in respect of the value of unrelated work carried out under the contract should generally rank as a suitable candidate for a set-off. Certainly, in the reverse situation where a contractor claims for the price of work carried out and an employer complains that the work is defective in some way, this is clearly capable of being brought into the equation, either as a set-off, or more probably as a straight defence of liability. Having said this, if a project was made up of units with a number of different purchasers or tenants, the result would be highly inconvenient if a claim were made against the contractor by a purchaser of a defective unit and he found himself faced with a set-off because the contractor had not been paid by the employer in respect of a different unit altogether. Clearly, the claim to set off here arises from or in connection with the contract but is it relevant to the term (in this example clause 5.4 of MPF) being enforced by the third party? Is the relevance of the term directed to a complaint concerning the sealant for the window system specifically, or is it related generally to workmanship and materials? The Law Commission Report (Privity of Contract: Contracts for the Benefit of Third Parties: Law Com No 242, 1996) formed the view that to include all defences and set-offs would cast the net too wide. The Report at paragraph 10.11 said:

> 'For where the third party is seeking to enforce a particular contractual provision, rather than the whole contract, it would seem that the defence or set-off should have to be relevant to the particular contractual provision. Otherwise a defence or set-off relating to an entirely separate clause, having no direct relevance to the particular contractual provision being enforced, could be used as a defence or set-off to the third party's claim. For example, if C seeks to enforce a payment obligation to him contained in, say, clause 20 of a construction contract between A and B, C's right should not be limited by a defence or set-off that A has against B in respect of, say, clause 5 which has nothing to do with clause 20.'

Then in paragraph 10.12:

> 'We therefore recommend that:
> ... the third party's claim should be subject to all defences and set-offs that would have been available to the promisor in an action by the promisee and which arise out of or in connection with the contract or, insofar as a particular contractual provision is being enforced by the third party, which arise out of or in connection with the contract and are relevant to that contractual provision.'

Also, in paragraph 10.13 of the Report :

> 'It was pointed out that allowing the promisor to raise defences (or set-offs or counterclaims) might have serious implications for the third party. For example, one of the advantages of reforming the third party rule is said to be to enable the original parties to a construction contract to grant subsequent owners or occupiers of the building contractual rights to repair of the premises, without the need for collateral warranties to be given to each owner or occupier. But the rights of, say, a subsequent owner, would be diminished if they could be met by a defence that the contractor had against the employer. The defence might be quite unknown to the subsequent owner.'

The situation is particularly acute where the project is in multi-occupation. While the requirement for the set-off to be relevant to the term being enforced must have some effect in restricting rights of set-off generally, it would seem to be too arbitrary to say that if there were to be one purchaser of the whole project the setting off for a failure to pay for work against a claim for defects would be possible, whereas if the project were divided among a number of purchasers in circumstances where the contractor's cross-claim for a failure to pay for work properly carried out was in relation to a different unit to that in respect of which a purchaser was claiming for defects, it would not be. The way in which the project is to be occupied might not have been decided at the time the contract was entered into. In either case the contractor probably has a right to raise the failure to pay as a set-off.

(4) The contractor has not been paid in respect of a loss and expense claim under clause 21. In particular, the contractor has been disrupted and delayed as a result of the activities on-site of the employer or his representative or advisors (see clause 21.2.1). Here there seems to be insufficient connection with the defects claim and it is unlikely that the contractor could use this cross-claim as a set-off.

Counterclaims

7-58 It should be noted that section 3(2) of the Act does not extend to counterclaims. This means that if the funder or purchaser or tenant brings an action or proceedings against the contractor, the contractor cannot raise a counterclaim which does not also amount to a defence or set-off in its own right. The Law Commission rejected the argument that section 3(2) should apply to counterclaims on the grounds that it would have been misleading and unnecessarily complex (see paragraph 10.10 of the Report). The reason for this is that a counterclaim can exceed the value of the third party claim and the effect of this would be to impose a burden on the third party as its claim against the promisor would be subject to a counterclaim against the promisee, which might exceed the value of the claim brought by the third party against the promisor. It is not the intention of the Act to alter the rule that a burden cannot be imposed on a third party without its consent. This exclusion of counterclaims makes it all the more important to distinguish between a set-off and a counterclaim which is not also a set-off.

However, though unlikely to occur often in practice, it can be noted that section 3(4) of the Act would enable the contractor to raise any matter by way of defence or set-off, or even if not arising from the contract, any counterclaim, against a funder that would have been available to him had the funder been a party to the MPF contract. In other words, if notionally it was the funder who was the other party to the contract rather than the employer, and for one reason or another, had it been the funder, the contractor would have had a defence set-off or counterclaim against the funder which would not have been available had it been the employer, the contractor can make use of this.

Contractor's obligations not affected by funder's enquiry into a relevant matter

F3 The obligations of the Contractor under or pursuant to clause 30.2 shall not be released or diminished by the appointment of any person by the Funder to carry out any independent enquiry into any relevant matter.

7-59 Clause 30.2 refers to the vesting in the funder of the rights contained in the schedule. To the extent that these rights impose obligations on the contractor, such obligations are not to be released or diminished because of the appointment by the funder of someone to carry out an independent enquiry into what is referred to as 'any relevant matter'. This effectively repeats clause 1(c) of MCWa/F 2001. However, in MCWa/F 2001, clause 1(c) relates only to clause 1 dealing with the contractor's obligation to comply with the building contract, whereas clause F3 of the schedule refers to all clauses in the schedule. Whether this will make much difference in practice is debatable as undoubtedly it is primarily aimed at the quality of the project. It is no doubt included *ex abundanti cautela* (out of an abundance of caution) as the chances of the appointment of any such person releasing or diminishing the contractor's obligation to comply with the contract in accordance with clause F1 or any other obligation in the schedule are highly unlikely. This is particularly so when clause F3 refers to the fact of the appointment rather than what the appointed person may say or do following the appointment.

The clause is presumably targeted essentially at the appointment of a specialist such as a surveyor, structural engineer, etc. who may examine the project or parts of it for some specific purpose.

The funder's step-in provisions

F4 The Funder has no authority to issue any direction or instruction to the Contractor in relation to the Contract unless and until the Funder has given notice under clause F6 or clause F10.

F5 The Funder has no liability to the Contractor in respect of amounts due under the Contract unless and until the Funder has given notice under clause F6 or clause F10.

F6 The Contractor agrees that in the event of the termination of the Finance Agreement by the Funder, the Contractor shall, if so required by notice in writing given by the Funder and subject to clause F11, accept the instructions of the Funder or its appointee to the exclusion of the Employer in respect of the Project upon the terms and conditions of the Contract. The Employer acknowledges that the Contractor shall be entitled to rely on a notice given to the Contractor by the Funder under this clause F6 as conclusive evidence for the purposes of the Contract of the termination of the Finance Agreement by the Funder and further acknowledges that such acceptance of the instructions of the Funder to the exclusion of the Employer shall not constitute any breach of the Contractor's obligations to the Employer under the Contract.

F7 The Contractor shall not exercise any right of termination of its employment under the Contract without having first:

.1 copied to the Funder any written notices required by the Contract to be sent to the Employer prior to the Contractor being entitled to give notice under the Contract that its employment under the Contract is terminated, and

.2 given to the Funder written notice that it has the right under the Contract forthwith to notify the Employer that its employment under the Contract is terminated.

F8 The Contractor shall not treat the Contract as having been repudiated by the Employer without having first given to the Funder written notice that it intends so to inform the Employer.

F9 The Contractor shall not:

.1 issue any notification to the Employer to which clause F7.2 refers, or

.2 inform the Employer that it is treating the Contract as having been repudiated by the Employer as referred to in clause F8

before the lapse of 7 days from receipt by the Funder of the written notice by the Contractor which the Contractor is required to give under clause F7.2 or clause F8.

F10 The Funder may, not later than the expiry of the 7 days referred to in clause F9, require the Contractor by notice in writing and subject to clause F11 to accept the instructions of the Funder or its appointee to the exclusion of the Employer in respect of the Works upon the terms and conditions of the Contract. The Employer acknowledges that the Contractor shall be entitled to rely on a notice given to the Contractor by the Funder under clause F10 and that acceptance by the Contractor of the instructions of the Funder to the exclusion of the Employer shall not constitute any breach of the Contractor's obligations to the Employer under the Contract. Provided that, subject to clause F11, nothing in clause F10 shall relieve the Contractor of any liability it may have to the Employer for any breach by the Contractor of the Contract or where the Contractor has wrongfully served notice under the Contract that it is entitled to terminate its employment under the Contract or has wrongfully treated the Contract as having been repudiated by the Employer.

F11 It shall be a condition of any notice given by the Funder under clause F6 or clause F10 that the Funder or its appointee accepts liability for payment of the sums due and payable to the Contractor under the Contract and for performance of the Employer's obligations including payment of any sums outstanding at the date of such notice. Upon the issue of any notice by the Funder under clause F6 or clause F10, the Contract shall continue in full force and effect as if no right of termination of the Contractor's employment under the Contract, nor any right of the Contractor to treat the Contract as having been repudiated by the Employer had arisen and the Contractor shall be liable to the Funder and its appointee under the Contract in lieu of its liability to the Employer. If any notice given by the Funder under clause F6 or clause F10 requires the Contractor to accept the instructions of the Funder's appointee, the Funder shall be liable to the Contractor as guarantor for the payment of all sums from time to time due to the Contractor from the Funder's appointee.

Summary

7-60 Clauses F4 to F11 inclusive (and clauses F12 and F17 are also relevant) contain step-in provisions which are commonly found in collateral warranties on behalf of funders, and sometimes purchasers or tenants. Clauses F6 to F11 inclusive are the equivalent of clauses 5 to 7 inclusive of the JCT Main Contract Warranty in Favour of Funders (MCWa/F 2001). The step-in can arise in two situations. Firstly under F6 where the finance agreement between the funder and the employer is terminated by the funder. In this instance, if the funder serves notice as required on the contractor, the employer acknowledges that the contractor is entitled to rely upon that notice as conclusive evidence of the termination of the finance agreement and also acknowledges that the acceptance thereafter by the contractor of instructions from the funder or the funder's appointee to the exclusion of the employer will not constitute a breach of the contractor's obligations to the employer under the contract.

Typical grounds for the termination of a finance agreement by the funder will be the employer's failure to fulfil preconditions or serious unremedied breaches by the employer as well as the employer's insolvency. The funder's right to step in is not dependent upon proof that there has been a termination of the finance agreement. The contractor is not concerned or required to enquire whether circumstances have arisen as between the employer and contractor entitling the contractor to step in. The funder's notice under clause F6 is conclusive evidence of the termination of the finance agreement.

Secondly, by clause F7, if the contractor wishes to terminate his employment under the contract or to treat it as having been repudiated by the employer, he must give notice to the funder of that intention and the funder then has 7 days in which to decide whether or not to operate the step-in provisions, during which time the contractor may not terminate his employment or bring the contract to an end. Further, if the contractor's proposed termination of his employment under the contract requires a warning notice, see for example clauses 33.1 and 34.1, a copy of that notice must be provided to the funder.

If the funder or his appointee does step in, the funder or his appointee may then issue instructions, though only to the extent permitted by the contract itself (see the first sentence of both clauses F6 and F10). The funder has no authority to give directions or issue instruction without an express right to do so. Clause F4 which expressly states this is probably inserted out of an abundance of caution to avoid any possibility that the funder could issue directions or instructions even without having given any notice under clauses F6 or F10.

Again, clause F5, providing that the funder has no liability to the contractor in respect of amounts due under the contract unless and until the funder has given notice under clauses F6 or F10, is no doubt the result of an abundance of caution to remove any possible, if remote, doubt there might otherwise have been.

The status of the step-in arrangement

7-61 The step-in arrangements on behalf of the funder are not easy to put into a known and accepted legal category. It does not appear to be a classic novation where a new contract replaces the existing contract which is discharged, and with the new contract being deemed to have existed from the outset. Possibly, the original contract coupled with a separate collateral warranty signed by employer, contractor and funder could suffice. Even then, a formal deed of novation is often called for. Under the MPF schedule it is more difficult to see how there is a tripartite agreement at all. The funder's notice does not create a new contract; instead the funder or his appointee steps into the employer's shoes and the original contract is clearly intended to continue in full force and effect. Further, the original contract is clearly not discharged. For example, in clause F10 it is provided that the contractor is not relieved of any liability he may have to the employer under the contract for any breach by the contractor, whether as a result of wrongly serving a notice that he is entitled to terminate or wrongly regarding the employer as having repudiated the contract or otherwise. In addition, clause F11 provides that on service of the notice by the funder, the contract is to continue in full force and effect.

A novation must be between the two original contracting parties and the third party, that is the funder or his appointee. Where these step-in arrangements are included in a collateral warranty, at least the funder is likely to be a party to it. In the MPF arrangement, the funder does not appear obviously to be a party to the MPF himself, nor of course does any appointee. This brings to light the issue of whether these step-in provisions can operate by means of section 1 of the Act at all.

The MPF contract is stated to remain in force on the basis that the employer and contractor give the funder or the funder's appointee the authority in such circumstances to take over the position of the employer and to issue instructions etc. The contractor remains liable to the employer for losses occasioned to the employer by the contractor's breaches prior to the step-in taking place, and then becomes liable to the funder or his appointee for future breaches. The funder or appointee becomes liable to the contractor for all sums owed to the contractor, whether by the employer in respect of sums becoming due and payable before the step-in arrangement, or thereafter for sums due following the step-in.

This arrangement appears to enable the funder or his appointee to take over the contract and in effect to receive all of the benefits available under that contract. Whether this is really what the Act envisages must be in some doubt. Section 1 talks about a third party obtaining the right to enforce a term which purports to confer a benefit on him. Throughout the Act it appears clear that the third party, even if entitled to all the benefits under the contract, never becomes the other party to the contract but is simply entitled to enforce those benefits. The Act does not appear to envisage the third party taking over liabilities, though it does envisage that the third party's right to enforce a term of the contract which is for his benefit is subject to and in accordance with any other relevant terms of the contract (section 1(4)) and presumably such other terms could impose onerous obligations as a condition of entitlement to the benefit. The Law Commission Report (Privity of Contract: Contracts for the Benefit of Third Parties: Law Com. No. 242, 1996) makes the point, when reaching the view that defences and set-offs should be allowed but counterclaims should not, that to allow them if they exceeded the value of the benefit received by the third party, would 'be an infringement of the principle that reform of the third party rule is intended to enable the enforcement of benefits by third parties not to impose burdens' (paragraph 10.10).

However, the Report makes it clear in paragraph 3.15 (referred to earlier – see paragraph 7-29) that such step-in provisions reflected in collateral warranties could equally well be achieved by the use of the granting of third party rights under section 1 of the Act.

The most favourable interpretation of these step-in provisions in the light of the Act is to say that the third party funder has an option which he can exercise in certain circumstances if he considers it to be beneficial to him. His exercise of that option is made subject to his taking on obligations which section 1(4) arguably allows the contract to impose. On this basis the Act can be used to achieve what for all practical purposes appears to be a novation.

Other aspects of the step-in provisions

RISKS FOR THE CONTRACTOR

7-62 Clause F9 provides that the contractor cannot terminate his employment or treat the contract as having been repudiated by the employer without having given 7 days' notice to the funder, during which time the funder has the option of stepping in. This is clearly a period of high risk for the contractor. During this time the contract does of course continue and the contractor has an obligation to proceed regularly and diligently with the project (see clause 9.2) and so will need to expend resources. However, if the contractor is unpaid in respect of sums due, he may well, if he has served the appropriate notice of proposed suspension as required by section 112 of the Housing Grants, Construction and Regeneration Act 1996, be able to suspend performance of his contractual obligations which could possibly protect him in this situation. Contractors should be mindful of this possibility. Clause F7 does not require the contractor to notify the funder if he intends to suspend performance of his contractual obligations. The contractor can suspend works for non-payment. The serious implications of work stopping because the contractor has not been paid means that a funder will want to be aware that suspension has taken place in order that he can take steps to protect his investment. An amendment requiring the contractor to provide a copy of his notice threatening suspension for non-payment pursuant to section 112 would seem to be sensible. However, some funder's warranties provide that during any notice period the contractor is expressly required to continue working. Depending upon the wording of such provisions, they may well be inconsistent with the contractor's statutory right under section 112, entitling the contractor on 7 days' notice to suspend performance of his contractual obligations while payment remains outstanding

Some funder warranties contain longer periods than 7 days, for example 21 days. This is often a specific requirement of funders to allow them sufficient time to assess whether to step in or not and in particular the financial implications of doing so. This clearly greatly increases the financial risks for the contractor in expending resources. In the event that the funder does not proceed, all this additional expense will be suffered by the contractor subject to his ability to claim it back from the employer. Even if the step-in is operated, the period during which the contractor is not permitted to exercise his legal right of termination etc. is in some funder's warranties left at the risk of the contractor, with the funder only being responsible for fulfilling payments and other liabilities incurred after the step-in.

Clause F10 provides that nothing in it relieves the contractor of any liability he may have to the employer. It could be that the funder steps in as a result of the contractor threatening to terminate his employment or treat the contract as repudiated by the employer. Following the step-in, as a result of which the employer is of course ousted, it may be that it can be shown that the contractor had no contractual right to do what he threatened to do. In those circumstances clause F10 expressly preserves the contractor's liability to the employer for any such breach. This is a useful saving provision for the employer to avoid any doubt as to the employer's rights in such a situation. It could clearly expose the contractor to a significant claim from an innocent employer.

This provision in clause F10 is made subject to clause F11 which provides that following the issue of the notice under clause F6 or F10, the contractor is to be liable to the funder or the funder's appointee under the contract in lieu of his liability to the employer. These two provisions may at first sight appear conflicting, but presumably for pre-step-in breaches of the contract the contractor remains liable to the employer, whereas for post-step-in breaches he is liable to the funder and any appointee in place of the employer.

It might be thought that there would also be a provision that in the event of the funder not stepping in, the failure by the contractor to terminate his employment or treat the employer's breach as a repudiation, while the notice period continued, should not be treated as a waiver of any breach of the employer which gave right to the contractor's entitlement to take the action proposed. No such provision is to be found in the schedule.

BENEFITS TO THE CONTRACTOR

7-63 By clause F11, if the step-in takes place, the contract is to continue in full force and effect as if no right of termination by the contractor etc. had arisen, and thereafter the contractor is liable to the funder or the funder's appointee instead of to the employer. The funder and any appointee of the funder become liable for sums due and payable to the contractor at the time of the step-in as well as subsequently. If it is the funder's appointee who steps-in, the funder is liable to the contractor on account of all sums due to the contractor from his appointee. The funder therefore acts as a payment guarantor of the appointee for sums outstanding at the time of step-in as well as for future payments. These are considerable benefits to the contractor. In practice, funders do not tend to favour such provisions. In some collateral warranties the funder is entitled to step in without any restriction and without guaranteeing the performance of the payment obligations under the contract by the replacement employer, who may well of course be a developer of the funder's choice. This would expose the contractor to considerable risk.

Copyright and design documents

> F12 Subject to the Employer having paid all monies due and payable under the Contract, or the Funder having made payment of such sums in accordance with clause F11, the Funder is granted the same rights in respect of the Design Documents as are granted to the Employer by clause 7 (*Copyright*) and any use made of the Design Documents is subject to the same conditions as are set out in clause 7.

7-64 Clause F12 is a simple provision giving the funder the same rights under clause 7 as the employer has in relation to 'Design Documents' (see commentary to clause 7 in relation to the employer's rights – paragraph 2-169 *et seq*.). This right is subject to either the employer or the funder having paid all sums due under the contract. Many funders are likely to find this proviso unacceptable.

It is surprising (and could well be a drafting slip) that these rights are not also granted to any appointee of the funder (see for example MCWa/F 2001 clause 8). On the other hand, if in one manner or another the contractor has been fully paid, any appointee having stepped in to the existing contract will be in the same position as the employer and so will be able to obtain the benefit of clause 7.

It might have been sensible in clause 7 for the employer to have been given an express right to call for design documents to enable copies for reproduction to be made rather than to leave this to arguable implication.

Contractor's professional indemnity cover

F13 Where clause 28 (*Professional indemnity*) applies the Contractor shall, upon request, provide to the Funder evidence that the insurance required by clause 28 is being maintained. At the time of its issue, the Contractor shall provide to the Funder a copy of any notification provided under clause 28.3.

7-65 Clause F13 provides that if clause 28 of the MPF applies, requiring the contractor to take out and maintain professional indemnity insurance on the terms stated in that clause, the funder can, in the same way as the employer, require the contractor to provide the funder with evidence that the insurance is being maintained. The contractor is also required to provide the funder with a copy of any notification provided to the employer under clause 28.3. The substantive provisions of clause 28 are commented on earlier (see paragraph 6-47 *et seq*.).

Assignment

F14 The rights contained in this schedule may be assigned without the consent of the Contractor by the Funder by way of absolute legal assignment, to another person (P1) providing finance or re-finance in connection with the carrying out of the Project and by P1, by way of absolute legal assignment, to another person (P2) providing finance or re-finance in connection with the carrying out of the Project. In such cases the assignment shall only be effective upon written notice thereof being given to the Contractor. No further or other assignment of the Funder's rights under the Contract will be permitted and in particular P2 shall not be entitled to assign these rights.

7-66 Clause F14 allows the funder to assign his third party rights without the consent of the contractor by way of absolute legal assignment to another funder. This second funder can also by way of absolute legal assignment again assign the third party rights to a third funder. The assignment is only effective upon written notice being given to the contractor and no further assignment is allowed. The definition of 'Funder' (see clause 39.2 and the appendix) should, provided the appendix does not include after the funder's identity some such words as 'including any assignees', restrict the assignments to two. If some such words are included, it could create a conflict between the definition of funder which does not include assignees, and clause F14 which talks about the funder (which would then include assignees) being able to assign twice. Contractors may wish to ensure that this possible difficulty does not arise when checking the appendix to the MPF.

While such restrictions on assignment may sometimes be acceptable to funding institutions, it may well be more of a problem to get investment purchasers to agree to the equivalent provision in the purchaser's or tenant's third party rights (see clause PT9).

It is important that any assignee ensures that the requirement for written notice to be given to the contractor is complied with to make the assignment effective.

The assignment can only be by way of an absolute legal assignment, which means that a funder could not assign part only of his interest in the project. It must be the whole of it.

Some assignment clauses provide that certain assignments, such as those by way of security or those between associated companies within the same group, should not count towards the two permitted assignments. Clause F14 does not do this.

Some assignment clauses also provide that the contractor should not be able to contend against any assignee that any loss which the assignee seeks to claim should be restricted by reason of the fact that the assignee was not named in the warranty. This is intended to enable the assignee to claim his own losses as a result of any breach by preventing the contractor from arguing that such loss is too remote and that it is only the losses which might have been suffered by the original funder which could be claimed. The assignee would of course argue to the contrary pointing out that the assignment to him was always envisaged and therefore provided his losses are those which would be foreseeable in respect of a typical assignee, they should be recoverable. Such a provision is not to be found in Part F of the schedule.

It would have been feasible not to expressly provide for assignment of the funder's rights at all, and instead to have relied upon including subsequent funders as being within the class of third parties who were intended to benefit. This would prevent any argument about foreseeable losses. This was envisaged in the Law Commission Report (Privity of Contract: Contracts for the Benefit of Third Parties: Law Com. No. 242, 1996) in paragraph 3.19 (see paragraph 7-29). As it is, there may be some risk here in relation to the level of damages recoverable by an assignee. However, the fact that the schedule expressly refers to assignees arguably makes their losses, if reasonably foreseeable at the time the contract is entered into, recoverable.

Notices

F15 Any notices required to be given by this schedule shall be given in accordance with clause 38.2.

7-67 As with other provisions in the schedule, the opportunity is taken to simply cross-refer to the MPF Conditions, in this case clause 38.2 which provides that any notice required to be given by the Schedule is to be by actual delivery, registered post (replaced by special delivery) or recorded delivery and that it shall take effect upon delivery. Clause 38.2 is commented upon in Chapter 9 when dealing with communications (see paragraph 9-10). This is not, however, without its difficulties. As clause 38.2 does not state the address to which the notice must be delivered or upon which address it must be served, presumably any notice served upon the funder by the contractor, e.g. under clause F7, needs to be served upon the registered office of the funder.

While it may be appropriate for notices such as those relating to a step-in situation or assignment to be subject to formality, it is not clear why the same reasoning has not been applied to assignment generally under clause 29, where there is no notice requirement of any kind to be found. Also, under clause 30.4 it would have been sensible for the employer's notice to the contractor in relation to the vesting rights of a purchaser or tenant, to have been subject to similar formality.

Time limit for commencing action or proceedings

F16 No action or proceedings for any breach of the rights contained in this schedule shall be commenced against the Contractor after the expiry of 12 years from the date of Practical Completion of the Project or the relevant Section (if applicable), whichever is the earlier.

7-68 The effect of this clause is to prevent the funder or any permitted assignee from commencing any action or proceedings for any breach of the funder's rights contained in the schedule after the expiry of 12 years from practical completion of the project. Where any potential action or proceedings in respect of a breach relates to any section of the project which has achieved practical completion, then the 12 years is measured from the date of practical completion of the section.

Unfortunately, clause F16 does not specifically address the situation where part of the project is taken over under clause 11 of the MPF, resulting in the taken over part of the project being treated as having achieved practical completion as from the date on which it was taken over. It might have been sensible to clarify the position in this situation. Presumably, as taking over part is not the same as practical completion of a section, the 12-year period will not begin to run until practical completion of the project or the section in which the part is contained.

The MPF assumes that the parties will execute it as a deed. On this assumption, the employer generally has 12 years from the date of the contractor's breach of contract in which to take action or commence proceedings, after which time the claim will become statute-barred, though, to obtain the benefit of a defence of limitation, the defendant must plead it specifically in his defence. Apart from defects remedied during the rectification period, the 12 years is likely to run from the date of practical completion of the project. A similar period of 12 years is therefore included in the schedule so far as the funder or any assignee is concerned.

Section 7(3) of the Act provides that the provisions of the Limitation Act 1980 relating to time limits in respect of actions based on either simple contracts (i.e. under hand) or upon a specialty (deeds and contracts executed as deeds) are extended to actions brought in reliance upon section 1 of the Act.

Contractor not liable to funder for delays

F17 Notwithstanding the rights contained in this schedule the Contractor shall have no liability to the Funder for delay under the Contract unless and until the Funder serves notice pursuant to clause F6 or clause F10. For the avoidance of doubt, the Contractor shall not be required to pay liquidated damages in respect of the period of delay where the same has been paid to or deducted by the Employer.

7-69 This clause mirrors that of clause 14 of MCWa/F 2001.

Clause F17 makes it clear that while the contractor may be liable to the employer in respect of delay under the contract, generally as delay damages is in the form of a liquidated amount, the contractor has no such liability to the funder prior to the funder operating the step-in provisions. A sentence is added for the avoidance of doubt, specifically stating that the contractor is not required to pay liquidated damages to the funder (or presumably its appointee) even following a step-in where this has been paid to or deducted by the employer. Delay to completion of the project prior to any step-in, and therefore its late delivery, will generally be dealt with between the employer and funder.

After the step-in has operated, the contractor will have a liability to the funder for delay under the contract. If the funder becomes the employer then this will no doubt be in the form of liquidated damages. However, it is by no means certain that the funder's loss will be the same as that of the original employer under the contract. The liquidated damages might therefore possibly be a penalty provision and unenforceable in this situation, but the funder could then seek general damages, though issues of remoteness of damage may be relevant. Clause F17 does not appear to prevent the funder claiming damages for delay which occurs under the contract where the funder's appointee has become the employer under it. This could therefore extend the contractor's liability for delay. It is unlikely that this was intended to be the situation.

Applicable law

> F18 This schedule shall be governed and construed in accordance with the laws of England and the English courts shall have jurisdiction over any dispute or difference between the Contractor and the Funder that arises out of or in connection with this schedule.

7-70 Disputes or differences between the contractor and the funder arising out of or in connection with the schedule are subject to the laws of England, and the English courts have jurisdiction. This mirrors to a large extent the position under the MPF as between employer and contractor (see clause 35.2 and the commentary on its substantive provisions – see paragraph 8-112).

It is clearly sensible to have the same provisions applying to the schedule as to the rest of the contract. It would probably be the case in any event (note section 1(4) of the Act which makes the enforcement of a term of a contract by a third party subject to any other relevant terms of the contract; this would include clause 35.2).

Third party rights from the contractor in favour of a purchaser or tenant (Part PT of the schedule)

7-71 This schedule provides for purchasers or tenants to have rights to enforce terms of the contract which are referred to in this part of the schedule. These rights are granted under section 1 of the Contracts (Rights of Third Parties) Act 1999. As there are no step-in provisions such as those in Part F of the schedule relating to funders (see clauses F6 to F11), the application of the Act to the schedule is more straightforward.

As with the funder's warranty, there is no net contribution clause and no separate prohibition against using certain materials, either or both of which are routinely found in collateral warranties in favour of purchasers or tenants. (See paragraphs 7-48 and 7-51 for a discussion of this in relation to Part F of the schedule, which is relevant also here.)

That section of the schedule dealing with the rights of purchasers and tenants is made up of 13 clauses, PT1 to PT13, which will now be considered in turn.

Contractor to comply with the contract: limits to any claim

PT1 The Contractor warrants as at and with effect from Practical Completion of the Project that it has carried out the Project in accordance with the Contract. In the event of any breach of this warranty and subject to clauses PT2 and PT3 the Contractor shall be liable for the reasonable costs of repair renewal and/or reinstatement of any part or parts of the Project to the extent that the Purchaser or Tenant incurs such costs and/or the Purchaser or Tenant is or becomes liable either directly or by way of financial contribution for such costs.

PT2 If the Appendix states that this clause is to apply the Contractor shall in addition to the costs referred to in clause PT1 be liable for any other losses incurred by a Purchaser or Tenant up to the maximum liability stated in the Appendix. Where no maximum liability is inserted in the Appendix then this clause shall not apply.

PT3 If the Appendix does not state that clause PT2 is to apply the Contractor shall be not liable for any losses incurred by a Purchaser or Tenant other than those costs referred to in clause PT1.

7-72 As the purchaser or tenant is assumed not to be providing finance for the project during construction, the contractor's warranty operates from practical completion. As from that date the contractor warrants that it has carried out the project in accordance with the contract.

Much of what has been said in the commentary to clause F1 of the schedule relating to funders is relevant so far as the contractor's obligations mirroring those it has under MPF are concerned. However, the contractor's liability in respect of any breach is subject to clauses PT2 and PT3. Clauses PT1, PT2 and PT3 deal with the extent of the liability of the contractor following a breach of his warranty to carry out the project in accordance with the contract. The position can be summarised as follows:

- The contractor is liable for the reasonable costs of repair renewal and/or reinstatement of any part of the project to the extent that the purchaser or tenant incurs such costs directly or alternatively becomes liable by way of a financial contribution for such costs, e.g. commonly the position of a tenant where there is multi-occupancy;
- PT2 applies if the appendix states that it does and if this is the case then in addition to the costs referred to above, the contractor is to be liable for other losses incurred by a purchaser or tenant subject to a maximum liability stated in the appendix. Accordingly, the appendix under the heading 'rights of a purchaser and/or tenant' has two parts requiring active completion before the contract is entered into. In relation to the schedule, the appendix provides as follows.

Rights of a purchaser and/or tenant

Clause PT2 is/is not to apply*.
Where no selection is made clause PT2 is not to apply.
Where clause PT2 applies the limit of the Contractor's liability to a Purchaser or Tenant for losses other than those referred to by clause PT1 is a total of

£ ______________
in respect of each breach/in total under the Contract*
Where clause PT2 does not apply or where no limit of liability is inserted the Contractor will only be liable to a Purchaser or Tenant in the manner provided by clause PT1.
*Delete as applicable

The first decision therefore is to decide whether PT2 applies or not and this is achieved by way of deletion. If no selection is made, then PT2 does not apply and the contractor's liability is limited to the costs referred to above in PT1.

If PT2 is stated to apply, then the next insertion is the limit of the contractor's liability for the purchaser's or tenant's losses. This itself is the subject of two monetary options:

(1) A financial limit in respect of each breach; or
(2) A total financial limit under the contract whatever the number of breaches.

Again, the selection is made by way of a deletion of that part which is not required. The appendix goes on to state that where PT2 does not apply or where no limit of liability is inserted, the contractor will only be liable for the costs referred to in clause PT1. Unfortunately, the appendix does not also provide, as it should, that if the option of a financial limit in respect of each breach or a total limit under the contract has not been selected, then again the contractor's liability should be limited to the costs referred to in clause PT1. As it stands, a figure could be inserted in the appendix with no indication given as to whether it is in respect of each breach or the total under the contract. This could cause the provision to fail totally due to a lack of certainty. It is imperative, therefore, particularly from the employer's point of view, to ensure that this appendix entry is carefully considered and effectively completed. Some employers will no doubt amend the appendix to provide for a suitable fall back to avoid such uncertainty.

Clause PT2 provides that the contractor's liability for these other losses is only to apply where a maximum liability is inserted in the appendix. This mirrors what is stated also in the appendix entry in this situation.

Clause PT3 provides that if the appendix does not state that PT2 is to apply, the contractor is not to be liable for losses other than those referred to in clause PT1. Again, this mirrors what is also stated in the appendix itself.

The level of liability which the contractor is prepared to accept will, of course, depend on a number of factors including the state of the construction and commercial property markets; the particular relationship between the employer and the contractor; and the restrictions under which the contractor may be required to operate, e.g. as set down by any holding company.

The kinds of loss for which the purchaser or tenant may seek recompense in addition to repair, renewal or reinstatement of any part of the project itself are, of course, likely to be consequential losses such as business interruption, including possible loss of profits, temporary relocation, storage, etc.

Equivalent rights in defence of liability

PT4 The Contractor shall be entitled in any action or proceedings by a Purchaser or Tenant to rely on any term in the Contract and to raise the equivalent rights in defence of liability as it would have against the Employer under the Contract.

7-73 The provisions of this clause in the light of the Act raise a number of important and interesting issues. These have been discussed earlier in relation to Part F of the Schedule and in particular in the discussion of clause F2 (see paragraph 7-52 *et seq.*). Similar points arise in relation to purchasers and tenants.

Contractor's obligations not affected by purchaser or tenant's enquiry into a relevant matter

PT5 The obligations of the Contractor under or pursuant to clause 30.4 shall not be released or diminished by the appointment of any person by a Purchaser or Tenant to carry out any independent enquiry into any relevant matter.

7-74 This clause mirrors clause F3 in Part F of the schedule dealing with funders, except that the appropriate reference is to clause 30.4 rather than clause 30.2. Similar comments to those made in relation to clause F3 are relevant here (see paragraph 7-59).

The purchaser or tenant has no authority to issue direction or instructions

PT6 A Purchaser or Tenant has no authority to issue any direction or instruction to the Contractor in relation to the Contract.

7-75 This expressly states what is in any event the position: that neither a purchaser nor a tenant has the authority under the MPF or otherwise to issue directions or instructions to the contractor so as to bind the employer.

Copyright and design documents

PT7 Subject to the Employer (or the Funder in accordance with clause F11) having paid all monies due and payable under the Contract, a Purchaser or Tenant is granted the same rights in respect of the Design Documents as are granted to the Employer by clause 7 (*Copyright*) and any use made of the Design Documents is subject to the same conditions as are set out in clause 7.

7-76 This clause mirrors clause F12 in Part F of the schedule dealing with funders, and similar comments apply where relevant (see paragraph 7-64).

Contractor's professional indemnity cover

PT8 Where clause 28 (*Professional indemnity*) applies the Contractor shall, upon request, provide to a Purchaser or Tenant evidence that the insurance required by clause 28 is being maintained. At the time of its issue the Contractor shall provide to any Purchaser or Tenant notified in accordance with clause 30.4 a copy of any notification provided under clause 28.3.

7-77 This clause is similar to clause F13 of Part F of the schedule in relation to the funder and similar comments can be made (see paragraph 7-65).

Assignment

PT9 The rights contained in this schedule may be assigned without the consent of the Contractor by a Purchaser or Tenant by way of absolute legal assignment, to another person (P1) taking an assignment of the Purchaser's or Tenant's interest in the Project and by P1, by way of absolute legal assignment, to another person (P2) taking an assignment of P1's interest in the Project. In such cases the assignment shall only be effective upon written notice thereof being given to the Contractor. No further or other assignment of a Purchaser's or Tenant's rights under the Contract will be permitted and in particular P2 shall not be entitled to assign these rights.

7-78 This clause mirrors clause F14 in Part F of the schedule dealing with the funder. Similar comments to those made earlier in respect of clause F14 apply here (see paragraph 7-66). In practice purchasers who are purchasing the project as an investment may seek an amendment with the intention of increasing the number of assignments possible. Much will depend on the state of the commercial property market.

Notices

PT10 Any notices required to be given by this schedule shall be given in accordance with clause 38.2.

7-79 This clause is identical to clause F15 of Part F dealing with the funder. Similar comments apply (see paragraph 7-67).

Time limit for commencing action or proceedings

PT11 No action or proceedings for any breach of the rights contained in this schedule shall be commenced against the Contractor after the expiry of 12 years from the date of Practical Completion of the Project or the relevant Section (if applicable), whichever is the earlier.

7-80 This clause is identical to clause F16 of Part F dealing with the funder. Similar comments apply (see paragraph 7-68).

Contractor not liable to purchaser or tenant for delays

PT12 For the avoidance of doubt, the Contractor shall have no liability to a Purchaser or Tenant under this schedule for delay in completion of the Project.

7-81 Late completion by the contractor does not entitle a purchaser or tenant to claim damages from the contractor. Any prejudice to the purchaser or tenant for which they want compensation must be dealt with in the agreement for sale or lease.

Applicable law

PT13 This schedule shall be governed and construed in accordance with the laws of England and the English courts shall have jurisdiction over any dispute or difference between the Contractor and a Purchaser or Tenant that arises out of or in connection with this schedule.

7-82 This clause is identical to clause F18 of Part F dealing with the funder. Similar comments apply (see paragraph 7-70).

Practical implications

7-83 We now turn to look at some of the practical implications of the issues raised earlier in relation to the schedule and consider how these might be dealt with in practice.

Issue

7-84 The need to amend the rights in the schedule as a result of negotiation by the employer with beneficiaries after the contract has been executed.

EXAMPLE

As discussed above, PT2 allows a limit on a contractor's liability to the purchaser/tenant to be introduced (see paragraph 7-72). The problem with a clause of this type appearing in the standard form is that psychologically it draws to the attention of the contractor negotiating a contract, that this type of limitation is a possibility. In reality caps of this kind in warranties to purchaser/tenant are by no means the market norm. If a cap is included in the schedule and subsequently a purchaser or tenant reviews the executed contract with a view to purchasing or taking a lease, the purchaser or tenant may find this or another provision unacceptable and request an amendment to the schedule.

SOLUTION

The best way to deal with this issue from the employer's perspective is to prevent it arising in the first place by taking a robust stance in negotiating the schedule and ensuring that the limits on liability are either deleted or kept to an acceptable minimum. The schedule can then be presented to a potential purchaser/tenant on a 'take it or leave it' basis.

Where some amendment is agreed this is often achieved quite simply because collateral warranties are annexed to the back of a building contract on the basis that the form annexed is subject to reasonable amendments requested by purchasers or tenants.

Clause 30 could be amended to state that rights that are in the schedule are subject to amendment at the request of the employer with the consent of the contractor not to be unreasonably withheld.

If the contractor subsequently consents and amendments are required as a result of negotiation with a purchaser or tenant, the schedule could be retyped as a separate document or photocopied and amended in manuscript and made the subject of a deed of variation which would vary the terms of the schedule in relation to that particular purchaser or tenant only. The beneficiary would need to execute this as would the contractor and the employer in order to ensure that all parties were bound by its terms. Another option is for the schedule to be amended by a simple letter circulated between the parties and signed by all of them. However, the beneficiary would be unlikely to find this acceptable.

Issue

7-85 What is to be handed over to the beneficiary in place of the collateral warranty?

EXAMPLE
A downside of collateral warranties is the need to produce copies for all of the parties to the warranties and circulate them for execution. This can often place a heavy administrative burden on parties to a contract, particularly when trying to get warranties from all sub-contractors involved in a project.

SOLUTION
The beneficiary under the schedule is not required to execute the schedule. However, the beneficiary would require at least a certified copy of the contract including the schedule in order to understand what rights he had and to keep for his records. This may appear onerous in terms of the amount of photocopying to be undertaken particularly if there are a large number of beneficiaries. However, currently a lawyer advising a prudent purchaser or tenant and reviewing the terms of a contractor's collateral warranty will ask for a copy of the building contract to see what, if any, limitations on liability it contains and to understand how they impact on the warranty. The contractor or employer may object to the contract sum being shown to a beneficiary and therefore this information may need to be deleted from the copy circulated to the beneficiaries.

Issue

7-86 How do you deal with the issue of different rights being given to different beneficiaries in the same class?

EXAMPLE
Different forms of the schedule may be acceptable to different beneficiaries. For example, a tenant who does not have a repairing obligation under a lease may be prepared to accept a limit on liability in relation to repair. Equally, any purchasers during the development may require step-in rights whereas a purchaser purchasing the development after the development has been completed may be content without step-in rights.

SOLUTION
This could be dealt with from the start by drafting different schedules for different types of tenants/purchasers/funders. Clauses that are usually thought to be acceptable in the market place for those particular classes of beneficiaries could be included in the same way as would have been done when negotiating collateral warranties. Alternatively, clauses could be placed in square brackets which may need to be introduced or deleted in particular circumstances with a footnote stating when and for what category of beneficiaries the clause is to be included/excluded. For example, step-in rights could be placed in the PT schedule in brackets with the footnote 'To be included for a purchaser at the request of the employer'.

Issue

7-87 What happens if different types of beneficiaries such as freeholders, funders to purchasers or first funders appear part way through the project?

EXAMPLE

If a funder named in the schedule or appendix was replaced part way through the project this would require a formal variation to the contract. The unamended contract provides for funders to be specifically named. The definition of 'funder' would not cover mortgagees of the completed project, institutions providing refinance after the project has been completed or institutions providing funding to a purchaser. All of these bodies will usually require warranties, and traditionally when amending construction contracts to make provision for warranties in favour of funders, the definition will be extended to cover these categories of funder.

SOLUTION

A solution to this is to extend the definition of funder so that it does not specifically mention funders by name but introduces categories of funders such as those mentioned above. This would allow rights to be given to the types of beneficiaries without the need to amend the schedule when a new beneficiary is introduced. The definition of 'Purchaser' and 'Tenant' could also be amended to cover all potential classes of beneficiary. This should be acceptable to the contractor as it is commonly introduced into building contracts. The vesting clause 30.2 would require amendment.

Issue

7-88 What happens if a beneficiary objects to the schedule and requests a collateral warranty as an alternative?

EXAMPLE

Beneficiaries and their advisors may feel that the schedule is an unknown quantity in the early days of its introduction, and may request a collateral warranty instead.

SOLUTION

A provision could be introduced allowing warranties as an alternative to the schedule. The other solution would be to delete the schedule entirely and replace it with provisions regarding the provision of warranties to the beneficiaries. This may become the norm in the short term while the schedule is gaining market acceptance.

Chapter 8
Default and dispute resolution

Content

8-01 This chapter looks at termination of the contractor's employment and dispute resolution under the MPF. The relevant clauses are:

- Clause 31 – Provisions applicable to termination generally (reservation of rights and remedies)
- Clause 32 – Termination by employer
- Clause 33 – Termination by contractor
- Clause 34 – Termination by either the employer or the contractor
- Clause 35 – Resolution of disputes
- Clause 36 – Mediation
- Clause 37 – Adjudication.

Termination of contractor's employment – clauses 31–34

Background

Introduction

8-02 Contractual obligations may come to an end in a number of ways. In the ordinary course of things, this will most commonly be brought about by the parties performing all their contractual obligations or promises.

8-03 However, contractual obligations can be discharged in a number of other ways, e.g. by accord and satisfaction (mutual discharge by agreement not to require performance of outstanding obligations); by the unilateral release by one party of the other's obligations; by frustration of the contract; by breach (a sufficiently serious breach of contract by one party might entitle the other party to treat himself as discharged from any further performance); or by contractual terms providing for discharge in certain events.

8-04 In this summary we are concerned only with the discharge of contractual obligations in accordance with the express conditions of the contract itself, and with the discharge of contractual obligations brought about by breach of contract by one of the parties.

Provision for discharge of the contract itself

8-05 Building contracts often make provision for one or both of the parties to have the right in certain situations either to bring the contract itself to an end or (and the distinction might be slight) to bring the employment of the contractor under the contract to an end.

8-06 At common law, a contracting party will only be able to regard himself as discharged from any further performance of his contractual obligations by a particularly serious breach of contract by the other party. The inclusion therefore of an express provision for termination has the advantage to the party exercising the right that it can permit such a discharge of further obligations for less serious breaches and even for an event which does not amount to a breach of contract at all (although its consequences might lead to a breach), e.g. administrative receivership or a voluntary arrangement for a composition of debts. However, in certain circumstances a term of this nature would have to be shown to be fair and reasonable by virtue of the provisions of the Unfair Contract Terms Act 1977 (section 3).

8-07 There might be very good reasons for terminating the employment of the contractor rather than bringing the contract as a whole to an end. It might be desirable for one or both parties that the contract survives and that express provisions can be called into play in the event of the employment of the contractor being terminated. Such express provisions may cover the right of the contractor to payment for work done but not paid for; rights over goods, materials, equipment and plant on site; and recovery of loss or damage incurred by the employer or the contractor as the case may be, as a result of the termination of the employment of the contractor.

8-08 If a breach of contract occurs which is sufficiently serious to be regarded as an intention by the party in breach no longer to be bound by the contract, i.e. a repudiation, and which is accepted as a repudiation by the other party, the effect of the parties' actions is to bring an end to their obligations as to future performance. On one view this might be seen as having brought the contract to an end, because the primary purpose of the contract is no longer to be achieved, but in fact the contract will remain in existence in the sense that a remedy in damages for breach of contract will survive, which will be calculated by reference to the terms of the contract, and certain express provisions of the contract (e.g. arbitration and, possibly, adjudication provisions) will remain operative for the purposes of regulating the parties' residual rights. In addition, the termination itself will not affect accrued rights and liabilities under the contract.

CONTRACTUAL DISCHARGE WITHOUT PREJUDICE TO COMMON LAW RIGHTS

8-09 It is perhaps a matter of debate whether or not the inclusion in the contract of express provision for termination takes away either party's right to treat the contract as discharged by reason of repudiation, in the absence of a clear statement to that effect in the contract. In the case of *Architectural Installation Services Limited* v. *James Gibbons Windows Limited* (1989), the question arose as to whether or not a party to the contract which provided a right to terminate subject to compliance with a comprehensive code retained his common law right to terminate, in circumstances where that code was not prefaced by the words 'without prejudice to other rights and remedies'. Judge Bowsher said:

> 'I would be sorry if the draftsmen of contracts felt it necessary to include such legal verbiage in order to avoid unintended results of their drafting. Construction contracts are already sufficiently complicated when the draftsmen seek to state what they do mean. They should not be burdened with the additional task of stating what they do not mean. When someone has obviously gone to a great deal

> of trouble to draft a contract, and two commercial parties have agreed to a contract in those terms, the court should be very reluctant to step in and suggest that those two parties also agreed something which was not written down in the agreement between them.'

It might, however, be argued that because 'two commercial parties' had agreed something which was written down, i.e. a right of termination subject to a comprehensive code, they would not expect the court to step in and conclude that the parties had not thereby intended to exclude any common law right to regard the contract as being terminated which was not subject to that same code. Indeed, in the case of *Lockland Builders Limited* v. *John Kim Rickwood* (1995), the Court of Appeal upheld the judge at first instance who had concluded that since the building contract in question had provided a comprehensive machinery for termination and this right had not been expressed to be without prejudice to the parties' common law rights, it followed that the machinery and the common law rights could co-exist only in circumstances where the contractor displayed the clear intention not to be bound by his contract. In the absence of such circumstances the machinery of the contract created the only effective way in which the agreement could be terminated. Of Judge Bowsher's comments in the *Architectural Installation Services* case, Lord Justice Russell in the Court of Appeal said:

> 'With all respect to the learned Judge, for my part I do attach significance to the absence of such words as "without prejudice to other rights and remedies", and I do not think that to include them would, as Judge Bowsher thought, involve verbiage in drafting.'

The leading texts appear to favour the view that in the absence of express provision, contractual terminations are not intended as a substitute for, or to exclude, common law rights: see *Hudson's Building and Engineering Contracts*, 11th edition, paragraph 12-006 *et seq.* and *Keating on Building Contracts* 7th edition, paragraph 6-83, though the latter notes the doubt created by the *Lockland* case.

8-10 The express right to terminate either the contract or the contractor's employment under it is often made subject to the service of one or more notices and any such notices must be clear and unambiguous in order to be valid. Further, a failure to comply with any procedural requirements, e.g. timetable or mode of service, might prevent the notice from taking effect. More is said about this topic later (see paragraph 8-48).

A wrongful termination of the contractor's employment by the employer might well itself amount to a repudiation of contract by the employer giving the contractor the right to treat the contract as at an end and to sue for damages for breach of contract.

A termination of the contractor's employment by the employer, which is challenged by the contractor, can create great difficulties if the contractor is unwilling to leave the site. The employer/landowner will no doubt want to engage another contractor to finish off the work. The continued presence on site of the original contractor will almost certainly make this impossible. The contractor's right to enter the site is likely to be by reason of an express or implied contractual licence to do so. The important question is whether or not, and in what circumstances, this licence is revocable. This issue has been considered when commenting on the contractor's rights of access to the site under clause 9.1 (see paragraph 3-03).

Under the MPF there is no express provision under which the contractor is required to give up possession of the site in the event of the employer determining the contractor's employment under the contract. However, clause 9.1 makes it clear that the contractor does not have an exclusive right to possession and indeed his licence to be on the site is limited to access. In these circumstances it is highly unlikely that the contractor, even following a disputed termination of his employment, would have any legal right to remain on the site.

Discharge by breach

8-11 One party to a contract may, by reason of the other party's breach, be entitled to treat himself as discharged from any obligation to further perform his contractual promises under the contract, and treat that other party's breach as being a repudiation by him of his contractual obligations. The innocent party will additionally have a right to claim damages consequent on the breach.

8-12 However, while any breach of contract can give rise to a claim for damages, it is not every breach of contract which will entitle the innocent party to a discharge from further liability to perform his own contractual promises. The default must be of a particularly serious or fundamental nature. Broadly speaking, it will be sufficiently serious to justify a discharge in three situations:

(1) In the case of a renunciation by one party of his contractual liabilities where that party by his words or conduct evinces an intention – whether due to unwillingness or inability – no longer to continue his part of the contract.
(2) Where by breach or default one party has rendered himself unable to perform his outstanding contractual obligations, e.g. if a contract requires the builder to be a member of the National House Building Council and to obtain an appropriate NHBC certificate and he is removed from the register kept by that body during the course of the contract.
(3) In the case of a very serious failure of performance by one party which will discharge the other. The failure must be of a fundamental nature going to the root of the contract.

The three situations referred to above may be described as repudiation or repudiatory events (sometimes perhaps less accurately as repudiatory breaches).

8-13 Once a repudiation has taken place, the innocent party must elect whether or not to accept the repudiation and this must be clear and unequivocal and must be communicated to the other contracting party. Once made, it cannot be withdrawn. The requirement that the election must be clear, unequivocal and communicated should not, however, be construed as a requirement that any particular form of election is required. In *Vitol SA* v. *Norelf Limited* (1996) the House of Lords recognised that non-performance of an obligation arising under a contract might suffice. In giving the unanimous opinion of the House, Lord Steyn gave the following example:

> 'Postulate the case where an employer at the end of a day tells a contractor that he, the employer, is repudiating the contract and that the contractor need not return the next day. The contractor does not return the next day or at all. It seems to me that the contractor's failure to return may, in the absence of any other explanation, convey a decision to treat the contract as at an end. Another example

may be an overseas sale providing for shipment on a named ship in a given month. The seller is obliged to obtain an export licence. The buyer repudiates the contract before loading starts. To the knowledge of the buyer the seller does not apply for an export licence with the result that the transaction cannot proceed. In such circumstances it may well be that an ordinary businessman, circumstanced as the parties were, would conclude that the seller was treating the contract as at an end.'

8-14 The innocent party may, if he wishes, treat the contract as continuing if this is still a possibility, despite the existence of repudiatory conduct. This will not prevent him from claiming damages while at the same time continuing with his contractual obligations.

8-15 If one party mistakenly treats an event as amounting to a repudiation by the other party and purports to accept it as such, this will often in itself create a repudiation which the wrongly accused party might be forced to accept. Indeed, in practice it is not unusual for both sides to argue that they have accepted a repudiation by the other.

However, if a party refuses to perform the contract, giving an inadequate or wrong reason, he may be able to justify his conduct if he subsequently discovers that at the time there existed another good reason entitling him to refuse further performance.

8-16 Once a repudiation has been accepted by the innocent party, both parties are excused from further performance of their primary obligations under the contract. Instead, secondary obligations are imposed on the guilty party, namely to pay monetary compensation for non-performance: see *Photo Production Ltd* v. *Securicor Transport Ltd* (1980).

8-17 Breach of a term of a contract which is not in itself sufficiently serious to amount to a repudiation could become so if it persists, especially after a notice from the innocent party requesting proper performance. Furthermore, the degree of wilfulness of a breach of contract might be a relevant factor in displaying an intention to no longer be bound by the contract terms.

8-18 Examples of serious breaches of contract amounting in the particular circumstances of the case to a repudiation include the prolonged failure by the employer to hand over the site to the contractor (*Carr* v. *J. A. Berriman (Property) Limited* (1953)); and a failure by the contractor to obtain the required performance bond (*Swartz & Son (Property) Limited* v. *Wolmaransstad Town Council* (1960)).

8-19 It has been suggested that a contractual term for due expedition by the contractor of the construction etc. of the contract works must be implied in building contracts, if not expressed, and that the repeated failure by the contractor to proceed with due expedition after notice by the employer will entitle the employer to treat the contractor's failure as a repudiation: see *Hudson's Building and Engineering Contracts*, 11th edition, paragraph 9.032 *et seq*. However, the decision at first instance of Mr Justice Staughton in the case of *Greater London Council* v. *Cleveland Bridge & Engineering Co Limited and Another* (1984) has thrown some doubt on the extent and nature of such a term. In the context of the MPF, however, this debate is of academic interest only because of the express obligation to proceed regularly and diligently (see clause 9.2). On the question of the construction of the express term to proceed regularly and diligently, see *West Faulkner Associates* v. *London Borough of Newham* (1994) (paragraph 3-14).

Remedies for breach of contract

INTRODUCTION

8-20 By far the more important remedy for either party to a building contract in the event of the other party's breach of that contract is a claim for damages. While it is possible in unusual circumstances to obtain an order for specific performance, whereby one party is compelled to honour his contractual obligations, or alternatively to obtain an injunction prohibiting a party from acting in breach of contract, damages remain the prime remedy.

DAMAGES FOR BREACH OF CONTRACT

8-21 The essential purpose of damages is to compensate the innocent party for loss or injury suffered through the other party's breach. The innocent party is, so far as money can achieve it, to be placed in the same position as if the contract had been performed. If no damage has been suffered or no damage can be proven, nominal damages will be awarded.

Difficulties in the assessment of damages do not prevent recovery. The fact that there are future losses involved might make assessment of damages very difficult, but nevertheless an assessment must be made and damages in respect of losses incurred and future likely losses must be awarded at the same time. It is not possible to have a series of actions claiming damages for breach of contract as and when the amount of future losses is ascertained.

8-22 However, it might in certain circumstances be possible for a contracting party to claim substantial damages even though no loss has been suffered. The House of Lords established in the cases of *St Martins Property Corporation Limited* v. *Sir Robert McAlpine Limited* and *Linden Gardens Trust Limited* v. *Lenesta Sludge Disposals Limited* (1993) that it might be possible for an employer under a building contract who sold the building which was the subject of the contract for full value and therefore in that sense would suffer no loss, to nevertheless claim substantial damages for breach of contract in respect of defective work. This has been discussed when dealing with assignment of the benefit of a building contract (see paragraph 7-07).

8-23 The Contracts (Rights of Third Parties) Act 1999 has now dramatically reformed the rule of 'privity of contract' and sets out the circumstances in which a third party (i.e. someone not a party to a contract) is to have a right nevertheless to enforce a term of a contract. This can enable someone who benefits from a contract to enforce beneficial terms in his own right and to claim damages. However, the contracting parties have the right to exclude the application of the Act to their contract. Until the MPF was published, all of the major standard forms of construction contract, including those of the JCT, expressly excluded the operation of the Act. The use of collateral warranties to fill the 'legal black hole' (see paragraph 7-07) has therefore continued. The MPF, however, makes use of the Act and includes an integrated schedule of third party rights in favour of purchasers, tenants and funders. A brief summary of the Act's provisions is included in the commentary on the schedule (see paragraph 7-20 *et seq.*).

CAUSATION AND REMOTENESS OF DAMAGE

8-24 Although causation and remoteness are two different concepts, they are often closely related and are therefore discussed together here. It is a matter of causation if the question which requires an answer is, 'was the loss caused by the breach of contract?' On the other hand, it is a matter of remoteness of damages if the question asked is, 'was the particular loss within the contemplation of the parties?' The leading case on remoteness of damage is *Hadley* v. *Baxendale* (1854).

FACTS

The plaintiff's mill was brought to a standstill by the breakage of their only crank shaft. The defendant carriers failed to deliver the broken shaft to the manufacturer at the time when they had promised to do so. They were sued by the plaintiff who sought recovery of the profits which they would have made had the mill been started up again without the consequent delay due to the late delivery of the broken shaft.

HELD

The facts known to the defendant carriers were insufficient to show that they were reasonably aware that the profits of the mill would have been affected by an unreasonable delay in the delivery of the broken shaft. The loss was therefore too remote. Baron Alderson said as follows:

> 'Where two parties have made a contract which one of them has broken, the damages which the other party ought to receive in respect of such breach of contract should be such as may fairly and reasonably be considered either arising naturally, i.e. according to the usual course of things, from such breach of contract itself, or such as may reasonably be supposed to have been in the contemplation of both parties, at the time they made the contract, as the probable result of the breach of it. Now, if the special circumstances under which the contract was actually made were communicated by the plaintiffs to the defendants, and thus known to both parties, the damages resulting from the breach of such contract, which they would reasonably contemplate, would be the amount of injury which would ordinarily follow from a breach of contract under these special circumstances so known and communicated. But, on the other hand, if these special circumstances were wholly unknown to the party breaking the contract, he, at the most, could only be supposed to have had in his contemplation the amount of injury which would arise generally, and in the great multitude of cases not affected by any special circumstances, from such breach of contract.'

The principles laid down in the passage quoted above have been interpreted and restated in more recent times in *Victoria Laundry (Windsor) Limited* v. *Newman Industries* (1949) and in *Koufos* v. *C. Czarnikow Limited (The Heron II)* (1967), and (in a construction context) in *Balfour Beatty Construction (Scotland) Limited* v. *Scottish Power Plc* (1994) (see below).

In the *Victoria Laundry* case it was stated that firstly, the aggrieved party is only entitled to recover such part of the loss actually resulting as was at the time of the contract reasonably foreseeable as liable to result from the breach; secondly, that what was reasonably foreseeable depends on the knowledge possessed by the parties at the time of entering into the contract; and thirdly, that for this purpose, knowledge 'possessed' is of two kinds: one imputed, the other actual. Everyone is taken to know the 'ordinary course of things' and consequently what loss is liable to result from a breach of contract in that ordinary course, but to this knowledge which the party in breach of contract is *assumed* to possess (whether he actually possesses it or not) there may have to be added in a particular case knowledge which he *actually* possessed of special circumstances outside the 'ordinary course of things', of such a kind that a breach in those special circumstances would be liable to cause a particular loss.

In the *Heron II* case Lord Upjohn stated the broad rule as:

> 'What was in the assumed contemplation of both parties acting as reasonable men in the light of the general or special facts (as the case may be) known to both parties in regard to damages as a result of a breach of contract.'

These two cases were applied by the House of Lords in the *Balfour Beatty* case, which is of interest on the question of the extent to which a party to a contract is presumed to know about the business activities of the other party.

FACTS

The plaintiffs were main contractors for the construction of the roadway and associated structures forming part of the Edinburgh city bypass. They installed a concrete batching plant near the site and entered into an agreement with the defendants for a temporary supply of electricity. Part of the plaintiffs' work was the construction of a concrete aqueduct to carry the Union Canal over the road. During this part of the works the batching plant ceased to work as a result of the rupturing of fuses provided by the defendants in their supply system. As a result of the failure and the resulting shut down of the batching plant, the construction of the aqueduct could not be completed by continuous pouring of concrete as required by the specification. The plaintiffs were obliged to demolish the partly constructed aqueduct and start again. The plaintiffs sued for damages.

HELD

The plaintiffs were not entitled to recover, the losses not being of a type which were within the defendants' contemplation. As Lord Jauncey put it:

> 'It must always be a question of circumstances what one contracting party is presumed to know about the business activities of the other. No doubt the simpler the activity of the one, the more readily can it be inferred that the other would have reasonable knowledge thereof. However, when the activity of A involves complicated construction or manufacturing techniques, I see no reason why B who supplies a commodity that A intends to use in the course of those techniques should be assumed, merely because of the order for the commodity, to be aware of the details of all the techniques undertaken by A and the effect thereupon of any failure of or deficiency in that commodity.'

MEASURE OF DAMAGES AND TIME OF ASSESSMENT

As a general rule, the assessment of damages will be based on the difference between the value of the subject matter of the contract in its defective or incomplete state, as compared to the value it would have had, had the contract been properly completed and fulfilled. However, in certain circumstances this may cause injustice or prove particularly difficult in terms of assessment, e.g. a contractor's failure to complete a building. In such a case the measure of damages is likely to be the difference between the contract price and the amount it actually costs the employer to complete the contract works substantially as it was originally intended.

The general rule is that the time at which the assessment of damages will take place will be the date when the cause of action accrues, i.e. in the case of breach of contract at the date of the breach, irrespective of when damage is suffered or known to be suffered as a result of the breach. However, in cases where the appropriate measure of damages is the cost of repair or completion of the outstanding work, damages are likely to be assessed at the time when the repairs or completion work ought reasonably to have been undertaken, and this may in appropriate circumstances be as late as the date of the hearing if the claimant was acting reasonably in not having the repairs carried out or the outstanding work completed before then.

The New Zealand case of *Bevan Investments* v. *Blackhall & Struthers* (1978) is an example of the flexible approach to the question of the most appropriate method for measuring damages in building contract cases and also as to the time at which such an assessment should be made. The learned authors of Building Law Reports summarised the court's findings in this case as follows (11 BLR 79):

'1. That the company was entitled to be put into the position it would have been in had the contract been performed;
2. That in building cases the loss should, prima facie, be measured by ascertaining the amount required to rectify the defects complained of and so give to the building owner the equivalent of a building on the land substantially in accordance with the contract;
3. That this rule was adopted unless the court was satisfied that some lesser basis of compensation could in all the circumstances be fairly employed;
4. That since the only practicable action in the present case was to complete according to the modified design, the cost of so doing was the reasonable measure of damages;
5. That in calculating those damages
   (a) to avoid an element of betterment credit should be given for the hypothetical additional cost of a proper initial design;
   (b) the assessment should be computed at the date of trial either because of the principle that damages were to be assessed by reference to the date when the reinstatement works could be reasonably carried out and, on the facts, it was reasonable to postpone the work until the issues of liability and damages were settled, or because such damage was not too remote in that it was foreseeable that the company might be unable to complete the works until a trial if the appellant failed to exercise the skill required of him.'

It can be seen therefore that the general principle of English law, which provides that, where the proper measure of damages is the cost of repairs etc, the cost is to be calculated on the basis of prices prevailing at or within a reasonable time after the discovery of the defective work, is unlikely to be applied when in all the circumstances it is reasonable for the innocent party to postpone the repairs etc. Indeed, in appropriate circumstances the time of assessment may be as late as the date of the hearing itself. There may be good commercial reasons for delaying repairs and the court is entitled to take these into account.

8-25 There are two important potential qualifications to the general rule referred to at the beginning of this section, one of which was referred to in the *Bevan Investments* case:

(1) The general rule will apply unless the court is satisfied that some lesser basis of compensation can fairly be employed; and

(2) In the context of defects, although a claimant can only recover as damages the costs which the defendant ought reasonably to have foreseen that the claimant would incur, reasonable costs do not mean the minimum amount which, with hindsight, it could be held would have sufficed. When the nature of the repairs is such that the claimant can only make them with the assistance of expert advice, the defendant ought, arguably, to have foreseen that the claimant would take such advice and be influenced by it.

On the first of these qualifications, reference must be made to the decision of the House of Lords in *Ruxley Electronics & Construction Limited* v. *Forsyth* (1995).

FACTS

The appellants entered into a contract whereby they agreed to construct for the respondent a swimming pool having a maximum depth at its deepest point of 7 ft 6 in. As constructed, the pool had a maximum depth of only 6 ft 9 in, and that depth was not achieved at the intended deepest point. At first instance, the judge made certain findings of fact: that the shortfall in depth did not decrease the value of the pool; that the cost of reconstruction would be £21 560; that the respondent had no intention of building a new pool. The judge found that the expenditure on reconstruction would be unreasonable since the cost would be wholly disproportionate to the advantage, if any, to be gained. The judge awarded a sum of £2500 as general damages for loss of amenity.

In the Court of Appeal the respondent's appeal was allowed, and an award was made of £21 500 being the estimated cost of rebuilding the pool. The appellants appealed to the House of Lords.

HELD

The proper application of the general principle that, where a party sustains loss by virtue of breach of contract, he is so far as money can do it to be placed in the same situation in respect of damages as if the contract had been performed, was not the monetary equivalent of specific performance but required the court to ascertain the loss the claimant had in fact suffered by reason of the breach. The cost of reinstatement was not the only possible measure of damage for defective performance under a building contract, and is not the appropriate measure where the expenditure would be out of all proportion to the benefit to be obtained – even if the alternative measure of value, diminution in value, would lead to only nominal damages.

The second qualification flows from the decision in the case of *The Board of Governors of the Hospital for Sick Children and Another* v. *McLaughlin & Harvey Plc and Others* (1987).

FACTS

Between 1977 and 1980 a new wing was constructed for the plaintiffs at Great Ormond Street Hospital. Soon after practical completion a walkway beam collapsed. As a result the plaintiffs commenced investigations which led to a detailed examination of the design and construction of the new wing, and, in due course, proceedings were commenced against the contractor, the architects and the structural engineers. The action was directed to be tried in a series of sub-trials, the first of which related to the question of the remedial work which the plaintiffs were entitled to carry out.

HELD

On the basis that the original design was negligent, the plaintiffs could recover the cost of remedial work which, by the date of the trial, they had carried out to the foundations of the building on the advice of their expert.

It is of course clear from this brief description of the Great Ormond Street case that there are limits to the building owner's right of recovery. Although it appears not to be open to the contract breaker to criticise the course honestly taken by the injured person on the advice of his experts, even though it appears by the light of after events that another course might have saved loss, it is still open to the contract breaker to argue that the scheme employed involved an element of betterment, i.e. the work carried out was more than was necessary properly to address the defects.

8-26 If the reason for building work being abandoned before completion is due to the contractor accepting a repudiation by the employer, this gives the contractor the right to sue for damages including his lost profit, if any. There has been some debate as to whether or not the contractor might, instead of suing for damages, bring an action in *quantum meruit* (as much as it is worth) for the work carried out by him; in other words to ignore the original contract sum and claim a reasonable sum instead: see *Lodder* v. *Slowey* (1904). This could be of considerable advantage to the contractor where his tender price was uneconomic and could be demonstrated to be less than the value of the work actually carried out. The difficulty with this argument is that the employer would in such a case be penalised by having to pay the contractor a sum in excess of the true damages suffered by the contractor as a result of the breach of contract. It is likely that a claim made on such a basis would fail.

MITIGATION

8-27 There is what is generally described as a duty on the part of the innocent party to take reasonable steps to mitigate the loss suffered as a result of the guilty party's breach of contract. It may be said very briefly that there are three rules:

(1) A claimant cannot recover for losses resulting from a defendant's breach of contract where the claimant could have avoided the loss by taking reasonable steps.
(2) If the claimant actually avoids or mitigates his loss, even by taking steps which went beyond what was reasonably necessary in all the circumstances, he cannot then recover such avoided loss.
(3) If the claimant incurs loss or expense by taking reasonable steps to mitigate the loss, he can claim such loss and expense incurred in taking the mitigating steps from the defendant, even if the result is that the losses flowing from the breach of contract have been exacerbated.

Strictly, the requirement to mitigate does not amount to a duty on the part of the innocent party, because there is no rule of law requiring such steps to be taken. However, there will be no right of recovery to the extent that it can be shown that any loss claimed could have been avoided by taking reasonable steps in mitigation.

Limitation of actions

8-28 It is a matter of public policy that there should be an end to litigation and that a time should arrive when stale demands can no longer be pursued. Accordingly, the Limitation Act 1980 prescribes the time limits after the expiry of which a claim will become 'statute-barred'. However, to obtain the benefit of a defence of limitation, the defendant must plead it specifically in his defence.

8-29 So far as contracts not under seal or not executed as a deed are concerned, by virtue of section 5 of the Act, no action founded on contract can be brought after the expiry of 6 years from the date on which the cause of action accrued. The cause of action accrues at the date of the breach of contract irrespective of when and if damage is suffered as a result of that breach. Typically, in building contracts, the last date therefore on which a cause of action for breach of contract will accrue will be the date of practical completion or substantial completion. If the contract is under seal or executed as a deed, then the period is 12 years instead of 6 years.

8-30 For actions arising out of tortious acts, the limitation period is 6 years from the date on which the cause of action accrued, except in relation to actions for personal injuries when that period is 3 years. In the tort of negligence, the cause of action does not accrue until damage is suffered. However, in relation to actions for damages for negligence where there are latent defects or damage other than in respect of personal injuries, the limitation period is now subject to the changes introduced by the Latent Damage Act 1986 which amends the 1980 Act.

As a result of the 1986 Act there is a special time limit for commencing proceedings where facts relevant to the cause of action were not known at the date that the cause of action accrued. It extends the present period of limitation to 3 years from the date on which the claimant knew or ought to have known the facts about the damage, where these facts became apparent later than the usual 6 years from the date on which the cause of action accrued, namely, in negligence actions, the date when the damage was suffered whether discovered or not.

However, there is an overriding long stop which operates to bar all negligence claims involving latent defects or damage brought more than 15 years from the date of the defendant's breach of duty.

In summary, the time limit runs out in respect of such actions at whichever is the later of the following:

- 6 years from the occurrence of damage;
- 3 years from the discovery of the damage following the expiry of the basic 6-year period, (but subject to a final time limit of 15 years from the breach of duty).

A right of action is given to anyone who acquires property which is already damaged where the fact of such damage is not known and could not be known to him at the time that he acquired the interest.

The 3-year period commences on the earliest date on which the claimant had both the knowledge required for bringing an action and the right to bring such an action. However, in relation to successive owners of buildings etc., if a predecessor in title knew or ought to have known of the damage, then the 3-year period begins at that time and does not start afresh when the new owner acquires his interest.

There are provisions dealing with what knowledge a claimant must have or ought to have had before the 3-year period begins to run. These cover such matters as the knowledge that the damage was attributable to negligence and, if a third party is involved, the identity of that party.

At the time of the passing of the 1986 Act into law, it was a distinct possibility that a duty of care in negligence could be owed by a builder or others concerned with construction, in favour of those who had or acquired an interest in the constructed building which suffered damage as a result of that negligence, and who had no contractual links with the wrongdoer. The possibility of such an action succeeding is now remote (see paragraph 7-16). The Act has little application to builders. It may still apply to professionals in those situations where the law imposes on them an independent duty to take care (see paragraph 2-186 dealing with negligent misstatement or advice).

8-31 Any deliberate concealment of defective work will prevent time from running until the defect is or ought to have been discovered. Section 32 of the Limitation Act 1980 provides that in certain cases such as fraud, concealment or mistake, time will not begin to run until the claimant has discovered or could with reasonable diligence have discovered the fraud, concealment or mistake. Fraud in this sense means that the defendant has behaved or acted in an unconscionable manner, for example, if a building contractor knowingly takes a risk by allowing bad workmanship or materials to be covered up: see *Applegate* v. *Moss* (1971).

The Latent Damage Act has no application to contractual negligence, i.e. breach of a contractually imposed duty of care: see *Iron Trades Mutual Insurance Co Limited* v. *J.K. Buckenham* (1989).

Effect of determination on the contractual provision for liquidated damage

8-32 Subject to what the express terms of the contract may say, the liability of the contractor to pay liquidated damages ends on the determination of the contractor's employment, whether under an express provision or under the general law. However, liquidated damages which have accrued to that date will probably still be deductible from or payable by the contractor.

We now turn to consider the relevant clauses of the MPF.

Summary of termination provisions

8-33 The termination provisions entitle the employer to terminate the contractor's employment under the contract based either on the contractor's default or his insolvency. Similarly, the contractor is entitled to terminate his own employment under the contract based on the employer's default or his insolvency. Both employer and contractor can also terminate the contractor's employment based on prolonged suspension as a result of certain non-fault events. In the case of the fault-based grounds, a 14-day warning notice is required to be given and only in the event that the default has not been remedied within the 14 days does the contract provide that the party not in default may give a termination notice. If the ground is insolvency, a notice of termination only is required.

The fault-based grounds are grouped together under the heading of 'Material Breach' and this is defined in clause 39.2. It is divided into defaults of the contractor, defaults of the employer and defaults of either party. The events themselves are discussed in detail below.

Following a termination, there is a financial code setting out what is to happen. Whether or not the employer intends to have the project finished by others affects the nature of the financial calculation to be made. This financial code is considered below.

It is expressly provided that the provisions of clauses 32 to 34 inclusive are without prejudice to any other rights or remedies which either party may possess.

Provisions applicable to termination generally – clause 31

Other rights or remedies

31.1 The provisions of clauses 32, 33 and 34 (*Termination*) are without prejudice to any other rights or remedies that the parties may possess.

8-34 Although the clauses commented on below identify the only grounds on the basis of which the employer or the contractor may terminate the contractor's employment under the contract, there is a potentially significant reservation made by clause 31, i.e. that the parties' rights under clauses 32 to 34 are without prejudice to any other rights or remedies which they may possess.

Whether these express words of reservation are necessary has been the subject of some debate. This issue has been considered earlier in this chapter (see paragraph 8-09).

Termination by employer – clause 32

Material breach and warning notice

32.1 If the Contractor commits a Material Breach of the Contract the Employer may give the Contractor a notice identifying the Material Breach and stating that it may terminate the Contractor's employment under the Contract if it fails to remedy the Material Breach within 14 days of the notice.

Material breach

8-35 The employer's right to terminate the contractor's employment is dependent upon the contractor committing a 'Material Breach'. This is defined in clause 39.2 as follows:

By the Contractor:

- failure to proceed regularly and diligently with the performance of its obligations under the Contract;
- failure to comply with an instruction;
- suspension of the Project or any part thereof, otherwise than in accordance with HGCRA 1996 or the circumstances described in clause 34.1;
- breach of the CDM Regulations;
- breach of the requirements of CIS;
- breach of any of the provisions of the Contract relating to Named Specialists or Pre-Appointed Consultants.

By the Employer:

- failure to issue a payment advice in the manner required by the Contract.

By either party:

- failure to make payment in the manner required by any payment advice issued under the Contract;
- failure to insure, as established in accordance with clause 27.3;
- any repudiatory breach of the Contract.

8-36 In the absence of any definition, it might be thought that a material breach would be either one which was sufficiently fundamental to amount to a repudiation of the contract by the party at fault, or at least a breach of some significance. This is certainly not how the term is defined in clause 39.2. As will be seen in relation to the specific matters which amount to a material breach, some of the failures or defaults in any given circumstances may not be such as to entitle the innocent party to treat the contract as at an end under the general law, and though they may indeed be breaches of contract, they do not have to be of a serious nature. This is all the more significant in the absence of a clause (c.f. WCD 98, e.g. clause 27.2.4) providing that a notice terminating the contractor's employment shall not be given unreasonably or vexatiously. It may well be that even in the absence of such qualification, it is implicit having regard to the MPF as a whole, that the material breach must at least be more than technical, insignificant or nominal. In the case of *Rice (trading as The Garden Guardian)* v. *Great Yarmouth Borough Council* (2003) the contractor was engaged on two contracts for ground maintenance and leisure services for a 4-year period based upon the Association of Metropolitan Authorities standard form for this type of contract. It involved the contractor in expending significant setting-up costs for the performance of the contract. It was alleged that the contractor was in breach and notice of termination was served. Clause 23 provided that if the contractor committed 'a breach of any of its obligations under the contract' the employer was entitled to terminate the contractor's employment. In other words, it was argued that the contract permitted termination for any breach. The Court of Appeal rejected this argument despite the literal wording of the contract. Lord Justice Hale said of the relevant clause:

> 'It appears to visit the same draconian consequences upon any breach, however small, of any obligation, however small.
>
> ... the notion that this term would entitle the council to terminate a contract such as this at any time for any breach of any term flies in the face of commercial common sense.'

Under the MPF, the contractor is likely to have incurred significant setting-up costs and will at almost any point in time have made a significant investment in the project. If the contractor's employment is terminated for some small breach of contract, he faces some potentially severe financial consequences under clauses 32.5 or 32.7. This is unlikely, realistically, to have been the intention of the parties. The same must be true for the employer's position under clause 33. However, unlike in the *Rice Case*, 'Material Breach' in clause 39.2 of the MPF expressly includes 'any repudiatory breach of the Contract' as a distinct ground for termination which, as a matter of interpretation, must mean that other listed matters need not amount to repudiatory breaches of contract. Breaches of contract falling short of a repudiation must therefore be contemplated as justifying a termination. Just where the line is to be drawn between a significant non-repudiatory breach and an insignificant breach will be a matter of argument. This definition of 'Material Breach' may well be amended in practice.

8-37 A further point to note is that, unlike for example WCD 98, clause 27.2, which, in identifying grounds of default which are similar in many respects to a material breach under the MPF, does so on the basis that the default relates to its occurrence only before the date of practical completion, there is no such limitation under the MPF. However, looking at the actions which follow a termination, e.g. clause 32.4.3 (the employer may make arrangements to complete the project), it will hardly be relevant in many situations after practical completion to terminate the contractor's employment.

8-38 Where it is the employer wishing to terminate, a material breach arises where the contractor is responsible for any of the following.

FAILURE TO PROCEED REGULARLY AND DILIGENTLY WITH THE PERFORMANCE OF CONTRACTUAL OBLIGATIONS

8-39 It should first be noted that it is not simply a failure to proceed regularly and diligently with construction work alone, or even with design and construction work; it will extend to any obligations under the contract, e.g. a requirement to insure pursuant to clause 27, or the failure to obtain an appropriate licence under the copyright provisions of clause 7.

By clause 9.2 the contractor is under an express duty to 'proceed regularly and diligently with the Project' and failure to do so is a ground for termination. The meaning of the words 'regularly and diligently' in the termination provisions in JCT 63 was considered by Mr Justice Megarry in the case of *London Borough of Hounslow* v. *Twickenham Garden Developments Limited* (1970) where he said (at 7 BLR 120):

> 'These are elusive words, on which the dictionaries help little. The words convey a sense of activity, of orderly progress, and of industry and perseverance: but such words provide little help on the question of how much activity, progress and so on is to be expected. They are words used in a standard form of building contract in relation to functions to be discharged by the architect, and in those circumstances it may be that there is evidence that could be given, whether of usage among architects, builders and building owners or otherwise, that would be helpful in construing the words.'

The meaning of the words 'regularly and diligently' was considered by the Court of Appeal in the case of *West Faulkner Associates* v. *London Borough of Newham* (1994). In that case, Lord Justice Brown said:

> 'My approach to the proper construction and application of the clause would be this. Although the contractor must proceed both regularly and diligently with the works, and although each word imports into that obligation certain discrete concepts which would not otherwise inform it, there is a measure of overlap between them and it is thus unhelpful to seek to define two quite separate and distinct obligations.
>
> What particularly is supplied by the word "regularly" is not least a requirement to attend for work on a regular daily basis with sufficient in the way of men,

> materials and plant to have the physical capacity to progress the works substantially in accordance with the contractual obligations.
>
> What in particular the word "diligently" contributes to the concept is the need to apply that physical capacity industriously and efficiently towards that same end.
>
> Taken together the obligation upon the contractor is essentially to proceed continuously, industriously and efficiently with appropriate physical resources so as to progress the works steadily towards completion substantially in accordance with the contractual requirements as to time, sequence and quality of work.'

Certainly the fact that the work is done defectively does not prevent it being carried out regularly: see *Lintest Builders* v. *Roberts* (1978). The case of *Greater London Council* v. *Cleveland Bridge & Engineering Co Limited and Another* (1984), particularly at first instance, may be relevant here (see paragraph 3-14).

FAILURE TO COMPLY WITH AN INSTRUCTION

8-40 If the contractor fails to comply with an instruction, this is a ground for the termination of his employment under the contract. There is no express qualification in relation to the seriousness or otherwise of any such failure. Mention has already been made earlier (see paragraph 8-36) as to whether nevertheless there is an implication that a notice of termination will not be given where the failure is insignificant in nature.

Instructions of course can be given under the MPF in many situations. Clause 2 deals with instructions generally (see paragraph 2-28 *et seq.*). It will be recalled that an alternative option for the employer where the contractor fails to comply with an instruction is to engage others to carry it out, in which case the contractor is liable for the employer's additional costs.

The determination provisions of some other JCT contracts, including WCD 98 (see clause 27.2.1.3), refer only to a failure to comply with instructions requiring the removal of non-complying work, materials or goods, with an added qualification that any refusal or neglect by the contractor to carry out such instruction must be such as to cause the works to be materially affected. The MPF reference to instructions generally is therefore significantly wider and does leave the contractor exposed to some degree of opportunism on the part of an unreasonable employer.

SUSPENSION OF THE PROJECT OR ANY PART THEREOF, UNLESS IN ACCORDANCE WITH THE HOUSING GRANTS, CONSTRUCTION AND REGENERATION ACT 1996 OR UNDER CLAUSE 34.1

8-41 The contractor is permitted to suspend performance of his obligations under the contract under section 112 of the Housing Grants, Construction and Regeneration Act 1996 for non-payment (see paragraph 3-71 *et seq.*). In addition, if there has been a suspension of the project or a substantial proportion of it as a consequence of *force majeure*, the occurrence of a specified peril, hostilities, or the use or threat of terrorism, either party can terminate the contractor's employment (see clause 34). Any other suspension by the contractor is a ground upon which the employer can terminate under clause 32.

The suspension does not have to be of the project as a whole. Indeed it can be any part of it. This is unlike WCD 98 which, in clause 27.2.1.1, provides 'without reasonable cause he [the Contractor] wholly or substantially suspends the carrying out of design or construction of the Works'. The MPF is more favourable to the employer than the WCD form in that the suspension can be in respect of any part of the project rather than having to be a substantial suspension. It may also be wider in that it refers to the 'Project' being suspended rather than the carrying out of design or construction. In the definitions clause 39.2, 'The Project' is defined as 'The works to be undertaken in accordance with the Contract, as defined in the Appendix'. The project description is the subject of an insertion into the appendix which is followed by the words 'as more fully described by the Requirements and the Proposals'. The definition of 'Contract' includes, among other documents, the requirements, and the proposals. The effect of this may be to enable the employer to contend that the suspension of obligations other than those relating directly to design and construction could fall within this heading, though this would be rare.

BREACH OF THE CDM REGULATIONS

8-42 If the contractor commits a breach of any of the CDM Regulations, this entitles the employer to terminate the contractor's employment. Again, the question arises as to whether it is implicit that such a breach must be other than insignificant (see paragraph 8-36).

By clause 1.2, the contractor is appointed the planning supervisor and principal contractor. He is required to notify the Health and Safety Executive of his appointment. By clause 1.3 the contractor warrants that he has the competence and will allocate the resources necessary to fulfil these roles and that of designer in the manner referred to in the CDM Regulations. Clause 3 deals generally with statutory requirements which are defined in clause 39.2 to include any Act of Parliament and any instrument, rule or order made under an Act of Parliament. This will include the CDM Regulations. Clause 3.1 expressly requires the contractor to comply with these statutory requirements.

The result would appear to be that any breach of the CDM Regulations, quite apart from those expressed in clauses 1.2 and 1.3, can provide grounds for the termination of the contractor's employment. There is no express provision in the MPF that the employer will comply with statutory requirements generally or with the CDM Regulations in particular. In the absence of such a provision, there is no reciprocal right in clause 33 for the contractor to determine his own employment if the employer is in breach of the CDM Regulations. This is to be contrasted with WCD 98 which includes a ground in clause 28.2.1.3 giving the contractor the right to terminate his own employment if the employer fails, pursuant to the contract conditions, to comply with the requirements of the CDM Regulations.

BREACH OF THE REQUIREMENTS OF THE CIS

8-43 If the contractor is in breach of the CIS, defined in clause 39.2 as:

> The Construction Industry Scheme established by the Income Tax (Sub-Contractors in the Construction Industry) Regulations 1993, as amended by the Income Tax (Sub-Contractors in the Construction Industry) (Amendment) Regulations 1998.

The right of the employer to terminate the contractor's employment for breach of the CIS is in addition to the employer's right to refuse to make further payments to the contractor until the contractor's failure to provide vouchers in accordance with the CIS in respect of payments received has been remedied (see clause 22.8).

BREACHES IN RELATION TO NAMED SPECIALISTS OR PRE-APPOINTED CONSULTANTS

8-44 If the contractor is in breach of any of the provisions of the contract relating to named specialists or pre-appointed consultants, this can be a ground for the employer to terminate the contractor's employment. Literally it includes any breach in relation to such provisions and not just serious or significant breaches (but see paragraph 8-36). The topic of pre-appointed consultants and named specialists is dealt with in clause 18 of the MPF (see paragraph 4-66 *et seq.* for a commentary). There can be many instances where breaches may be of little real consequence, e.g. the late supply by the contractor to the employer of a copy of a contract entered into with a named specialist (see clause 18.3). Mention has already been made (see paragraph 8-36) concerning the question whether a termination can legally be made in respect of such breaches.

FAILURE TO MAKE PAYMENT IN THE MANNER REQUIRED BY ANY PAYMENT ADVICE ISSUED UNDER THE CONTRACT

8-45 If the contractor fails to make payment by the final date for payment, namely 14 days after the issue of the payment advice, this will provide a ground for the employer to terminate the contractor's employment. Occasionally there can be a negative payment advice issued by the employer which requires the contractor to pay money. An example would be the discovery of a significant amount of defective work which had been valued and included in an earlier payment advice on the assumption that it complied with the contract; another example would be where the contractor was liable to pay the employer liquidated damages in excess of the value of work carried out, both of which have been included in a payment advice.

On occasions, the contractor may be able to discharge his obligation to make payment by means of a set-off, provided any applicable notice of withholding has been duly given (see paragraph 5-163 *et seq.*).

FAILURE TO INSURE WHERE REQUIRED BY THE CONTRACT TO DO SO

8-46 Clause 27 of the MPF deals with insurances. It provides that policies of insurance are to be provided and maintained as indicated in the appendix to the contract. The appendix item has spaces for completion covering the types of insurance, the document in which they are described, whether it is the employer or the contractor who is to insure, and the amount of any excess. If the contractor is obliged to take out and maintain a particular policy but fails to do so, or if he fails to provide documentary evidence of having provided and maintained the required insurance, entitling the employer to make the assumption that there has been a failure to insure (see clause 27.3), this amounts to a ground for termination of the contractor's employment. It should be noted that clause 27.3 in this situation also entitles the employer to take out the appropriate insurance, with the contractor being liable to pay any costs incurred in providing and maintaining it. The exercise of this option by the employer would not appear to deprive him of the right to nevertheless terminate the contractor's employment, subject to the issue already discussed (see paragraph 8-36) regarding the significance of any breach.

ANY REPUDIATORY BREACH OF THE CONTRACT

8-47 This is an overall saving provision under which, if the contractor commits any breach of the contract which entitles the employer to treat the contractor's breach as a repudiation of it, the employer can terminate the contractor's employment. This is without prejudice to the employer's right to treat the contract itself as at an end as a result of the contractor's repudiation. Breaches of contract sufficiently serious to be treated as a discharge of the innocent party from his obligations under the contract have been discussed earlier (see paragraph 8-11 *et seq.*) as has its the effect on the interpretation of the other listed matters (see paragraph 8-36).

Warning notice

Under clause 32.1, if the contractor commits a material breach of the contract the employer may give the contractor a notice identifying the breach and stating that the employer may terminate the contractor's employment if the contractor fails to remedy the breach within 14 days of the notice. The warning notice is a pre-condition to any further notice purporting to terminate the contractor's employment.

METHOD OF SERVICE

8-48 By clause 38.2, the notice must be given by actual delivery, registered post (replaced by special delivery) or recorded delivery. In a matter of such importance, it would have been sensible for clauses 32.1 and 32.2 to have expressly cross-referred to clause 38.2. It takes effect upon delivery. Unlike many of the JCT forms, e.g. WCD 98, clause 27.1, there is no deemed service. It is extremely important therefore that evidence of delivery is obtained whether from a courier or through Royal Mail records.

The procedural requirements of clauses 32.1 and 38.2 should be followed precisely, although there is some authority in English law for a commonsense business construction to be given to commercial contracts (*Goodwin & Sons* v. *Fawcett* (1965)), in which a notice required by the contract to be served by registered post which was in fact served by recorded delivery was held to be valid on this basis). Other jurisdictions have not been so benign: in the case of *Central Provident Fund Board* v. *Ho Bock Kee* (1981) before the Singapore Court of Appeal, a purported determination was held to be invalid when delivery was effected by hand rather than (as required by the contract) by registered post. In the case of *J. M. Hill & Sons Limited* v. *London Borough of Camden* (1980), Lord Justice Ormrod resisted an unduly strict construction of the requirements for the timing of a notice of determination, in the following terms:

> 'Nothing is more distasteful to me than to construe a business contract in this formalistic sense. I think it makes a mockery of the law. To construe this contract and to treat what happened in this case between the contractors and the local authority in the way which (counsel submits it should be treated) is to make, I think, a farce of such contractual provisions . . .
>
> There can be no doubt here that it never occurred to any sensible person that there was any defect whatever in the operation of (the notice provision). Everybody concerned knew perfectly well what was happening. No one was in the very slightest degree prejudiced by it or pretends that they were prejudiced by it or would be believed if they did. So that it is the most purely formal point.'

Having said this, it would be open to a party seeking to challenge the validity of such an important notice, to point to a failure to comply with a clearly expressed contractual requirement, particularly when the grounds for the possible termination would not amount to repudiatory conduct at common law. Strict compliance is, therefore, advisable.

Where the service of the notice is by actual delivery, whether this necessarily means some form of personal service is unclear. The words in isolation do not require such a narrow construction. However, taking clause 38.2 as a whole, there would be little point in specifying registered post (replaced by special delivery) or recorded delivery as acceptable means of service if actual delivery could be effected by ordinary post or electronic transmission.

As with other communications, the notice must be sent either to the address notified from time to time by the contractor for the purposes of communications, or if there is no such address notified, it will be the address given at the beginning of the MPF, which provides for the insertion of the registered office address for both parties. Accordingly, unless the contractor has agreed that notices should be delivered to his principal place of business, service at such an address (if different from the registered office) will not be an option.

The warning notice operates from the time of delivery and the contractor has 14 days from the delivery of the notice to remedy the material breach.

CALCULATING TIME FOR CONTRACTOR'S RESPONSE

8-49 The first question to consider is whether clause 39.1 of the MPF applies to a warning notice served pursuant to clause 32.1. Clause 39.1 provides that where, under the contract, an act is required to be done within a specified period of days (but not weeks, months or years) after or from a specified date, the period is to begin immediately after that date. This reflects the provisions of section 116 of the Housing Grants, Construction and Regeneration Act 1996 Act except that the application of clause 39.1 is restricted to situations where the contract refers to 'days', whereas section 116 refers simply to 'a specified period' which could therefore include a period of weeks, months or years. Like section 116, clause 39.1 provides that the period begins immediately after the specified date. If clause 39.1 applies, it means that in relation to the warning notice the day of delivery is expressly excluded so that the contractor then has a period of 14 days after the date of delivery in which to remedy the breach. Even if clause 39.1 does not apply (see next paragraph), the day of delivery will almost certainly be excluded from the calculation of the 14 days. Clause 39.1 (like section 116) goes on to provide that if any such period includes Christmas Day, Good Friday or a day which under the Banking and Financial Dealings Act 1971 is a bank holiday in England and Wales, that day shall be excluded.

There must, however, be some doubt as to whether clause 39.1 applies at all to the warning notice. The warning notice does not actually require the contractor to take action to remedy the breach; it simply states that if the contractor fails to remedy the breach the employer may proceed by a further notice to terminate the contractor's employment. Clause 39.1 probably does not therefore apply, though the matter is not free from doubt. In these circumstances, the employer would be wise to give the contractor the benefit of the doubt by excluding Christmas Day, Boxing Day and bank holidays when calculating the 14-day period. However, if the employer chooses to make such an allowance, when it is not in fact necessary because clause 39.1 does not apply, the effect will be that the employer will be reducing the time in which a notice of termination of the contractor's employment must be given, namely within a further 14 days from the expiry of the first 14 days (see clause 32.2).

It is imperative therefore that the employer's notice of termination of the contractor's employment is issued within 28 days of the delivery of the warning notice, calculated by including any such bank holidays occurring during the period of operation of the warning notice, but excluding any bank holidays occurring during the 14-day period during which the termination notice may be served (see paragraph 8-53).

CONTENTS OF THE NOTICE

8-50 Clause 32.1 requires the notice both to identify the material breach and to state that the employer may terminate the contractor's employment if the contractor fails to remedy it within 14 days. Great care should be taken to expressly include these matters within the notice. So far as identifying the material breach is concerned, the notice should refer to the relevant matter from the list in the clause 39.2 definition of 'Material Breach'. Clearly the alleged material breach should be particularised so far as reasonably possible so that the contractor is aware of what steps need to be taken to remedy it. The information required may depend on what communications have previously passed between the parties in relation to the matter. On the other hand, the notice certainly does not need to be framed in the manner of a formal legal pleading.

In the case of *London Borough of Hounslow* v. *Twickenham Garden Developments* (1970), the architect, in giving the notice required under clause 25(1) of JCT 63, said:

> 'I therefore hereby give notice under clause 25(1) of the contract dated ... that in my opinion you have failed to proceed regularly and diligently with the works.'

Of this notice Mr Justice Megarry said as follows (7 BLR 115):

> 'I do not read the condition as requiring the architect, at his peril, to spell out accurately in his notice further and better particulars, as it were, of the particular default in question. All that I think the notice need do is to direct the contractor's mind to what is said to be amiss: and this was I think done by this notice.'

The termination notice

> 32.2 If the Material Breach has not been remedied within 14 days of the Employer's notice under clause 32.1 the Employer may by a further notice issued at any time within the subsequent 14 days terminate the Contractor's employment under the Contract.
>
> 32.3 In the event that the Contractor becomes Insolvent the Employer may at any time by notice to the Contractor terminate the Contractor's employment under the Contract.

Termination notice following failure to remedy material breach

8-51 If the contractor has failed to remedy the material breach within 14 days of delivery of the employer's warning notice (see paragraph 8-49 for the calculation of this) then, under clause 32.2, the employer may issue a further notice at any time within the subsequent 14 days to terminate the employment of the contractor. If the breach is remedied within the 14-day period but it is subsequently repeated, a further warning notice is still required. This is unlike the position under WCD 98 where there is a facility for dispensing with the warning notice if the default is subsequently repeated (see clause 27.2.3).

Clause 32.2 itself does not say anything about the contents of the termination notice. It would be highly advisable for the employer to ensure that the termination notice cross-referred to the warning notice and, though perhaps out of an abundance of caution, also restated the material breach complained of, added to which should be the statement that it has not been remedied and that accordingly the employment of the contractor is terminated.

THE TERMINATION NOTICE AND CLAUSE 38.2

8-52 Clause 38.2 deals with the giving of notices under clause 32. Any such notice must be given by actual delivery, registered post (replaced by special delivery) or recorded delivery and takes effect upon delivery. It should be noted however that under clause 32.2, it is the 'issue' of the termination notice within the subsequent 14 days which is important. It is unfortunate that clause 32.2 does not expressly require the notice not only to be issued but also to be 'given' (per clause 38.2) so as to make it clear beyond doubt that the formalities of clause 38.2 as to the method of service apply. Nevertheless it is clear that these formalities are intended to apply to such a notice and the reference to issuing the notice within 14 days is intended to ensure that the ability of the employer to comply with clause 32.2 is within his own control. It is implicit that the notice must also be given. It will therefore take effect on delivery. It is probably the case that the notice would be issued when it is put into circulation by being sent by registered post (replaced by special delivery) or recorded delivery or, in the case of actual delivery, by being handed to the courier with adequate instructions to be able to effect actual delivery.

The termination itself will not take effect until actual delivery of the notice.

TIMING OF THE NOTICE

8-53 Possibly unlike the calculation of the contractor's 14-day period in which to remedy the material breach following the warning notice (see paragraph 8-49), clause 39.1 will apply as the issuing of a notice of termination is certainly 'an act [which] is required to be done within a period of days ... after or from a specified date'. The calculation of the 14-day period within which the employer must issue the notice of termination will therefore not only commence on the 15th day after the warning notice is delivered (see paragraph 8-49 for the calculation of this period), but will also exclude Christmas Day, Good Friday or a day which under the Banking and Financial Dealings Act 1971 is a bank holiday in England and Wales.

NOTICE AFTER BELATED REMEDY?

8-54 Can a notice of termination be issued if, after the expiry of the period of operation of the warning notice, the contractor remedies the breach? There is some difficulty of meaning in clause 32.2. For example, supposing the material breach is not remedied within 14 days of the warning notice so the employer decides, perhaps on the 15th or 16th day that a termination notice should be given. Can this still be given if before it is issued the breach has been remedied? If there was every reason to believe that the remedying of the breach was a genuine and permanent action on the part of the contractor, it would be highly artificial and technical to proceed with a termination notice, particularly without the saving provision to be found in other JCT contracts that a notice must not be given unreasonably or vexatiously.

EFFECT OF EMPLOYER'S FAILURE TO ISSUE TIMELY NOTICE OF TERMINATION

8-55 Another problem with the operation of the clause, and a potential difficulty for the employer, is that the termination notice must be issued within 28 days of the delivery of the warning notice. Supposing therefore that a warning notice is delivered and that the contractor continues to fail to remedy the breach for a further, say, 6 weeks during which the employer has not issued a notice of termination; the employer cannot now issue a notice of termination as it will not be possible to issue the notice within 14 days of the expiry of the 14 days of the warning notice. The employer would therefore have to give a further warning notice, giving the contractor the further 14 days from its delivery in which to remedy the breach.

Termination notice following insolvency

8-56 In addition to the employer's right to terminate the contractor's employment where the contractor is guilty of a material breach under clause 32.1, the employment of the contractor may also be terminated under clause 32.3, at any time, by notice to the contractor in the event of the contractor becoming 'Insolvent'. This is defined under clause 39.2 as follows:

> Either party is Insolvent when it makes a composition or arrangement with its creditors, or becomes bankrupt, or, being a company:
>
> - makes a proposal for a voluntary arrangement for a composition of debts or scheme of arrangement to be approved in accordance with the Companies Act 1985 or Insolvency Act 1986 as the case may be; or
> - has a provisional liquidator appointed; or
> - has a winding up order made; or
> - passes a resolution for voluntary winding up (except for the purposes of amalgamation or reconstruction); or
> - under the Insolvency Act 1986 has an administrator or an administrative receiver appointed.

These insolvency events are identically worded to those in many other JCT contracts such as WCD 98 (clause 27.3.1). However, the consequences of a termination based on insolvency are treated very differently under the MPF. In many other JCT contracts, certain of the insolvency events give rise to automatic determination, whereas the others give rise to a right on the part of the employer to determine by notice. In addition, other JCT contracts also deal with such matters as requiring the contractor to notify the employer if it has made a composition or arrangement with creditors etc.; provision for a specific agreement dealing with the continuation or novation or conditional novation of the contract; as well as more detailed provisions in relation to the financial code which comes into play in connection with the employer completing the outstanding works, or alternatively deciding not to do so.

The employer's notice to the contractor terminating the contractor's employment due to the contractor's insolvency can be given at any time. Having said this, if following the contractor becoming insolvent, an agreement is reached with the contractor or those acting for the contractor in his insolvency, under which the contractor is able to continue on the same or different terms, this would amount to a waiver of the employer's right to terminate. To be sure, in such a situation, any agreement to continue should specifically provide that the employer will not serve a termination notice in respect of the happening of that insolvency event.

The formalities in connection with the giving of the termination notice under clause 32.3 are governed by clause 38.2 (see paragraphs 8-48 and 9-10).

No corruption termination provision

Many JCT contracts, e.g. WCD 98, clause 27.4, entitle the employer to end the employment of the contractor if the contractor has offered or given or agreed to give any gift or consideration as an inducement or reward in connection with the obtaining or execution of the contract or if the contractor has committed an offence under the Prevention of Corruption Acts 1889 to 1916, or, where the employer is a local authority, if the contractor shall have given any fee or reward, receipt of which is an offence under section 117(2) of the Local Government Act 1972 or any amendment or re-enactment of it. There is no such anti-corruption provision in the MPF. Such clauses have been little used and in any case it may well be the position that any employer would be entitled to rescind the contract on the occurrence of certain of these events in view of the serious nature of the matters referred to.

Consequences of a termination of the contactor's employment by the employer

32.4 If the Contractor's employment is terminated under clause 32:
.1 the Contractor shall not remove any materials plant or equipment from the Site unless expressly permitted to do so by the Employer;
.2 the Contractor shall provide to the Employer all Design Documents prepared in connection with the Project;
.3 the Employer may make such other arrangements as it considers appropriate to complete the Project;
.4 the Employer shall not be obliged to make any further payment to the Contractor other than in accordance with clause 32.

8-57 Clause 32.4 deals with the consequences of a termination of the contractor's employment under the contract on any of the grounds referred to in clauses 32.1 or 32.3. The MPF has much less detail with significantly reduced wording in connection with the consequences of termination. This for the most part appears to be achieved without much disadvantage.

An initial point to note is that unlike many JCT contracts, e.g. WCD 98, clause 27.6.2, there is no provision for the employer to make use of any huts, plants, tools or site materials on the site belonging to the contractor, or to require their removal. Neither is there any provision under which the employer can call for an assignment to him without payment of the benefit of any supply or sub-contracts which the contractor may have in place (clause 27.6.3.1). Similarly, there is no provision under which the employer may pay a supplier or sub-contractor in respect of materials or goods and then deduct that sum from sums otherwise due to the contractor under the termination provisions, or otherwise claim it as a debt (clause 27.6.3.2). Such provisions as these are rarely of any significant use in practice. Any arrangements of this nature tend to be made on a specific ad hoc basis as and when appropriate.

Following termination under clause 32, the following provisions apply.

Contractor not to remove materials, plant or equipment from the site without express permission (clause 32.4.1)

8-58 This is an express prohibition on the contractor who is not to remove materials, plant or equipment from the site without express permission. As an express permission would be a communication, it would need to be in writing or in accordance with some other agreed procedure (see clause 38.1). Accordingly, the contractor ought not to rely upon any verbal permission. Unlike other JCT contracts, e.g. WCD 98, see clause 27.6.4, as mentioned in the previous paragraph, there is no provision enabling the employer to require the contractor to remove huts, plant, tools, equipment, goods and materials from the site with the right, if this does not happen, to remove and sell them, holding the proceeds of sale less costs to the credit of the contractor. These are in any event steps which would be open to the employer without such an express provision if the contractor fails within a reasonable time of a request to remove such items from the site. The contractor's failure to remove materials, plant or equipment etc. would put the employer in the position of an involuntary bailee and as such he may, though the position is not free from doubt, be able to avail himself of the wide powers of sale conferred by sections 12 and 13 of the Torts (Interference with Goods) Act 1977. However, the rights available under the Act would appear not to apply where the employer is aware that the items are not owned by the contractor, and the employer will need to be careful to ensure that materials, plant or equipment etc. belongs to the contractor and not to a third party such as a sub-contractor or leasing company; otherwise, the employer will risk committing the tort of Interference with Goods.

Though it could be the case that materials on the site belong to the employer, e.g. as a result of their being included in the definition of a stage where stage payments apply, the clause is clearly aimed at materials, plant and equipment etc. on site which belongs to the contractor. If the purpose behind this is to enable the employer to claim some form of lien upon them against any resulting sums due from the contractor to the employer, then this may face problems. If the contractor is insolvent and the effect of any such purported lien is to favour the employer over the position of other creditors, it could fall foul of the statutory provisions as to the mandatory *pari passu* discharge of liabilities which cannot be contracted out of: see *British Eagle International Airlines* v. *Compagnie Nationale Air France* (1975).

Contractor is to provide to the employer all design documents prepared in connection with the project (clause 32.4.2)

8-59 This is clearly a vitally important provision to enable the employer to satisfactorily arrange for the completion of the project by others. 'Design Documents' is defined (clause 39.2) to include drawings, specifications, details, schedules of levels, setting out dimensions and the like for explaining the requirements or proposals, and which are necessary to enable the contractor to execute the project or which are required by any provision in the requirements. This may be compared with WCD 98, clauses 27.6.1 and 5.5, which make clear that it includes documentation relevant to the maintenance and operation of the works. In the MPF, it is not at all certain that the definition of design documents is sufficiently wide to include any operating or maintenance manuals and the like which may exist in relation to plant or equipment already installed. It might be sensible therefore for clause 32.4.2 to be amended to expressly incorporate all available 'as built' information and operating and maintenance information otherwise required by the contract to be delivered at practical completion (see definition of 'Practical Completion' under clause 39.2), to the extent that it is available. In a different respect clause 32.4.2 is wider than clause 27.6.1 of WCD 98. Under the MPF, the contractor must provide all design documents whether used or yet to be used for the project. Under WCD 98 it is only design information in relation to completed work which is to be provided.

Employer may make such other arrangements as it considers appropriate to complete the project (clause 32.4.3)

8-60 It is appropriate having regard to the financial accounting provisions in clause 32.5 that there should be an express provision entitling the employer to make other arrangements for the completion of the project, though such a right would clearly exist in any event. This provision is generously and simply worded. However, the employer's room for manoeuvre must of course take into account a general duty to mitigate loss and act reasonably in the way in which he incurs additional costs in completing the project.

Typically, if the employer intends to see the project through to completion, he is likely, if at all possible, to engage some or all of the contractor's consultants (if any). Indeed, if the contractor has taken over the employer's pre-appointed consultants (see clause 18), it is possible that any novation agreement between the employer, pre-appointed consultants and the contractor, will contain provisions dealing with this eventuality whereby the employer steps back in, in place of the contractor. Alternatively, there may be a warranty between the pre-appointed consultants and the employer which contains appropriate step-in provisions. Similar provisions may also be included in any warranty between the employer and a named specialist appointed as a sub-contractor by the contractor pursuant to clause 18. In both cases such provisions may also deal with the situation where the employer wishes, in effect, to transfer the pre-appointed consultants or named specialists to a replacement contractor.

The employer may choose not to finish off the project, e.g. by selling it on, incomplete, to another developer. In this case, or if the employer does not begin to make arrangements for the completion of the project within six months of the termination, the MPF provides for financial closure under clause 32.7 (see paragraph 8-69).

Employer is not obliged to make any further payment to the contractor except as provided in clause 32 (clause 32.4.4)

8-61 At the point of termination, the employer ceases to be under any obligation to make further payments to the contractor except as provided for in clause 32 and in particular clauses 32.5 to 32.8. There is no express qualification to this provision. It could therefore mean that an interim payment advice which had not been honoured by its final date for payment and which therefore was overdue in circumstances where the employer was in breach of contract, would cease to be payable as a result of the termination. This can be contrasted with other JCT contracts, e.g. WCD 98, clause 27.6.5.1, which provides in such circumstances that the contractor can still seek payment in respect of amounts properly due and which the employer has unreasonably not paid. Under the MPF, all sums are held in suspense pending the completion of the project by others, or alternatively until the failure of the employer to commence making such arrangements for completion of the project within six months from the termination.

If however, at the time of the termination, the amount due at that point in time has been the subject of an adjudicator's decision, the courts will enforce payment. In the Court of Appeal decision in the case of *Ferson Contractors Limited* v. *Levolux A.T. Limited* (2003), it was held that a similar type of provision (clauses 29.8 and 29.9 of GC/Works/Sub-Contract) must be read as not applying to monies due by reason of an adjudicator's decision. The adjudicator's decision took precedence over such a provision; otherwise a main purpose of the adjudication provisions, namely to provide a rapid means of enforcing the payment of money due, would be defeated. This was the position in the *Ferson* case even though the contract provided that upon the determination, monies which were then due ceased to be due. It is all the more likely therefore that clause 32.4.4, which does not expressly state that existing sums due will cease to be due, will be construed as not affecting an adjudicator's decision that payment should be made.

On the other hand, if a termination comes about as a result of liquidation, or possibly if it follows a termination, the effect of Rule 4.90 of the Insolvency Rules 1986 requiring a determination of the balance of account between the parties in such circumstances would mean that summary judgment enforcing the payment by the employer to the contractor would serve no useful purpose and judgment would not therefore be given. See *Bouygues (UK) Ltd* v. *Dahl Jensen (UK) Ltd* (2000). If, on the other hand, the contractor was in receivership, summary judgment would be likely to be given but with the judgment being stayed. See *Rainford House Limited (in Administrative Receivership)* v. *Cadogan Limited* (2001), a case in which Judge Seymour QC said:

> 'Whereas in the case of a company in liquidation it is inevitable that the process contemplated by Rule 4.90 of the Insolvency Rules 1986 will be undertaken, that is not the position in a case in which the claimant is a company in administrative receivership. In the latter case one cannot tell what the outcome of the receivership will be. In a case in which there is not, inevitably, a need for a determination more or less as matters then stand between the parties of the net state of accounts, in which process the correctness of the decision of the adjudicator must be evaluated, and there is not otherwise any defence to a claim to enforce the award of an adjudicator it seems to me that the factors which led *Chadwick LJ in Bouygues (UK) Ltd* v. *Dahl Jensen (UK) Ltd* to consider that it would not be appropriate to give summary judgment at all are not present. However, if there is credible evidence that the claimant is insolvent, in my judgment that is a highly material matter for the court to consider in relation to any application for a stay of execution of the judgment in favour of the claimant.'

Financial accounting following termination

8-62 32.5 Subject to clause 32.7, when the Project has been completed and the defects rectification provisions of the Contract fulfilled the Employer shall issue a payment advice setting out:

.1 the additional costs incurred by the Employer in undertaking the Project compared with the costs that would have been incurred had the Project been completed by the Contractor in accordance with the Contract; and

.2 any loss and/or damage suffered by the Employer and for which the Contractor is liable, whether arising as a consequence of the termination or otherwise.

32.6 The Employer may issue to the Contractor interim payment advices in respect of amounts that are due to it under clause 32.5 as and when they are incurred.

32.7 If the Employer does not commence to make other arrangements for the completion of the Project within a six-month period commencing on the date of termination the Employer shall issue a payment advice setting out:

.1 the amount that the Employer was liable to pay the Contractor at the date of termination, calculated in accordance with the Contract as if the Contractor's employment had not been terminated; and

.2 any loss and/or damage suffered by the Employer and for which the Contractor is liable, whether arising as a consequence of the termination or otherwise.

Allowance shall be made for payments made to date in order to determine the amount payable by one party to the other.

Following completion of the project

8-63 Clause 32.5 provides that, subject to an exception where the employer decides not to complete the project at all (see clause 32.7), when the project has been completed and all defects rectification provisions of the contract have been fulfilled, the employer is to issue a payment advice setting out the financial position. Payment advices under MPF can of course be a positive or negative sum, requiring payment to or by the contractor. In the case of termination, however, the payment advice issued under clause 32.5 will be negative in the sense that it will show a sum due to the employer from the contractor (but see paragraph 8-66 for a possible exception).

The timing of this payment advice is therefore linked to the rectification provisions of the contract having been fulfilled. Those provisions are to be found in clause 17, which presumably means that the payment advice will follow whichever of the following is applicable:

- The issue of a statement under clause 17.2 after the expiry of the rectification period when all defects that the contractor has been instructed to remedy have been remedied; or
- The issue of a statement under clause 17.3 where the contractor has failed to rectify defects and the employer either intends to engage others to rectify (in which case a proper estimate of the costs is to be included), and/or does not intend to rectify the defects (in which case an appropriate deduction which the employer intends to make is to be provided).

An unresolved difficulty exists to a greater or lesser extent with provisions of this type where the contractor's employment is terminated rather than the contract itself. One reason for this is to ensure that use can then be made of express contractual provisions dealing with the financial position following the termination. The difficulty comes in tying in provisions such as those in clause 17, which have to be fulfilled before the issue of the payment advice following completion of the project, with the fact that the contractor is no longer acting under the contract. Clause 17 assumes that the contractor is still employed under the contract, whereas clause 32 is predicated on the basis that he is not. It is not possible for the contractor to fulfil his obligations under clause 17.2, and clause 17.3 does not operate unless the contractor has been instructed to remedy the defects. The contractor clearly will not be so instructed if his employment has been terminated. The only way in which these provisions could fit together is if the reference to 'Contractor' could be treated as a reference to the replacement contractor. However, the definition of 'Contractor' in clause 39.2 refers to the 'party to the contract named as such or any assignee to whom the Employer has consented in accordance with clause 29'. This definition does not therefore include a replacement contractor.

Another difficulty with the payment advice issued under clause 32.5 is its relationship with the final payment advice issued under clause 22.6. Presumably, following the termination of the contractor's employment, no final advice can or is required to be issued under this contract. Any replacement contractor may or may not enter into a contract to complete the outstanding work under an identical MPF. If a final payment advice was still to be issued it would effectively do the job of the clause 32.5 payment advice (see clause 22.6 for the calculation which is to be carried out).

Clauses 32.5.1 and 32.5.2 provide for what must be set out in the payment advice. This is dealt with below. It is, however, worth pointing out at this stage that as it may become clear before the provisions of the contract relating to rectification have been fulfilled, that the employer has already incurred sums over and above anything which may still be owed to the contractor, clause 32.6 accordingly enables the employer to issue interim payment advices in respect of any such amounts due as and when they are incurred.

THE PAYMENT ADVICE UNDER CLAUSE 32.5

The additional costs in undertaking the project compared with the costs that would have been incurred had the project been completed by the contractor in accordance with the contract (clause 32.5.1)

The way in which this calculation will generally be made is to add together the sums which the employer has already paid to the contractor and any further additional sums which the employer will have to pay in order to get the project completed. This total will be compared with the sum which would have been payable to the original contractor had his employment not been terminated and had he completed the project in accordance with the contract. The amount already paid to the contractor should not be too difficult to calculate. It would be the amount actually paid or otherwise properly discharged, e.g. by the exercise of a valid set-off.

In respect of the additional costs incurred by the employer in completing the project, there would be included any payments made in purchasing materials and goods or payments made to others, such as to the replacement contractor or to the consultants in order to carry out and complete the project. The amount of additional costs incurred by the employer in this connection will be recoverable only to the extent that such expenditure was reasonably incurred. The employer has a duty to take reasonable steps to mitigate his losses. However, the approach of the courts – and so also therefore of an arbitrator – is unlikely to be a strict one and what is reasonable in all the circumstances is likely to give the employer reasonable room for manoeuvre. It will not be sufficient for the contractor simply to establish that the outstanding work could have been completed at less cost. He must also show that the employer's actions were positively unreasonable.

8-64 The costs that would have been incurred had the employment of the contractor not been terminated and had he completed in accordance with the contract, need to be assessed. This is a notional final account. It might of course include the value of the work not even included in the original contract sum (e.g. for variations carried out in the course of completing the works). If it is the case (as is likely) that the employer will have had to pay the substitute contractor more – both for the original and the varied works – than he would have had to pay the original contractor, then there would be a claim for the additional cost of both.

8-65 How would an assessment be made of loss and expense, if circumstances arise during the completion works entitling the replacement contractor to reimbursement of loss and expense that would also have given the original contractor an entitlement to claim? There would in fact be no logical basis on which the original contractor's theoretical loss could be assessed, and the likelihood is that any reimbursement to the substitute contractor in respect of such a claim would have to be excluded from the comparison. If a strict comparison was attempted, there could be difficulty in relation to loss and expense payable to the substitute contractor over and above that which would have been payable to the original contractor. Further, as the causing of the loss and expense, with the exception of variations, is likely to be attributable to a default or act of prevention on the employer's part, it is unclear whether the original contractor would be required to meet this particular loss.

The potential for disagreement in carrying out this theoretical comparison is great, all the more so because the exercise will be carried out by the employer himself.

If the sum of the additional costs incurred, plus payments made (or properly discharged) exceeds the total amount which would have been payable to the original contractor had his employment not been terminated, the difference will be reflected in the payment advice. If it is less, then there will be a balance in favour of the contractor. There is a problem with this scenario (see next paragraph). Whatever this figure, the total of the payment advice must also allow for any sum in favour of the employer as a result of the operation of clause 32.5.2.

8-66 An important point to note is that clause 32.5.1 refers only to 'additional costs' so it assumes an outcome in which there is a greater net cost to the employer as a result of the termination. In other words, it is assumed that the employer will pay more to have the remaining works carried out than would have been the case had the contractor's employment not been terminated. It could of course happen that the cost of completing is actually less than would have been payable for that work, had the contractor's employment not been terminated. Clause 32.5.1 does not provide for the comparison to be made where the costs turn out to be less rather than more. In this case it would appear that no sum is to be included. Any reduction is not part of the calculation. This prevents the contractor getting the benefit of this saving which in effect therefore becomes a windfall for the employer. It is unlike, for example, the position under WCD 98 where the contractor gets the benefit of any such saving (clause 27.6). However, in any such situation, though it is difficult to see on the wording of clause 32.5 how the payment advice will include it, any outstanding payment advice in the contractor's favour and the value of any work or other matter covered by clause 22.5 must surely be included and could therefore, exceptionally, result in payment to the contractor.

Any loss and/or damage suffered by the employer for which the contractor is liable, whether arising as a consequence of the termination or otherwise (clause 32.5.1)

8-67 This is widely worded. Firstly, unlike some other JCT contracts, e.g. WCD 98, see clause 27.6.6.1, the reference here is to loss and/or damage rather than being limited to 'direct' loss and/or damage. Further, in clause 27.6.6.1, the loss or damage must be caused as a result of the determination, whereas under the MPF it is loss or damage whether arising as a consequence of the termination or otherwise. It will certainly therefore include losses equivalent to those recoverable as damages for breach of contract at common law which are not already picked up under clause 32.5.1. As has been stated earlier (see paragraph 8-32), once a termination of the contractor's employment has taken place, further liability for liquidated damages will cease. This will open up to the employer the ability to claim as damages sums which might otherwise have been included as part of the liquidated damages calculation had the termination not taken place. This might include loss of profits, e.g. as a result of the loss of rental income, or delay in a production facility coming on stream. It could also include payments made by the employer as damages or compensation, e.g. where other direct contractors who were to follow the completion of the project, such as fitting out contractors, are unable to gain access at the appropriate time; or compensation to purchasers or tenants of the development. The nature and extent of common law damages has been dealt with briefly earlier in this chapter (see paragraph 8-21 *et seq*.).

Any amount stated as payable under the payment advice, assuming it shows an amount due to the employer, will become due on its issue and has a final payment date of 14 days after the due date (clause 32.8 and clause 22.4).

INTERIM PAYMENT ADVICES (CLAUSE 32.6)

8-68 If, at any time before the interim payment advice under clause 32.5 is issued, the employer can establish that amounts have already become due during the period prior to the rectification of defects obligations being fulfilled, these can be included in an interim payment advice showing sums due from the contractor to the employer. Clearly these will have to be taken into account in the calculation of the final payment advice under clause 32.5. The employer does not therefore have to wait until the rectification of defects obligations have been complied with under clause 17 before seeking recovery of sums due from the contractor.

Employer not commencing arrangements for completion of the project within six months from the date of termination

8-69 By clause 32.7, if the employer does not start to make arrangements for the completion of the project within 6 months from the termination of the contractor's employment, the employer must issue a payment advice based upon the assumption that the employer will not be completing the project. The fact that the advice will not need to contain particulars of the cost of completion will, in many cases, make the preparation of the account a rather more straightforward and therefore quicker process. The matters to be set out in the payment advice are:

- The amount the employer is liable to pay the contractor at the date of termination, calculated in accordance with the contract as if the contractor's employment had not been terminated; and

- Any loss and/or damage suffered by the employer for which the contractor is liable, whether arising as a consequence of termination or otherwise.

Allowance is then to be made for payments made to date so that the net amount payable can be calculated.

The calculation of the amount for which the employer is liable will include the items set out in clauses 22.5.1 to 22.5.4. The overall objective of this clause is clear even if the wording is not perhaps perfect, e.g. the contractor may have carried out work prior to the termination, but at the date of termination, the employer may not be liable to pay anything for it if the termination is between the dates for the issue of interim payment advices. Nevertheless the value of such work must clearly be reflected in the payment advice. This includes not only a proportion of the contract sum and the value of any changes, but also the amount of any reductions under clauses 6.7 (employer not liable to pay for work executed otherwise than in accordance with approved design documents); clause 18.5 (employer not liable to pay the contractor in respect of services provided by pre-appointed consultants or work undertaken by named specialists in the absence of a Model Form of Novation or the appointment of a named specialist in the manner provided by clause 18.3); and also any amounts that either party is liable to pay to the other in accordance with the provisions of the contract.

As with the payment advice referred to in clauses 32.5 and 32.6, the issue of the payment advice under clause 32.7, if it shows a sum due to the employer, renders the amount due immediately, with a final date for payment of 14 days thereafter.

Due and final date for payment

32.8 Any amount stated as payable by a payment advice issued under clause 32 shall become due and shall have a final date for payment established in the manner provided by clause 22.4.

8-70 Clause 32.8 provides that any amount stated as payable by a payment advice issued under clause 32 becomes due and has a final date for payment in the same manner as in clause 22.4 (see paragraph 5-133).

Termination by contractor – clause 33

8-71 Clause 33 provides for the contractor to determine his own employment under the contract on the grounds stated in clauses 33.1 and 33.2 (material breach by the employer) and in clause 33.3 (insolvency of the employer). Clause 33.4 deals with the practical and financial consequences and clause 33.5 with the payment mechanism in respect of any sum due from one party to the other.

8-72 Material breach, warning notice and termination notice

33.1 If the Employer commits a Material Breach of the Contract the Contractor may give a notice to the Employer identifying the Material Breach and stating that the Contractor may terminate its employment under the Contract if the Employer fails to remedy the Material Breach within 14 days of the notice.

33.2 If the Material Breach has not been remedied within 14 days of the Contractor's notice under clause 33.1 the Contractor may by a further notice issued at any time within the subsequent 14 days terminate its employment under the Contract.

33.3 In the event that the Employer becomes Insolvent the Contractor may at any time by notice to the Employer terminate its employment under the Contract.

Material breach

8-73 The contractor's right to terminate his own employment is dependent upon the employer committing a 'Material Breach'. This is defined in clause 39.2 as follows:

By the Contractor:

- failure to proceed regularly and diligently with the performance of its obligations under the Contract;
- failure to comply with an instruction;
- suspension of the Project or any part thereof, otherwise than in accordance with HGCRA 1996 or the circumstances described in clause 34.1;
- breach of the CDM Regulations;
- breach of the requirements of CIS;
- breach of any of the provisions of the Contract relating to Named Specialists or Pre-Appointed Consultants.

By the Employer:

- failure to issue a payment advice in the manner required by the Contract.

By either party:

- failure to make payment in the manner required by any payment advice issued under the Contract;
- failure to insure, as established in accordance with clause 27.3;
- any repudiatory breach of the Contract.'

The nature of a material breach has in general terms been discussed when looking at the employer's right to terminate the contractor's employment on this ground (see paragraph 8-35 *et seq.*). Similar points can be made here, though the comparative references to WCD 98 clause 27 will be to various provisions in clause 28 of that contract.

8-74 Where it is the contractor wishing to terminate, a material breach arises where the employer is responsible for any of the following.

FAILURE TO ISSUE A PAYMENT ADVICE IN THE MANNER REQUIRED BY THE CONTRACT

8-75 It is obvious that both the issue and payment of payment advices are vital to the cash flow of the contractor and therefore vital to his ability to continue trading. It is therefore appropriate that the contractor should be entitled to terminate his employment if the employer fails to issue or pay a payment advice. However, as already mentioned above (see paragraph 8-36) there is no general qualification such as the inclusion of a proviso that a termination notice will not be issued if it is unreasonable or vexatious. There could be numerous, what might be called technical breaches of contract by the employer in relation to the issue of a payment advice. It may be issued a little later than it ought to be; or it may not include all that it should. If these are properly regarded as insignificant or nominal in nature, it may be that despite the literal wording of clause 33.1 in relation to material breaches (and its definition in clause 39.2), it would be regarded as not amounting to a ground for termination (see paragraph 8-36). To be sure therefore, the contractor should be satisfied that the failure to issue a payment advice can be objectively regarded as serious in nature.

Further, if the employer has a genuinely arguable defence to any allegation of an alleged failure to issue a payment advice as required by the contract, the contractor would be strongly advised to take appropriate legal advice as to the strength of each party's case as, if the contractor terminates his employment in circumstances where the employer's position was justified under the contract, this will place the contractor in breach (almost certainly in repudiatory breach) with the contractor likely to suffer the very serious consequences flowing from such a breach. In any situation where the contractor is not absolutely sure of his ground, he may be well advised to consider following the adjudication route to obtain a prompt decision in relation to the dispute or difference which has arisen.

FAILURE TO MAKE PAYMENT IN THE MANNER REQUIRED BY ANY PAYMENT ADVICE ISSUED UNDER THE CONTRACT

8-76 If the employer fails to make payment by the final date for payment, namely 14 days after receipt by the employer of a VAT invoice for the due amount, this will provide a ground for the contractor to terminate his own employment. The same point as made above in relation to insignificant or non-serious breaches, applies here also.

On occasions, the employer may be able to discharge his obligation to make payment by means of a set-off, provided any applicable notice of withholding has been duly given. For the most part, matters which under other JCT contracts would be the subject of a set-off, e.g. liquidated damages, will under the MPF be included in the calculation of the payment advice (clause 22.5), so avoiding the need to set off or provide a withholding notice under clause 23 or under section 111 of the Housing Grants, Construction and Regeneration Act 1996 (e.g. paragraphs 2-40, 3-34, 5-111 and 5-163).

FAILURE TO INSURE WHERE REQUIRED BY THE CONTRACT TO DO SO

As this is a ground on which either party may terminate the contractor's employment, it has already been discussed when considering the employer's rights (see paragraph 8-46). The same points are relevant here also.

ANY REPUDIATORY BREACH OF THE CONTRACT

8-77 This is an overall saving provision under which, if the employer commits any breach of the contract which entitles the contractor to treat the employer's breach as a repudiation of it, the contractor can terminate his own employment. This is without prejudice to the contractor's right to treat the contract itself as at an end as a result of the employer's repudiation. Breaches of contract sufficiently serious to be treated as a discharge of the innocent party from his obligations under the contract have been discussed earlier (see paragraph 8-11 *et seq.*).

Also, as this is a ground on which either party may terminate the contractor's employment, it has already been discussed when considering the employer's rights (see paragraph 8-47). The same points are relevant here also.

Notices

8-78 Clauses 33.1 and 33.2, dealing with the giving by the contractor of a warning notice and termination notice respectively, mirror the provisions of clause 32.1 and clause 32.2 relating to the serving of such notices by the employer. The commentary given in respect of the method of service, the timing and content of such notices in relation to clauses 32.1 and 32.2 (see paragraphs 8-48 to 8-53) apply also *mutatis mutandis* (with necessary changes) to clauses 33.1 and 33.2 and reference should be made to that commentary.

Insolvency

8-79 In addition to the contractor's right to terminate his own employment where the employer is guilty of a material breach under clause 33.1, the contractor can also terminate his own employment at any time, by notice to the employer, in the event of the employer becoming 'Insolvent'. This is defined under clause 39.2 as follows:

> Either party is Insolvent when it makes a composition or arrangement with its creditors, or becomes bankrupt, or, being a company:
>
> - makes a proposal for a voluntary arrangement for a composition of debts or scheme of arrangement to be approved in accordance with the Companies Act 1985 or Insolvency Act 1986 as the case may be; or
> - has a provisional liquidator appointed; or
> - has a winding up order made; or
> - passes a resolution for voluntary winding up (except for the purposes of amalgamation or reconstruction); or
> - under the Insolvency Act 1986 has an administrator or an administrative receiver appointed.

These insolvency events are identically worded to those in many other JCT contracts such as WCD 98, clause 28.3.1. Such an insolvency gives rise to a right on the part of the contractor to terminate his own employment by notice.

Notice on insolvency

8-80 The contractor's notice to the employer terminating his own employment due to the employer's insolvency can be given at any time. Having said this, if following the employer becoming insolvent, an agreement is reached with the employer or those acting for the employer in his insolvency, under which the employer is able to continue on the same or different terms, this would amount to a waiver of the contractor's right to terminate. To be sure, in such a situation, any agreement to continue should specifically provide that the contractor will not serve a termination notice in respect of the happening of that insolvency event.

8-81 The formalities in connection with the giving of the termination notice under clause 33.3 are governed by clause 38.2 (see paragraphs 8-48 and 9-10).

Consequences of a termination of the contractor's employment by the contractor

> 33.4 In the event that the Contractor's employment is terminated under clause 33 the Contractor shall:

.1 remove all its materials plant or equipment from the Site without delay;
.2 prepare an account setting out its valuation of its entitlements under the Contract at the date of termination (including any entitlements in respect of Changes and other amounts for which the Employer is liable) together with its reasonable costs of removal from the Site as a consequence of the termination and any loss and/or damage suffered by the Contractor and for which the Employer is liable, whether arising as a consequence of the termination or otherwise;
.3 issue a statement that compares the total amount included in the above account with the total payments previously received by the Contractor in order to determine the balance that is to be paid by one party to the other.

8-82 Clause 33.4 deals with the consequences of a termination by the contractor of his own employment under the contract on any of the grounds referred to in clauses 33.1 or 33.3. The MPF has less detail with reduced wording in connection with the consequences of termination than most other JCT contracts. This for the most part appears to be achieved without much disadvantage. The contractor is to remove his materials, plant or equipment from the site without delay. An interesting difference between this clause and clause 32.4, which applies where it is the employer terminating the contractor's employment, is that there is no equivalent to clause 32.4.2 under which the contractor is to provide to the employer all design documents prepared in connection with the project. Such a provision does feature in WCD 98 in both clause 27.6.1 (employer determining contractor's employment) and clause 28.4.1 (contractor determining his own employment). Employers may well seek to amend clause 33.4 to include a similar provision to that set out in clause 32.4.2.

So far as financial accounting matters are concerned, the contractor under clause 33.4.2 is to prepare an account setting out his valuation of his entitlements under the contract at the date of termination which are to expressly include entitlements in respect of changes and also other amounts for which the employer is liable. The account will therefore include any loss and expense entitlements of the contractor prior to the termination. In addition to this, the contractor must include in the account his reasonable costs of removal from the site as a consequence of the termination and also any loss or damage suffered and for which the employer is liable, whether arising as a consequence of the termination or otherwise.

The loss or damage suffered by the contractor for which the employer is liable will include matters such as the payment of damages to sub-contractors or suppliers whose contracts have been consequentially breached by the contractor or which have been terminated; or the contractor's loss of contribution to head office overheads and loss of profit in connection with the lost work. The contractor's claim in this regard will be equivalent to that which he could make in a claim for damages for breach of contract at common law: see *Wraight Limited* v. *P.H. & T. (Holdings) Limited* (1968).

Once the account has been prepared the contractor, under clause 33.4.3, is to issue a statement comparing the total amount included in the account with the total payments previously received. This will determine the balance that is to be paid by one party to the other. It is highly likely of course that this will result in a sum being due to the contractor. The most likely exception to this would be where work has been found to be defective having earlier been paid for on the basis that it complied with the contract.

Due and final dates for payment

33.5 Any amount identified by the statement issued under clause 33.4.3 as properly payable to the Contractor shall become due for payment upon the receipt of a VAT invoice by the Employer and any amount identified as payable to the Employer shall become due upon the issue of the statement. The final date for payment shall be 14 days after the amount to be paid becomes due.

By clause 33.5, any amount identified by the statement under clause 33.4.3 as properly due for payment will be due, where it is due from the employer to contractor, on receipt by the employer of a VAT invoice from the contractor. Where the amount properly due is due from the contractor to the employer, it becomes due upon the issue of the statement itself. The final date for payment is then 14 days after the amount to be paid becomes due.

As it is the contractor who is preparing the account and statement, clause 33.5 refers to an amount identified by the statement which is 'properly payable to the Contractor'. This is clearly intended to protect the employer so that it cannot be argued that an amount becomes actually due simply because it is contained in the statement. This avoids the risk that the contractor can obtain an adjudicator's decision in his favour, notwithstanding that the employer can establish that the sum claimed is incorrect, but where the employer has failed to serve a withholding notice in accordance with section 111 of the Housing Grants, Construction and Regeneration Act 1996.

Termination by either employer or contractor – clause 34

Prolonged suspension or removal of insurance cover for terrorism – grounds and notice of possible termination

34.1 If the carrying out of the Project or a substantial proportion of the Project is suspended for the period stated in the Appendix as a consequence of force majeure, the occurrence of a Specified Peril or hostilities involving the United Kingdom, or the use or threat of terrorism as defined by the Terrorism Act 2000 then either party may issue to the other a notice specifying the circumstances of the suspension and stating that if the circumstances continue for a further 14 days it may terminate the Contractor's employment under the Contract.

8-83 Where the project or a substantial portion of it is suspended for the period stated in the appendix to the contract as a consequence of certain events, this entitles either party to issue a notice that if the circumstances causing the suspension to continue for a further 14 days, that party may terminate the contractor's employment under the contract. If no period is stated in the appendix, the default period will be 13 weeks' suspension. Clause 34.1 does not expressly say whether the suspension has to be continuous or could be made up of lesser periods which, when added together, equal or exceed the qualifying period. Considering the wording of clause 34.1 and the appendix item, the suspension is probably intended to be one continuous period.

A common feature of the grounds giving rise to the right to terminate (with a possible exception relating to specified perils – discussed later) is that they do not involve fault on the part of either party.

The grounds for a qualifying suspension giving rise to the possibility of a termination of the contractor's employment by either party worth further discussion are as follows.

8-84 ***Force majeure***

This is of uncertain meaning (see the discussion of it in paragraph 3-56). What is certain is that its interpretation in this contract will be affected by the context in which it appears. Accordingly, its meaning in clause 12 (Extension of time) (see clause 12.1.1) may be somewhat different to its meaning in clause 34.1. For instance, in the former (see clause 12.1.4) there is a separate event which refers not only to the use or threat of terrorism, but also to the activities of relevant authorities in dealing with such a threat. Clause 34.1 also includes a separate event which refers to the use or threat of terrorists but it does not include a reference to such other activities and it may be argued therefore that they would be covered by *force majeure* in clause 34.1 but not covered by *force majeure* in clause 12. Also, it has already been suggested in the commentary to clause 12.1.1 (see paragraph 3-57) that exceptionally adverse weather conditions may possibly fall within the meaning of *force majeure* so that a prolonged suspension due to exceptionally adverse weather could give rise to a termination of the contractor's employment.

8-85 ***The occurrence of a specified peril***

'Specified Peril' is defined in clause 39.2 as:

> Fire, lightning, explosion, storm, tempest, flood, escape of water from any water tank, apparatus or pipe, earthquake, aircraft or other aerial devices or articles dropped therefrom, riot or civil commotion.

The occurrence of certain of the specified perils could be attributable to the negligence of either the employer or the contractor, e.g. fire or explosion. In other JCT contracts, e.g. WCD 98, clauses 28A.1.2 and 28A.7, this possibility is dealt with. In that contract, where the specified peril is the result of the contractor's negligence, he is not entitled to terminate his own employment. Where the specified peril is the result of the employer's negligence, while the employer may still be able to terminate the contractor's employment, the contractor is entitled to recover any direct loss or damage caused by the termination, which is not the case if the specified peril is not caused by the employer's negligence. In clause 34, there are no such qualifications. Finally, it is important to note that the specified peril does not have to directly relate to the site or the project itself. It could, for instance, affect the supply of essential materials which cause a qualifying suspension to the project.

8-86 ***The use or threat of terrorism as defined by the Terrorism Act 2000***

This was considered in relation to the same words used in clause 12.1.4 dealing with adjustments to the completion date (see paragraph 3-65), and reference should be made to the commentary on it. There is in clause 34.3 a further ground upon which the employer can terminate the contractor's employment if terrorism cover, as defined in clause 39.2 , is no longer available.

8-87 ***Notice of possible termination***

Clause 34.1 provides that if there is a qualifying suspension, either party may issue to the other a notice specifying the circumstances of the suspension and stating that if those circumstances continue for a further 14 days (from delivery of the notice – see clause 38.2) the party serving the notice may terminate the contractor's employment under the contract.

The notice must of course clearly specify the circumstances of the suspension, which should include reference to the ground relied upon, e.g. *force majeure* or a specified peril, and sufficient information about the occurrence to enable the employer or contractor, as the case may be, to appreciate the situation and how he may, if at all possible, bring the suspension to an end within the 14 days. The notice must also expressly state that if the circumstances do continue for a further 14 days, then the party serving the notice may terminate the contractor's employment.

As to the method of service of the notice, reference should be made to the commentary to clause 32.1 under the heading 'Method of service' (see paragraph 8-48).

Termination notice

34.2 If the circumstances continue for a further 14 days the party that issued the notice may by a further notice at any time within the subsequent 14 days terminate the Contractor's employment under the Contract.

8-88 By clause 34.2, if the circumstances referred to in the notice under clause 34.1 continue for a further 14 days, the party issuing the notice may by a further notice at any time within the subsequent 14 days terminate the contractor's employment under the contract. Any such notice should cross-refer to the earlier notice and, for good measure, restate the circumstances. As to the method of service of the notice, reference should be made to the commentary to clause 32.2 (see paragraph 8-52 *et seq.*).

Determination by the employer where terrorism cover is no longer available

8-89 34.3 Where Terrorism Cover is no longer available the Employer may, by notice, terminate the Contractor's employment under the Contract.

8-90 'Terrorism Cover' is defined in clause 39.2 as:

Cover under any policy required to be provided by the Contract against the physical loss or damage to work executed or site materials caused by an act of terrorism as defined by the Terrorism Act 2000.

The definition of terrorism in the Terrorism Act 2000 has been set out in relation to the commentary to clause 12.1.4 (see paragraph 3-65). Clause 27.8 provides that if terrorism cover ceases to be available, then on the later of the date that it ceases or (where it is the contractor's responsibility to provide and maintain the cover) the date notified by the contractor to the employer that it or any part of such cover has ceased, the risk of any loss that would otherwise have been covered by the policy rests with the employer. Because of this, the employer is given the right to terminate the contractor's employment as an alternative to taking on this risk.

8-91 **Consequences of a termination of the contractor's employment under clauses 34.2 or 34.3**

> 34.4 If the Contractor's employment is terminated under clause 34 the Contractor shall:
> .1 remove all of its materials plant or equipment from the Site without delay;
> .2 provide to the Employer all Design Documents prepared in connection with the Project;
> .3 prepare an account setting out its valuation of its entitlements under the Contract at the date of termination (including any entitlements in respect of Changes and other amounts for which the Employer is liable) together with its reasonable costs of removal from the Site as a consequence of the termination;
> .4 issue a statement that compares the total amount included in the above account with the total payments previously received by the Contractor in order to determine the balance that is to be paid by one party to the other.

Clause 34.4 provides for what is to happen if the contractor's employment is terminated. It provides that the contractor is to:

- Remove all of his materials, plant or equipment from the site without delay (clause 34.4.1).
- Provide to the employer all design documents prepared in connection with the project (clause 34.4.2). Virtually identical wording in clause 32.4.2 was considered earlier (see paragraph 8-59).
- Prepare an account setting out his valuation of his entitlements under the contract at the date of termination (including those relating to changes and any other amounts for which the employer is liable), together with his reasonable costs of removal from the site as a consequence of the termination (clause 34.4.3). This is similar to the wording in clause 33.4.2 except that as, generally speaking, the termination under clause 34 is not based upon an employer default, there is no provision for the contractor to seek loss or damage. With this exception the commentary on clause 33.4.2 is applicable (see paragraph 8-82).
- Issue a statement comparing the total amounts included in the account referred to above with the total payments previously received by the contractor, so that the balance payable by one party to the other can be determined (clause 34.4.4). This is identical to the wording of clause 33.4.3 and reference can be made to the commentary on that clause (see paragraph 8-82).

Due date and final date for payment

> 34.5 Any amount identified by the statement issued under clause 34.4.4 as properly payable to the Contractor shall become due for payment upon the receipt of a VAT invoice by the Employer and any amount identified as payable to the Employer shall be due upon the issue of the statement. The final date for payment shall be 14 days after the amount to be paid becomes due.

8-92 Clause 34.5 provides that the amount identified by the statement issued under clause 34.4.4 as properly payable to the contractor shall become due for payment upon receipt of a VAT invoice by the employer. If it is the employer who seeks payment, the amount is due upon the issue of the statement itself. The final date for payment in either case is 14 days after the amount to be paid becomes due. As it is the contractor who is preparing the account and statement, clause 34.5 refers to an amount identified by the statement which is 'properly payable to the Contractor'.

This is clearly intended to protect the employer so that it cannot be argued that an amount becomes actually due simply because it is contained in the statement. This avoids the risk that the contractor can obtain an adjudicator's decision in his favour, notwithstanding that the employer can establish that the sum claimed is incorrect but where the employer has failed to serve a withholding notice in accordance with section 111 of the Housing Grants, Construction and Regeneration Act 1996.

Resolution of disputes – clauses 35–37

Background

8-93 When a contract is entered into no one intends that there should be disputes or differences. Furthermore, most contracting parties do not expect to have disputes and differences. Nevertheless they can and do arise from time to time. The field of dispute resolution has been transformed in recent years. The following developments have taken place.

Litigation

8-94 Litigation through the courts is now subject to the Civil Procedure Rules 1998 (CPR) which came into effect from 26 April 1999. They apply to civil court proceedings.

It is not intended to deal here in any detail with the CPR, though the following points can be noted:

- Rule 1.1 contains the overriding objective which is to enable the court to deal with cases justly. It includes so far as practicable:
  - (a) Ensuring that the parties are on an equal footing;
  - (b) Saving expense;
  - (c) Dealing with the case in ways which are proportionate to the:
    - ○ amount involved
    - ○ importance of the case
    - ○ complexity of the issues
    - ○ financial position of each party;
  - (d) Ensuring that it is dealt with expeditiously and fairly;
  - (e) Allotting to it an appropriate share of the court's resources, while taking into account the need to allot resources to other cases.
- The court will expect the parties to co-operate with each other in the conduct of proceedings
- The parties are required to help the court to further the overriding objective (Rule 1.3).
- By Rule 1.4, the court is to further the overriding objective by actively managing cases and this includes:
  - (a) Encouraging the parties to co-operate with each other in the conduct of the proceedings;
  - (b) Identifying the issues at an early stage;
  - (c) Deciding promptly which issues need full investigation and trial and accordingly disposing summarily of the others;
  - (d) Deciding the order in which issues are to be resolved;

(e) Encouraging the parties to use an alternative dispute resolution procedure if the court considers that appropriate, and facilitating the use of such a procedure;
(f) Helping the parties to settle the whole or part of the case;
(g) Fixing timetables or otherwise controlling the progress of the case;
(h) Considering whether the likely benefits of taking a particular step justify the cost of taking it;
(i) Dealing with as many aspects of the case as it can on the same occasion;
(j) Dealing with the case without the parties needing to attend at court;
(k) Making use of technology; and
(l) Giving directions to ensure that the trial of a case proceeds quickly and efficiently.

8-95 Rule 1.4(e) is particularly noteworthy and it is very clear that the courts will be keen to see that the parties investigate the possibility of alternative ways of resolving their dispute short of litigation. Two cases provide an indication of the courts' approach to this issue. In *Cable & Wireless plc* v. *IBM United Kingdom Ltd* (2002), Mr Justice Coleman went out of his way to support a clause in a contract between the parties that they would in good faith attempt to resolve any dispute by using an alternative dispute resolution procedure. This was so even though the clause was not as certain as it could have been. The clause was contained in a global framework agreement relating to information technology. Clause 41 provided:

> '(1) The parties shall attempt in good faith to resolve any dispute or claim arising out of or relating to this agreement or any local services agreement promptly through negotiations between the respective senior executives of the parties who have authority to settle the same...
> (2) If the matter is not resolved through negotiation, the parties shall attempt in good faith to resolve the dispute or claim through an alternative dispute resolution (ADR) procedure as recommended to the parties by the Centre for Dispute Resolution. However, an ADR procedure which is being followed shall not prevent any party or local party from issuing proceedings.'

The obligation on the parties therefore was to attempt in good faith to adopt an alternative dispute resolution procedure as recommended by the Centre for Dispute Resolution (now the Centre for Effective Dispute Resolution).

The judge distinguished this clause from other provisions, which have been held to be a mere promise to negotiate. Mr Justice Coleman said:

> 'However, the English courts should nowadays not be astute to accentuate uncertainty (and therefore unenforceability) in the field of dispute resolution references. There is now available a clearly recognised and well-developed process of dispute resolution involving sophisticated mediation techniques provided by trained mediators.'

Accordingly, where one party had sought to litigate without following any alternative dispute resolution procedure, the court was prepared to stay the court proceedings.

This approach was supported by CPR 1.4 and also by pronouncements of the Court of Appeal, e.g. see *Dunnett* v. *Railtrack Plc* (2002), in which the Court of Appeal refused to make a costs order in favour of Railtrack and against the unsuccessful claimant in the light of Railtrack's refusal to consider arbitration or mediation.

In the second case, *Hurst* v. *Leeming* (2002) Mr Justice Lightman in the Chancery Division considered the issue of a refusal by a party to mediate as a relevant factor in the exercise of the court's discretion when considering an award of costs. In the instant case, exceptionally, as mediation would have offered no realistic prospect of resolving the dispute, the defendant was not penalised for having refused it. This case concerned a solicitor who brought proceedings against his former partners. The judge made the point that the decision of a party to ignore a mediation clause on the basis that it would not have achieved a settlement is a high-risk strategy. The judge said:

> 'The critical factor in this case, in my view, is whether, objectively viewed, a mediation had any real prospect of success. If mediation can have no real prospect of success a party may, with impunity, refuse to proceed to mediation on this ground. But refusal is a high-risk course to take, for if the court finds that there was a real prospect, the party refusing to proceed to mediation may, as I have said, be severely penalised. Further, the hurdle in the way of a party refusing to proceed to mediation on this ground is high, for in making this objective assessment of the prospects of mediation, the starting point must surely be the fact that the mediation process can and does often bring about a more sensible and more conciliatory attitude on the part of the parties than might otherwise be expected to prevail before the mediation, and may produce a recognition of the strengths and weaknesses by each party of his own case and that of his opponent, and of willingness to accept the give and take essential to a successful mediation.'

8-96 Prior to the commencement of proceedings, the court will expect the parties to act reasonably in exchanging documents relevant to the claim and generally in trying to avoid the necessity for the start of proceedings. This will include following any relevant pre-action protocols such as the Pre-Action Protocol for Construction and Engineering Disputes (August 2000).

8-97 More is said about mediation below (see paragraph 8-108 *et seq.*).

8-98 Construction cases are generally dealt with in the Technology and Construction Court and such proceedings are covered by Part 60 of the CPR and the Practice Direction for TCC claims.

8-99 There is also available The Technology and Construction Court Guide (December 2001).

Details and more information can be found on the Court Service website http://www.courtservice.gov.uk.

Arbitration

8-100 The MPF is unique among JCT contracts (outside its consumer range) in that it does not expressly provide for arbitration as a dispute resolution method. To take a dispute or difference to arbitration would therefore require the parties to agree to this either by amending the contract to include an arbitration agreement, or by agreeing at some later time to take part in an ad hoc reference to arbitration. In either case the parties may wish to also agree to the application of a particular set of arbitration rules such as those included in other JCT contracts which do provide for arbitration, namely the JCT Edition of the Construction Industry Model Arbitration

Rules (CIMAR). Any agreement to arbitrate must, if the Arbitration Act 1996 is to apply, be an 'arbitration agreement' within section 6 of the Act.

Arbitration is a very common and long standing means of resolving disputes in the construction industry. It is now generally governed by the Arbitration Act 1996 replacing, in particular, the Arbitration Acts 1950, 1975 and 1979.

The 1996 Act has sought to make arbitration a more efficient and cost-effective means of dispute resolution than it has been in the recent past, particularly in relation to construction disputes. The Act is in part a consolidation of earlier legislation and common law and in part a modification of it. Its essential thrust is to help facilitate speed and cost effectiveness in the arbitral process, and to increase the scope for the parties to control the proceedings, with however, fall back provisions to apply where the parties have not agreed what should happen. The provisions of the Act have been significantly influenced by the Model Law on Arbitration, drafted by UNCITRAL, the International Trade Law Committee of the United Nations.

8-101 An interesting and unusual feature of the Act is that it contains a statement of overriding objectives; a kind of mission statement. Section 1 provides:

> '(1) The provisions of this Part are founded on the following principles, and shall be construed accordingly:
> (a) the object of arbitration is to obtain the fair resolution of disputes by an impartial tribunal without unnecessary delay or expense;
> (b) the parties should be free to agree how their disputes are resolved, subject only to such safeguards as are necessary in the public interest;
> (c) in matters governed by this Part the court should not intervene except as provided by this Part.'

8-102 The general duty of the arbitrator is set out in clause 33 which provides:

> '(1) The Tribunal shall
> (a) act fairly and impartially as between the parties, giving each party a reasonable opportunity of putting his case and dealing with that of his opponent, and
> (b) adopt procedures suitable to the circumstances of the particular case, avoiding unnecessary delay or expense, so as to provide a fair means for the resolution of the matters falling to be determined.
> (2) The Tribunal shall comply with that general duty in conducting the arbitral proceedings, in its decisions on matters of procedure and evidence and in the exercise of all other powers conferred on it.'

8-103 In passing it is worth noting that the duty of the arbitrator is to act 'fairly' as well as impartially. The duty of the adjudicator – see section 108(2)(e) of the Housing Grants, Construction and Regeneration Act 1996 – is to act impartially. There is no reference to the adjudicator acting fairly. This difference may be significant. Having regard to the very tight time-scales which govern adjudication proceedings, even for the most complex cases, it may be that the adjudicator will not be able to act fairly although he will still be able to act impartially. The adjudicator has a duty to reach a decision within a stated time. Within that framework he must treat the parties equally in the sense that he must not show partiality.

It is not possible in a book of this kind dealing with the MPF generally, to deal in detail with the subject of arbitration. Reference should be made to standard works on the subject such as that of Mustill & Boyd *Commercial Arbitration*, 3rd edition.

STAY OF COURT PROCEEDINGS

8-104 However, one final point is worth particular mention. Should one party seek to litigate a dispute rather than to arbitrate it in accordance with an arbitration agreement entered into, the courts will upon application stay the court proceedings. The court has no discretion to allow the proceedings to continue: see section 9 of the Act; *Halki Shipping Corporation* v. *Sopex Oils Ltd* (1997); *Davies and Middleton & Davies Ltd* v. *Toyo Engineering Corporation* (1997); and *Ahmad Al-Naimi* v. *Islamic Press Agency Incorporated* (1998). In this last case, Judge Bowsher QC sitting in the Technology and Construction Court, held that a section 9 stay should be granted even where there was a dispute between the parties as to whether their contractual dispute was covered by an arbitration agreement at all. However, in *Birse Construction Ltd* v. *St David Ltd.* (1999), Judge Humphrey Lloyd QC regarded the *Ahmad Al-Naimi* case as one on its particular facts. If on a section 9 application the question is raised as to whether there is a valid arbitration agreement, this is likely to be determined by the court itself.

The matter can only continue in the courts if the defendant has taken an appropriate step in the proceedings to answer the substantive claim; or where the court is satisfied that the arbitration agreement is null and void, inoperative or incapable of being performed.

Adjudication

8-105 In the construction industry, adjudication is essentially a rapid decision-making process producing a temporarily binding decision. It is especially important in relation to the question of cash flow as its history in the construction industry demonstrates.

HISTORY

8-106 Adjudication was introduced into the construction industry to provide contractual control of the common law right to set off, which allegedly was being abused by main contractors to the prejudice of their sub-contractors. It was introduced in the 1970s in relation to forms of sub-contract for use with JCT main contacts (popularly known as the 'Blue Form' (domestic) and the 'Green Form' (nominated)) in order to obtain a quick decision in connection with the vital question of cash flow where contractors sought to set off sums against amounts which sub-contractors said were due to them.

Adjudication at that time had a very narrow role to play, being specific to the situation in which the main contractor set off from sums otherwise due to his sub-contractor losses allegedly caused by delay or disruption on the sub-contractor's part. The adjudicator's jurisdiction was limited to:

- Upholding the deduction; or
- Ordering the contractor to pay some or all of the amount to the sub-contractor; or
- Ordering the sum or any part of it to be paid to a stakeholder.

If either party did not accept the decision and they could not reach agreement, then the matter went to arbitration.

Later, adjudication was introduced for a wider range of issues in the supplementary provisions in the WCD 81 form of contract.

In 1994 the New Engineering and Construction Contract (now the Engineering and Construction Contract) introduced adjudication provisions intended to apply to all disputes.

8-107 On the coming into force of Part II of the Housing Grants, Construction and Regeneration Act 1996 (and the statutory instrument containing the Statutory Scheme for Construction Contracts) on 1 May 1998, both the JCT and ICE introduced adjudication provisions aimed at complying with the requirements of section 108 of the Act. If the adjudication procedures set out in these contracts fail in any respect to meet the requirements of section 108, then the Statutory Scheme for Construction Contracts will apply in full. In addition, if a construction contract such as the MPF is used which does not contain any adjudication provisions, other than in relation to the appointment of an adjudicator, then again, the Statutory Scheme will apply.

The two essential features of adjudication in the construction industry can be seen to be (1) a tight time-scale within which an adjudicator must make a decision, and (2), while the decision itself must be complied with, nevertheless an aggrieved party is entitled to refer the underlying dispute or difference to arbitration or litigation as the case may be. These two attributes are closely connected. If tight time-scales are in operation, particularly where the dispute or difference raises complex issues and involves consideration of a considerable amount of relevant material, it is important that if the adjudicator's decision is unacceptable to one of the parties, that party should be able to refer the dispute for resolution under a fuller and more considered method of dispute resolution such as arbitration or litigation.

Rapid adjudication can, on occasions, prejudice a party in a number of ways. For example, not all relevant points may be able to be made; not all relevant documentation may be available within the time-table allowed; it may not be possible in the time permitted to be represented by someone of a party's choice, whether a lawyer or other suitably trained representative; further, it may be that in the time permitted, it is not possible, even though it may be desirable, to have an oral hearing, for example, where there is a dispute as to a factual matter which might best be resolved by having the benefit of oral examination and cross-examination of witnesses.

It is not proposed to deal in any detail with adjudication in this book. It is an interesting and rapidly developing area. Cases come before the court on a regular basis and reference can be made to the many periodicals published in this area as well as to textbooks dealing specifically with this topic, e.g. *Construction Adjudication* by John L. Riches and Christopher Dancaster published by LLP Reference Publishing.

Mediation

8-108 An increasingly effective and therefore popular alternative dispute resolution mechanism is where the parties agree to mediate with the assistance of an independent mediator. The courts will encourage the parties to follow this route wherever possible (see paragraph 8-95). There are dedicated bodies who provide a mediation service, e.g. Centre for Effective Dispute Resolution (CEDR).

The JCT has in Practice Note 28 provided a guide to mediation as a method of resolving disputes. It gives a fair picture of the nature of mediation and how it can be used in the construction industry.

SUMMARY OF JCT PRACTICE NOTE 28

8-109 Disputes can be resolved either by litigation, arbitration, adjudication, or by agreement between the parties involved. Mediation is a way in which the negotiation of such an agreement may be facilitated.

This involves the appointment of an independent person. Their qualifications will depend upon the nature of the dispute. The intention is not to impose a solution but to try to steer the parties themselves towards a settlement.

Mediation may not be suitable if one or more of the following factors are present:

- A legal precedent is required;
- A party needs an injunction;
- A party wants a public hearing;
- Where a party does not genuinely want to reach an agreement.

If the parties consider that a mediator could assist in achieving a settlement, the use of a Mediation Agreement is advised. Such an agreement is provided by way of 'Example A' of the Practice Note.

The main matters to be covered are:

- A clear and precise statement of the issues in dispute;
- A declaration that the parties wish to resolve the dispute with the help of a mediator;
- A period during which the mediation is to take place, leaving parties free to withdraw without reasons prior to a binding settlement being achieved;
- The name and qualifications of the mediator;
- A provision that the mediation is to be confidential and without prejudice and that no party in any other proceedings will call the mediator to give evidence;
- A date for meeting the mediator to discuss how the mediation is to be conducted;
- The question of how costs are to be dealt with as well as the apportionment of the mediator's fees;
- A provision that if the mediation results in settlement the parties will execute a binding agreement setting out its terms.

The effect on progress of work of agreeing to mediation should be considered and an agreement reached as to how that effect is to be dealt with. If a rapid binding decision is required, then adjudication may sometimes be more appropriate.

The agreement to try mediation should not prevent any party commencing or continuing other proceedings, unless a binding settlement is reached.

It is recommended that unless and until agreement is reached the parties should proceed wherever possible with the contract as if there was no dispute. If the dispute involves more than two contracting parties special arrangements may be needed to allow for this.

If any settlement affects any decision or opinion already made or expressed by the contract administrator, if any, under the contract, a copy of the decision should be provided as it may affect any such person's professional responsibilities under their appointment.

Lists of mediators are maintained by the RIBA, RICS, ACE, CEDR and CIARB.

The Practice Note sets out a sample of the following:

- Mediation agreement;
- Agreement appointing a mediator;
- Agreement following the resolution of the dispute by mediation.

Choice of law and jurisdiction clauses

8-110 The purpose of a jurisdiction clause is generally to determine the choice of law, i.e. which country's law is to apply to the contract wherever the litigation may take place, and also to determine where any litigation will take place. It is better to do this than to litigate about where to litigate. What cannot be controlled, of course, is how any given system of background law before which such a clause comes to be construed, will interpret and give effect to it. What is within the draftsman's control is to achieve the maximum certainty possible so far as drafting can do so.

Where a standard form of contract is intended to be used between parties from England, Wales and Northern Ireland, it is obviously appropriate to produce a jurisdiction clause which seeks to ensure that litigation takes place in those parts of the UK. Any such clause also needs to be drafted well enough to enable injunctions to be sought if proceedings are brought by one party against the other elsewhere. Well drafted clauses may also prevent a foreign court from considering that the agreement could be impeached in legal proceedings wrongly brought before that court. The clause may also be effective in providing a defence to the recognition and enforcement of a foreign judgment which is obtained in breach of the jurisdiction clause. To guarantee, as far as possible, that any dispute is litigated within the stated jurisdiction, it is generally appropriate to provide that the jurisdiction will be 'exclusive'. If the clause is non-exclusive, then clearly a party could seek to litigate elsewhere. In a standard form this may be an acceptable position to take. However, there may be some employers who will seek to provide that the jurisdiction clause either gives exclusive jurisdiction to the English courts, or alternatively allows only the employer to litigate in either the English courts or some other court of his choice.

We now turn to consider what the MPF provides in relation to the above matters.

Choice of dispute resolution method

35.1 Should any dispute or difference arise between the parties in relation to this Project:
- .1 where the parties agree to do so, the dispute or difference may be submitted to mediation in accordance with the provisions of clause 36 (*Mediation*);
- .2 the dispute or difference may be referred to adjudication in accordance with the provisions of clause 37 (*Adjudication*);
- .3 the dispute or difference may be resolved by legal proceedings.

8-111 Clause 35.1 provides that disputes or differences between the parties in relation to the project are to be dealt with in accordance with sub-clauses 1 to 3, namely mediation, adjudication and litigation. There is no express provision for arbitration. If the parties wish to provide for arbitration, this can be done by amending the contract, before it is signed, to provide for an arbitration agreement or, alternatively, the parties can agree to refer a dispute or difference to arbitration on an ad hoc basis at the time that it requires resolution. Arbitration has been discussed briefly earlier (see paragraph 8-100 *et seq.*).

If the parties agree to submit to mediation, this is to be in accordance with clause 36. If either party wishes to refer the dispute or difference to adjudication, this is done in accordance with the provisions of clause 37.

If either party wishes to commence proceedings in the courts to resolve the dispute or difference, they are free to do so.

Choice of law and jurisdiction

35.2 The Contract shall be governed and construed in accordance with the laws of England and the English courts shall have jurisdiction over any dispute or difference that may arise.

8-112 This is the choice of law and jurisdiction clause. Such clauses have been briefly discussed in general terms earlier (see paragraph 8-110). By means of this clause, it is provided firstly that the contract is to be governed and construed in accordance with the laws of England. Accordingly, in whichever jurisdiction proceedings are commenced, the courts of that territory will be required to apply English law.

Secondly, this clause provides that the English courts shall have jurisdiction over any dispute or difference that may arise. This is a non-exclusive jurisdiction clause. In other words, it does not provide that the English courts are the only courts who could have jurisdiction. This issue has been discussed earlier (see paragraph 8-110). It means that if proceedings are commenced in another jurisdiction, for example where the project is situated, there may be an issue as to whether the proceedings should take place in the English courts or those of the territory in which the project is being built. It is not intended to discuss jurisdictional matters in this book. Readers are recommended to consult appropriate textbooks such as Dicey and Morris *The Conflict of Laws*, 13th edition, (pages 424–51), published by Sweet & Maxwell. As to the drafting of such clauses, a useful analysis of the problems, and drafting suggestions can be found in Appendix VII of Briggs and Rees *Civil Jurisdiction and Judgments*, 3rd edition, published by Informa Professional.

One issue of considerable importance in relation to jurisdiction is to ensure that the jurisdiction clause is drawn sufficiently wide to include not only actions for breach of contract but also issues which arise before the contract is entered into, e.g. misrepresentation or even challenges to the validity or existence of the contract. The fact that clause 35.1 refers to 'any dispute or difference . . . in relation to the Project' should have the effect of giving the English courts a wide jurisdiction as to the types of claim it can deal with.

Mediation

8-113 36.1 Either party may identify to the other any dispute or difference as being a matter that it considers to be capable of resolution by mediation and, upon being requested to do so, the other party shall within 7 days indicate whether or not it consents to participate in mediation with a view to resolving the dispute. The objective of mediation under clause 36 shall be to reach a binding agreement in resolution of the dispute.

8-114 36.2 The mediator or selection method for the mediator shall be determined by agreement between the parties.

Agreeing to mediate

8-115 Clause 36 deals in simple and basic terms with the possibility of the parties agreeing that the dispute or difference between them should be referred to mediation - see clause 35.1.

For the most part this clause is self-explanatory. The decision to mediate in order to, if possible, resolve a dispute or difference is a matter of agreement between the parties. The only obligation contained in this clause is that, one party having identified the dispute or difference as being capable of resolution by mediation, the other party upon request must indicate within 7 days of the request whether it consents to participate. If there is no such indication within the 7-day period, that may technically be a breach of the contract, but it is difficult to see how it could really cause any loss. The last sentence referring to the objective of mediation being to reach a binding agreement in resolution of the dispute while written in peremptory terms, is probably of little contractual significance. Its value is likely to be in conditioning and focusing the minds of the parties before mediation starts.

Mediation is undoubtedly an increasingly efficient and cost-effective way of resolving many disputes and differences. This topic was discussed in more detail earlier (see paragraph 8-108 *et seq.*).

It may be thought that if the parties have been unable themselves to resolve a dispute or difference which exists between them, it is unlikely that they will readily agree to seek mediation which, after all, is a process which is intended to get the parties to settle their differences. It may be that some would prefer a clause which requires the parties to submit to the process, at least by providing for the appointment of a mediator if one cannot be agreed and also providing that both parties attend at least the first meeting before the mediator. This can have the effect of changing people's attitudes.

Even with the purely consensual approach reflected in the drafting of clause 36.1, the approach of the courts in encouraging mediation, even to the point of penalising a party in costs if he has unreasonably refused to take part in the mediation process (see cases mentioned earlier at paragraph 8-95), is likely to cause both parties to seriously consider the possibility of mediation.

Appointment of mediator

8-116 It is for the parties to agree either a named mediator or alternatively the method by which a named mediator can be ascertained. Clearly if there is no agreement on either of these things then the mediation will not take place. It is very much a consensual process. If, under clause 36.1, the parties have agreed to seek mediation in order to resolve their dispute or difference, it is extremely unlikely that they will fall out in respect of obtaining the appointment of a mediator.

Adjudication

37.1 Either party may at any time refer any dispute or difference arising under the Contract to adjudication in accordance with the provisions of The Scheme.

37.2 The adjudicator shall be the person named in the Appendix. Where no person is named or where the named adjudicator is unable to act the adjudicator shall be selected in the manner set out in the Appendix.

Right to refer dispute or difference to adjudication

8-117 Clause 37 is the enabling clause in the event that either party to the contract wishes to refer a dispute or difference to adjudication.

Almost without exception, any project constructed using the MPF will be a construction contract under Part II of the Housing Grants, Construction and Regeneration Act 1996. The most likely exceptions will be where the exclusion order, Construction Contracts (England and Wales) Exclusion Order 1998, applies. There is similar legislation for Northern Ireland and Scotland.

8-118 A party to a 'construction contract' (see section 104 of the Act) is entitled as of right to refer any dispute or difference arising under the contract to adjudication. A particular point to note here is that while in clause 35.1 reference is made to 'any dispute or difference ... in relation to the Project', in clause 37 itself the reference is to 'any dispute or difference arising under the Contract'. These are significantly different phrases. The wording in clause 37.1 reflects the wording of section 108(1) of the Act. Accordingly, it is only disputes or differences which *arise under* the contract which are covered by the clause. This is to be contrasted with the wording of clause 35.1. The latter would encompass matters such as misrepresentation or negligent misstatement and even possibly disputes or differences which could vitiate the contract itself, such as common mistake. It is likely that none of these matters will be covered by the adjudication clause.

8-119 The MPF does not provide for an internal adjudication procedure. Unlike any other major JCT contract to date, the MPF instead refers to 'The Scheme' which is defined in clause 39.2 as:

> The Scheme for Construction Contracts made in accordance with the provisions of section 114 of the HGCRA 1996.

Part 1 of the Statutory Scheme deals with adjudication.

Adjudication has been briefly discussed earlier (see paragraph 8-105 *et seq.*) but has not been dealt with in any detail in this book. However, it is worth stressing here that the adjudicator's decision is not permanently binding unless the parties wish it to be. Even so, if the dispute is subsequently referred to arbitration or litigation, the adjudicator's decision relating to that dispute must nevertheless be complied with in the meantime, unless there is a successful jurisdictional challenge.

Appointment of adjudicator

8-120 Under clause 37.2, the options for the appointment of an adjudicator are either for a person to be named in the appendix to the MPF, or if there is no name or if the named adjudicator is unable to act, then the adjudicator is to be selected in the manner set out in the appendix. The appendix therefore provides a space for the name of an adjudicator to be inserted. It is clearly sensible for an approach to be made to the adjudicator intended to be named to ensure that he is willing to act and also that the terms upon which he is prepared to act, particularly his fees, are agreed in advance. This could save problems later in connection with any difficulties over the terms of the adjudicator's appointment.

The appendix also provides space for the name of an appointing body to be inserted in the event that no adjudicator is named or the named adjudicator is not able to act. The appointing body selected is likely to be an 'adjudicator nominating body'. The Statutory Scheme in paragraph 2(3) gives a definition of an 'adjudicator nominating body' as:

> 'In this paragraph, and in paragraphs 5 and 6 below, an "adjudicator nominating body" shall mean a body (not being a natural person and not being a party to the

> dispute) which holds itself out publicly as a body which will select any adjudicator when requested to do so by a referring party.'

It is a wide definition. The most popular adjudicator nominating bodies are the Royal Institution of Chartered Surveyors; the Royal Institute of British Architects; the Chartered Institute of Arbitrators; the Construction Industry Council; the Technology and Construction Solicitors Association; and the National Specialist Contractors Council..

If no adjudicator is named or the named adjudicator is unable to act and there has been no insertion in the appendix of an appointing body, the appendix goes on to provide that in this situation the adjudicator is to be appointed by the President of the Royal Institution of Chartered Surveyors.

Naming an adjudicator at the outset has the advantage that, as he is already appointed, he should be in a position to quickly deal with disputes or differences referred to him as no time will be lost in going through the appointment process. Against this, however, many parties will prefer to wait until a dispute has arisen before seeking an adjudicator, in order to make sure that the appointee is someone appropriately qualified and experienced to deal with the particular area of dispute which has arisen. It is worth pointing out, however, that the adjudicator is entitled to obtain such technical or legal advice as he considers necessary.

Chapter 9
Communications definitions and attestation

Content

Clause 38 deals with communications between the parties and clause 39 contains the definition clause for the MPF. Following clause 39 is the attestation section of the MPF.

Communications – clause 38

Summary

9-01 Clause 38 provides that communications are to be:

- In writing, or alternatively in accordance with the procedure (if any) specified in the appendix to the MPF for electronic communications, or by any other means agreed in writing between the parties;
- Sent to either the address notified by the parties for receiving communications, or if no address is provided then to the address included at the head of the MPF conditions which will generally be the registered office of each party.

Any notice required to be given by either the third party rights schedule or under the termination provisions in clauses 32 to 34 must be by actual delivery, registered post (replaced by special delivery) or recorded delivery, and is effective once delivered.

Methods of communication and special situations

9-02 38.1 Save as provided in clause 38.2 all communications required to be made by one party to the other in accordance with the Contract shall be in writing or by the procedure specified in the Appendix for electronic communications (if applicable) or by such other means as shall have been agreed in writing by the parties. All communications shall be sent to either the address notified from time to time by a party for the purposes of communications or, if no address has been notified, the address given in these Contract Conditions.

38.2 Any notice required to be given by the Third Party Rights Schedule or by clauses 32, 33 and 34 (*Termination*) shall be given by actual delivery, registered post or recorded delivery and shall take effect upon delivery.

The methods of communication

9-03 Clause 38 can be compared with section 115 of the Housing Grants, Construction and Regeneration Act 1996 dealing with the service of notices and other documents under a construction contract. It provides:

'(1) The parties are free to agree on the manner of service of any notice or other document required or authorised to be served in pursuance of the construction contract or for any of the purposes of this Part.
(2) If or to the extent that there is no such agreement the following provisions apply.
(3) A notice or other document may be served on a person by any effective means.
(4) If a notice or other document is addressed, pre-paid and delivered by post
(a) to the addressee's last known principal residence or, if he is or has been carrying on a trade, profession or business, his last known principal business address, or
(b) where the addressee is a body corporate, to the body's registered or principal office,
it shall be treated as effectively served.
(5) This section does not apply to the service of documents for the purposes of legal proceedings, for which provision is made by rules of court.
(6) References in this Part to a notice or other document include any form of communication in writing and references to service shall be construed accordingly.'

Communications 'required' to be made

9-04 The HGCRA refers to notices and other documents that are 'required or authorised' to be served; Clause 38.1 refers to communications that are 'required' to be made. There is no mention in clause 38.1 of communications authorised rather than required. It does not obviously therefore cover communications that are permitted rather than required, e.g. instructions for a change. It could be argued that such an instruction is not 'required' to be made at all, though the better view must be that if the employer is issuing an instruction as permitted under the contract, then the communication of that instruction is required to be made in writing etc. The position must be similar in other permitted communications.

Communications in writing

9-05 It may be noted that the Act refers to notices and other documents, as indeed do other JCT contracts, e.g. WCD 98 clause 1.5. Clause 38 does not mention these but instead refers to all communications. In one sense clause 38 may therefore appear to be wider than the Act. However, it then goes on to restrict its scope by requiring these communications to be in writing unless otherwise agreed. As a result it is not obvious that this includes communications in drawn form, e.g. operating the design submission procedure in relation to design documents under clause 6. Compare this with section 115(6) of the Act which provides that references to notices or documents includes communications in writing but is not limiting, so that communication by drawn information is not prevented. There is no doubt that clause 38 is intended to include drawn information and this is likely to be the way in which it is interpreted but not without some straining of the meaning of 'in writing'.

The requirement for communications to be in writing can cover letters, faxes (at any rate if received and legible) and possibly emails if actually printed out.

Electronic communications

9-06 Any means of electronic communication are to be selected by specifying them in the appendix. This can of course include email. It also introduces the possibility of using some form of electronic data interchange.

Electronic data interchange (EDI)

9-07 Some JCT contracts, e.g. WCD 98, clause 1.8, provide for the possibility of the parties using EDI. In those contracts, there are 'Supplemental Provisions for EDI' annexed to the conditions of contract. Something similar can be adapted for use if required under the MPF. Provision can then be made for the parties to agree, by entering into an Electronic Data Interchange (EDI) Agreement, to the electronic exchange of communications under the contract. A choice is then made whether the EDI Agreement is to be either the EDI Association Standard EDI Agreement or alternatively the European Model EDI Agreement.

The use of such means of communication is likely to become increasingly popular, and properly regulated can reduce paperwork and save time. In the JCT supplemental provisions for EDI, Annex 2 to WCD 98, the following points are particularly noteworthy:

- Nothing within the EDI Agreement selected is to override or modify the application or interpretation of the contract unless it is expressly provided for within the supplemental provisions for EDI.
- The contract documents are to state the types and classes of communications and the persons between whom the data is to be exchanged. If nothing is contained in the contract documents then it can subsequently be agreed in writing between the parties.
- If the contract documents require a type or class of communication to which the chosen standard form of EDI Agreement applies, to be in writing it shall be validly made if exchanged in accordance with the EDI Agreement, but there are exceptions to this where the communication must be in writing specifically in accordance with the relevant provisions of the contract:
    - any determination of the employment of the contractor;
    - any suspension by the contractor of performance of his obligation under the contract;
    - the final account and final statement;
    - any invoking by either party of dispute resolution procedures;
    - any agreement between the parties amending the contract conditions or the EDI supplemental provisions.
- The contract procedures applicable to the resolution of disputes or differences will also apply to any dispute or difference regarding the supplemental provi-

sions for EDI or the exchange of any data under the selected EDI Agreement. Accordingly any dispute resolution provisions which may be in the selected standard form of EDI Agreement shall not apply to such disputes or differences.

It will be appreciated that while careful note must be taken of the exceptions to communicating by electronic data interchange, nevertheless it covers a wide range of communications including: interim payment advices; the warning notice prior to the determination of employment notice itself; communications in connection with procedures applicable to the resolution of disputes or differences, with the exception of the notice of intention to refer to adjudication; and the notice commencing arbitration proceedings (see Rule 2.1 of the JCT 1998 Edition of the Construction Industry Model Arbitration Rules). The supplemental provisions for EDI, if incorporated, need therefore to be considered alongside relevant clauses in respect of adjudication and rule 14 of the Construction Industry Model Arbitration Rules in connection with arbitration, both of which deal with the method of serving documents. (The MPF provides for the Statutory Scheme to apply to adjudication and does not provide for arbitration at all.)

Where the parties choose to litigate disputes or differences, the provisions of the EDI Agreement cannot override the provisions as to service contained in the Civil Procedure Rules 1998 which may apply, unless those rules themselves permit the parties to agree on the means of service.

The JCT has produced a comprehensive book explaining the system and how to implement it. It also provides a code of practice, glossary of terms, and appendices containing the JCT Supplemental Provisions for EDI and specimen forms of the EDI Association: Model Form of EDI Agreement and the European Model EDI Agreement.

Other means of communication agreed in writing

9-08 Clause 38.1 provides a third alternative means of communications: such other means as shall have been agreed in writing by the parties. Theoretically at least this could include an agreement in writing that future communications shall be made orally. This is clearly not a sensible course to take even in relation to simple or routine matters. In any event, it would be inapplicable in relation to clauses where documents are required to be sent. It would, however enable the parties to agree to change the method of communication during the contract from, for example, letters or faxes to emails.

Address for communications

9-09 All communications have to be sent to the address notified from time to time by a party for the purposes of communications. In default of notification of an address, communications must be served at the address given at the beginning of the MPF Conditions, which will be the registered office of each party. For letters or documents sent through the postal system, there will of course need to be a postal address. For other forms of communication it may be different. For example, for faxes it will be a fax number and for electronic communications it will be an electronic address such as an email address. There seems no reason why different addresses cannot be used for different purposes, e.g. interim applications for payment being directed to a specific address; drawings being sent directly to the address of a professional advisor as part of the design submission procedure, etc.

Third party rights schedule and termination

9-10 Clause 38.2 provides for a special means of service or delivery in respect of notices required under the third party rights schedule or in respect of termination of the contractor's employment under clauses 32, 33 and 34. In these cases, any such notice must be given by either actual delivery, registered post (replaced by special delivery) or recorded delivery and they are to take effect upon delivery. Proof of delivery whether through Royal Mail or a signed receipt in connection with actual delivery is important. Unlike many of the JCT forms, e.g. WCD 98, clause 27.1, there is no deemed service provision. It is extremely important therefore that evidence of delivery is obtained.

These procedural requirements should be followed precisely. As to notices under the third party rights schedule, see paragraphs 7-67 and 7-79. For notices under the termination provisions, see paragraphs 8-48, 8-52, 8-78, 8-81, 8-87 and 8-88.

Definitions and meanings – clause 39

9-11 Clause 39 deals firstly with the measurement of time under the MPF, where it provides that an act must be done within a stated period of time; secondly, it deals with the application of the definitions section of the MPF and contains the list of definitions.

Time within which an act must be done

39.1 Where under the Contract an act is required to be done within a specified period of days (but not weeks, months or years) after or from a specified date, the period shall begin immediately after that date. Where the period would include Christmas Day, Good Friday or a day which under the Banking and Financial Dealings Act 1971 is a bank holiday in England and Wales, that day shall be excluded.

Measurement of time where contract specifies a number of days within which an act is to be done

9-12 Clause 39.1 provides that where the contract (that is the conditions, appendix, third party rights schedule, requirements, proposals and the pricing document – see clause 39.2, Definitions), requires an act to be done within a specified period of days (weeks, months or years are expressly excluded) after or from a specified date, that period is to be begin immediately after that date and it excludes Christmas Day, Good Friday or bank holidays.

This can be contrasted with section 116 of the Housing Grants, Construction and Regeneration Act 1996 which provides:

'(1) For the purposes of this Part periods of time shall be reckoned as follows.
(2) Where an act is required to be done within a specified period after or from a specified date, the period begins immediately after that date.
(3) Where the period would include Christmas Day, Good Friday or a day which under the Banking and Financial Dealings Act 1971 is a bank holiday in England and Wales or, as the case may be, in Scotland, that day shall be excluded.'

It can be seen that clause 39.1 and section 116 are very similar. However, while section 116 applies only to periods of time for the purposes of Part II of the Act, which concerns specific or limited matters, clause 39.1 applies to the whole of the MPF but, unlike section 116, restricts its application to situations where the contract refers to a period of days.

In respect of both the Act and the clause, it only applies where an act is required to be done within a specified period. This period begins immediately after the relevant date from which the period of days is calculated. For example, where under clause 22.4 the employer must pay the contractor within 14 days from the date the sum becomes due, namely, on receipt of the contractor's VAT invoice, the date of receipt of the invoice is excluded in making that calculation.

Although the Act and clause both talk about an act being 'required to be done', in this context it almost certainly extends to situations where an act is merely authorised but where, if that authority is exercised, its exercise is required to be done within a specified period. A good example of this is in clauses 32.2, 33.2 and 34.2, where the employer or contractor as appropriate are not bound to serve a termination notice but if they choose to, they have a 14-day period in which to do it.

This requirement for an act to be done is bound to give rise to difficulties. Firstly, so far as the contract is concerned, this provision extends to all of the documents making up the contract and so will cover anything contained in contract documentation as well as in the conditions themselves. Secondly, the contract in many instances refers to periods of days where it is difficult to be sure whether *an act is required to be done* within the specified time. Those instances where it is clear that clause 39.1 applies, are to be found in clauses 6.3, 6.8 (twice), 12.4, 12.6 (twice), 18.8, 20.3, 20.7, 20,8, 20.9 (twice), 21.5, 21.6 (twice), 32.2, 33.2 and 34.2. Those instances where it is unclear whether clause 39.1 applies are to be found in clauses 27.3, 32.1, 33.1 and 34.1.

If the act is required to be done before rather than after or from a specified date the calculation does not apply.

Definitions and meanings

39.2 The definitions and meanings in clause 39 apply throughout the Contract Conditions, the Appendix and the Third Party Rights Schedule. Any reference within these definitions to a statute, statutory instrument or other subordinate legislation ('legislation') is to the legislation as amended and in force from time to time, including any re-enactment or consolidation of it, with or without amendment.

9-13 Clause 39.2 applies the definitions contained within clause 39.2 throughout the contract conditions, appendix and third party rights schedule. It does not apply them to other contract documents such as the requirements, proposals or the pricing document. The failure to include the formal printed parts of the pricing document is unfortunate. It contains a number of words or phrases beginning with capital letters and clearly intended to have the meaning given in the definitions section of clause 39.2: examples are employer, contractor, requirements, contract sum, the pricing document itself, changes, practical completion. Where else can the meaning of these words be found? They must be intended to have their clause 39.2 meanings. Some of these words are crucially important in the context of the pricing document, e.g. the contract sum, changes and practical completion.

Clause 39.2 also provides, typically, that any reference within the definitions to a statute, statutory instrument or other subordinate legislation includes that legislation as it may be amended and enforced from time to time, including re-enactments or consolidations.

Figure 9.1 shows the definitions and meanings of words or phrases in clause 39.2 which are used throughout the conditions, appendix and third party rights schedule. These definitions are discussed where appropriate in relation to the clauses in which they appear. The third column of the table indicates the appropriate paragraph(s) of this book in which the particular definition is discussed.

Term:	**Definition or meaning:**	**Paragraph number**
Base date	The date identified in the appendix.	2-193
Base rate	The rate set from time to time by the Bank of England's Monetary Policy Committee, or any successor.	5-174 5-175
CDM Regulations	The Construction (Design and Management) Regulations 1994.	2-22
Change	• Any alteration in the requirements and/or proposals that gives rise to an alteration in the design, quality or quantity of anything that is required to be executed in accordance with the contract; or • any alteration by the employer of any restriction or obligation set out in the requirements and/or proposals as to the manner in which the contractor is to execute the project, or the imposition of additional restrictions or obligations; or • any matter that the contract requires to be treated as giving rise to a change Provided always that the alteration or matter referred to above is not required as a result of any negligence or default on the part of the contractor.	5-09; 5-10
CIS	The Construction Industry Scheme established by the Income Tax (Sub-Contractors in the Construction Industry) Regulations 1993, as amended by the Income Tax (Sub-Contractors in the Construction Industry) (Amendment) Regulations 1998.	5-162 8-43

Fig. 9.1 Table of definitions and meanings.

Term:	Definition or meaning:	Paragraph number
Completion date	The completion date stated in the appendix or fixed from time to time in accordance with clause 12 (*Extension of time*). Where the appendix identifies that there is more than one section then references to the completion date are to the completion date of the relevant section.	3-15
Contract	The contract conditions, the appendix, the third party rights schedule, the requirements, the proposals and the pricing document.	2-04 7-35
Contractor	The party to the contract named as such or any assignee to whom the employer has consented in accordance with clause 29 (*Assignment*).	2-03 5-19
Contract sum	The amount stated in the appendix.	3-45; 5-107
Defect	Any fault in the project that arises as a consequence of a failure by the contractor to comply with his obligations under the contract, together with the consequences of that fault.	4-52
Design documents	Drawings, specifications, details, schedules of levels, setting out dimensions and the like which are required to be prepared by the contractor for the purposes of explaining and amplifying the requirements and/or proposals, which are necessary to enable the contractor to execute the project or which are required by any provision in the requirements.	2-120 2-170
Employer	The party to the contract named as such or any assignee permitted by clause 29 (*Assignment*).	7-11
Finance agreement	The agreement between the funder and the employer for the provision of finance for the project.	7-35
Funder	The person or syndicate providing funding for the purposes of the project, as identified in the appendix.	7-13 7-34
HGCRA 1996	The Housing Grants, Construction and Regeneration Act 1996.	
Insolvent	Either party is insolvent when it makes a composition or arrangement with its creditors, or becomes bankrupt, or, being a company: • makes a proposal for a voluntary arrangement for a composition of debts or scheme or arrangement to be approved in accordance with the Companies Act 1985 or Insolvency Act 1986 as the case may be; or • has a provisional liquidator appointed; or • has a winding up order made; or • passes a resolution for voluntary winding up (except for the purposes of amalgamation or reconstruction); or • under the Insolvency Act 1986 has an administrator or an administrative receiver appointed.	8-56 8-79
Joint Fire Code	The edition of the Joint Code of Practice on the Protection from Fire of Construction Sites and Buildings Undergoing Renovation, published by the Construction Confederation and the Fire Protection Association, that is current at any particular time.	6-32

Fig. 9.1 *(Contd).*

Term:	Definition or meaning:	Paragraph number
Material breach	By the contractor: • failure to proceed regularly and diligently with the performance of his obligations under the contract; • failure to comply with an instruction; • suspension of the project or any part thereof, otherwise than in accordance with HGCRA 1996 or the circumstances described in clause 34.1; • breach of the CDM Regulations; • breach of the requirements of CIS; • breach of any of the provisions of the contract relating to named specialists or pre-appointed consultants. By the employer: • failure to issue a payment advice in the manner required by the contract. By either party: • failure to make payment in the manner required by any payment advice issued under the contract; • failure to insure, as established in accordance with clause 27.3; • any repudiatory breach of the contract.	8-35 to 8-38 8-73
Model form	Where applicable, the Model Form of Novation Agreement that forms a part of the requirements.	4-82
Named specialist	A sub-contractor or consultant that is either identified by name in the requirements or that is to be selected by the contractor from a list of specialists contained in the requirements.	4-77
Others	Persons whose presence on the site has been authorised by the employer, other than the contractor, his sub-contractors and suppliers and any other persons under the control and direction of the contractor.	3-70 6-10
Planning supervisor	The person appointed for the project in accordance with regulation 6(1)(a) of the CDM Regulations.	
Practical completion	Practical completion takes place when the project is complete for all practical purposes and, in particular: • the relevant statutory requirements have been complied with and any necessary consents or approvals obtained, • neither the existence nor the execution of any minor outstanding works would affect its use, • any stipulations identified by the requirements as being essential for practical completion to take place have been satisfied, and • the health and safety file and all 'as built' information and operating and maintenance information required by the contract to be delivered at practical completion has been so delivered to the employer. Where the appendix identifies that there is more than one section then, unless stated otherwise, references to practical completion are to be read as references to the practical completion of the relevant section.	2-100 3-21
Practical completion of the project	Practical completion of the project occurs upon practical completion or, when there is more than one section, when all the sections have achieved practical completion.	2-100

Fig. 9.1 *(Contd).*

Term:	Definition or meaning:	Paragraph number
Pre-appointed consultant	A consultant identified in the requirements as having been appointed by the employer with the intention that the appointment be novated to the contractor in accordance with clause 18 (*Pre-Appointed Consultants*).	4-75 4-82
Pricing document	The document identified in the appendix containing the contract sum analysis and particulars of the manner in which the contract sum is to be paid to the contractor.	3-45 5-179
Principal contractor	The person appointed for the project in accordance with regulation 6(1)(b) of the CDM Regulations.	
Project	The works to be undertaken in accordance with the contract, as defined in the appendix.	3-12 8-41
Proposals	The documents identified in the appendix that have been prepared by the contractor in order to set out the manner in which he intends to satisfy the requirements.	1-07
Purchasers	Any and all first purchasers of all or any part of the project.	7-41
Rectification period	The twelve-month period commencing on the date practical completion of the project occurs.	4-51
Requirements	The documents identified in the appendix that have been prepared by the employer in order to set out its requirements for the project and identify the boundaries of the site.	1-06
Section	Where the appendix identifies more than one section, the parts of the project so defined by the requirements.	
Site	The area where the project is to be constructed and whose boundaries are defined in the requirements.	3-11
The Scheme	The Scheme for Construction Contracts made in accordance with the provisions of section 114 of the HGCRA 1996.	8-119
Specified peril	Fire, lightning, explosion, storm, tempest, flood, escape of water from any water tank, apparatus or pipe, earthquake, aircraft or other aerial devices or articles dropped therefrom, riot or civil commotion.	3-60 8-85
Statutory requirements	In relation to the project: • any Act of Parliament and any instrument, rule or order made under any Act of Parliament; • any regulation or byelaw of any local authority or of any statutory undertaker which has jurisdiction with regard to the project or with whose systems those of the project are or will be connected; and • any directive of the European Community having the force of law.	2-42
Tenants	Any and all first tenants of all or any part of the project.	7-41
Terrorism cover	Cover under any policy required to be provided by the contract against the physical loss or damage to work executed or site materials caused by an act of terrorism as defined by the Terrorism Act 2000.	6-39 8-90
VAT	Value Added Tax.	5-178

Attestation

The attestation part of the MPF immediately following clause 39.2 provides as follows:

EXECUTED AS A DEED BY THE EMPLOYER

Either:

By affixing hereto its common seal
In the presence of:

Or:

Acting by a director and its secretary*/two directors* whose signatures are here subscribed:
Namely ....
[Signature] .... Director
and ....
[Signature] .... Secretary/Director*

EXECUTED AS A DEED BY THE CONTRACTOR

Either:

By affixing hereto its common seal
In the presence of:

Or:

Acting by a director and its secretary*/two directors* whose signatures are here subscribed:
Namely ....
[Signature] .... Director
and ....
[Signature] .... Secretary/Director*

*delete as applicable

Commentary

9-14 The MPF provides for each party to execute the contract as a deed either by affixing its common seal or acting by a director and its secretary or two directors.

Since the coming into force on 31 July 1990 of section 1(1)(b) of the Law of Property (Miscellaneous Provisions) Act 1989, deeds may be executed by individuals without the need for a seal to be affixed. Companies incorporated under the Companies Act can affix a common seal if they wish. Alternatively they need not affix a common seal. Any document signed by a director and secretary of the company or by two directors and expressed to be executed by the company as a deed has the same effect as if executed under the common seal of the company (see section 36A Companies Act 1985).

The two main characteristics of a deed are that it does not require consideration in order to be enforceable and that under limitation legislation, a right of action on a contract under seal is, in general, barred 12 years after its accrual, whereas in respect of a contract under hand, the period is 6 years.

Parties to the MPF will be well used to the formalities of executing contracts as deeds and these are not discussed further here.

Chapter 10
The appendix to the Major Project Form

10-01 The MPF is a standard form of contract. For it to work in respect of any individual project, certain variables have to be provided for, e.g. date for access to the site, date for completion, level of liquidated damages, etc. In addition, it is often desirable, if options are to be provided, for the appropriate option to be selected in a systematic and obvious way. This is achieved by means of an appendix to the MPF conditions. It is a way of tailoring the contract to the individual project. Sensibly, where the appendix provides an option, a default answer is often provided in the event that no option is selected.

The various items referred to in the appendix are discussed as appropriate in commenting upon the clauses to which they relate (the relevant clauses are identified on the left hand side of the appendix against each relevant item). Accordingly it is not intended to discuss them further here.

THE APPENDIX

THE CONTRACT CONDITIONS

Definitions

The Project is:

____

as more fully described by the Requirements and the Proposals

The Contract Sum is:

____

The Requirements are:

____

The Proposals are:

____

The Pricing Document is:

____

The Planning Supervisor previously appointed by the Employer (if any) is:

____

The Funder (if any) is:

___

The Base Date is:

___

Clause etc.	*Subject*
6	**Design Documents**
	Design Documents shall be submitted to the Employer for review in the following quantities and format:
	___
8	**Ground conditions**
	Clause 8.2 is to apply/is not to apply*
	Where an alternative is not selected clause 8.2 *is not to apply;*
9	**Commencement and completion**
	The Contractor will be given access to the Site on ___
	The Completion Date is:
	*Project/Section 1 ___
	*Section 2 ___
	*Section 3 ___
	Or such other date as may be established by the operation of clause 12 *(Extension of time).*
10	**Damages for delay**
	The daily rate of liquidated damages is:
	*Project/Section 1 £ ___
	*Section 2 £ ___
	*Section 3 £ ___
14	**Bonus**
	The daily rate of bonus for early Practical Completion is:
	*Project/Section 1 £ ___
	*Section 2 £ ___
	*Section 3 £ ___
	Where no rate is specified the rate shall be NIL.

*Delete as applicable.

18 **Pre-Appointed Consultants and Named Specialists**

The provisions of clause 18 in relation to Pre-Appointed Consultants are/are not* to apply.

Where no selection is made, the provisions shall not apply.

19 **Cost savings and value improvements**

The proportion of any benefit to be paid to the Contractor is ___%.

Where no proportion is specified the proportion shall be 50%.

22 **Payments**

Prior to Practical Completion of the Project the Employer shall issue an interim payment advice each month on the:

________________________________ of the month.

Where no date is stated, the payment advice shall be issued on the 28th of each month.

The Employer shall not be obliged to issue an interim payment advice after Practical Completion of the Project where the amount stated as due to either party is less than:

£________________________________

Where no amount is stated the amount is to be £10 000.00.

27. **Insurances**

The policies of insurance to be provided and maintained in accordance with the Contract are those defined by the documents listed in the following table, copies of which documents are attached to the Contract. The party responsible for providing and maintaining each policy of insurance is identified below.

Type of insurance: ________________________________

As detailed in attached documents reference: ________________

Insurance to be provided and maintained by: ________________

Amount of excess (Clause 27.6)[c] ________________________

Type of insurance: ________________________________

As detailed in attached documents reference: ________________

Insurance to be provided and maintained by: ________________

Amount of excess (Clause 27.6)[c] ________________________

Type of insurance: ________________________________

*Delete as applicable.
[c] See Guidance Note.

As detailed in attached documents reference: ___________________

Insurance to be provided and maintained by: ___________________

Amount of excess (Clause 27.6)[c] ___________________

Type of insurance: ___________________

As detailed in attached documents reference: ___________________

Insurance to be provided and maintained by: ___________________

Amount of excess (Clause 27.6)[c] ___________________

Type of insurance: ___________________

As detailed in attached documents reference: ___________________

Insurance to be provided and maintained by: ___________________

Amount of excess (Clause 27.6)[c] ___________________

28 **Professional indemnity**

Clause 28
does/does not* apply.

Where no selection is made, clause 28 *shall not apply.*

The limit of indemnity shall be not less than:

£ ___________________

for any one claim or series of claims arising out of one event/in aggregate for any one year.*

34 **Termination by either the Employer or the Contractor**

The period of suspension before a notice may be issued in accordance with clause 34.1 is:

Where no period is stated the period is to be 13 *weeks.*

37 **Adjudication**

The adjudicator is:

____

Where no adjudicator is identified or where the identified adjudicator is not able to act then the adjudicator shall be appointed by:

____

Where neither of the above are completed or where the named appointer fails to act the adjudicator shall be appointed by the President of The Royal Institution of Chartered Surveyors.

[c] See Guidance Note.

*Delete as applicable.

38 **Communications**

The communications that may be made electronically, and the format in which those communications are to be made are as follows:

If none are identified, all communications are to be in writing, unless subsequently agreed otherwise.

THE THIRD PARTY RIGHTS SCHEDULE

Rights of a Purchaser and/or Tenant

Clause PT2
is/is not to apply*.

Where no selection is made clause PT2 is not to apply.

Where clause PT2 applies the limit of the Contractor's liability to a Purchaser or Tenant for losses other than those referred to by clause PT1 is a total of:

£__________

in respect of each breach/in total under the Contract.*

Where clause PT2 does not apply or where no limit of liability is inserted the Contractor will only be liable to a Purchaser or Tenant in the manner provided by clause PT1.

THE PRICING DOCUMENT

The applicable rule for the determination of the manner in which the Contractor is to receive payments in respect of the Contract Sum is Rule A/B/C/D*, as set out in the Pricing Document.

If no rule is selected, Rule A shall apply.

---

*Delete as applicable.

Table of cases

The following abbreviations of law reports are used

AC or App Cas	Law Reports Appeal Cases
ALJR	Australian Law Journal Reports
All ER	All England Law Reports
All ER (Comm)	All England Law Reports (commercial cases)
BCL	Building and Construction Law (Australia)
BLM	Building Law Monthly
BLR	Building Law Reports
CA	Court of Appeal
Ch	Chancery
CILL	Construction Industry Law Letter
CLC	Commercial Law Cases
CLR	Construction Law Reports
Com Cas	Commercial Cases
Const LJ	Construction Law Journal
DLR	Dominion Law Reports (Canada)
EG	Estates Gazette
EGLR	Estates Gazette Law Reports
EWCA Civ	England & Wales Court of Appeal (Civil Division)
EWHC	England & Wales High Court
Ex	Exchequer
Ex D	Law Reports Exchequer Division
HBC	Hudson on Building Contracts
HL	House of Lords
ICR	Industrial Cases Reports
KB	King's Bench
LGR or LGLR	Local Government Law Reports
Lloyd's Rep	Lloyd's Reports
LR CP	Law Reports Common Pleas
LSG	Law Society Gazette
M & W	Meeson and Welsby's Exchequer Reports
NZCA	New Zealand Court of Appeal Reports
NZLR	New Zealand Law Reports
PC	Privy Council (New Zealand)
QB or QBD	Queen's Bench Division
RPC	Reports of Patent, Design and Trade Mark Cases
SALR	South African Law Reports
SC	Session Cases
SCLR	Scottish Civil Law Reports

SJ	Solicitors Journal
SJLB	Solicitors Journal Lawbrief
SLT	Scottish Law Times
TCLR	Technology and Construction Law Reports
TLR	Times Law Reports
Term Rep	Term Reports
Tr LR	Trading Law Reports
WLR	Weekly Law Reports

Table of statutes

MPF clause number index to text

Paragraph numbers in bold indicate detailed discussion

MPF and WCD 98 clause comparisons

MPF clause	WCD 98 clause	Book paragraph number
2.1	4.1.1	2-30, 2-33
2.1	4.2, 4.3	2-29, 2-33
2.3	4.1.1, 4.1.2, 4.2	2-37
4.2	2.4.1	2-58
5	2.5.1	2-75
5.3	2.1, 2.5.1, 8.4, 8.5, 12.1, 16.1, 23.1.1	2-100
5.5	8.1.2, 8.1.3	2-119
9.1	23	3-02
9.4	16.1	3-21, 3-26
10.1	24	3-33
12.1.2	25.4.3	3-62
12.4	25.3.3.1	3-87
12.8	25.3.2	3-105
15.1	Article 3	4-04
16	8.3, 8.4	4-14
16	8.5	4-15
16.1	8.3	4-17, 4-18, 4-19, 4-21
16.2.3	8.4.2	4-35
16.2.4	8.4.3	4-40
17	16.2	4-46, 4-47, 4-49
17.1	16.2	4-52
18.3	S4.2	4-84
18.5	S4.2.2	4-96
20	12.5	5-09
20	12.4.2	5-09
20	4.3.2	5-09
20	12.1.1	5-10
20	12	5-12
20	12.1.2	5-14
20.6	12.4.1	5-47
21	26	5-67
21	26.4	5-68
21	26	5-70
21	28.4.4.4	5-70
21.3	26.1, 26.1.1	5-83
21.3	26.1	5-89
21.4	26.1.2	5-92
22	30	5-102
22.1	30.3.3	5-121

MPF clause	WCD 98 clause	Book paragraph number
22.2	30.3.1, 30.3.2, 30.3.3	5-125
22.2	30.3.1.2	5-127
22.3	30.3.3, 30.3.4, 30.3.5	5-131
22.6	30.5, 30.6, 30.7, 30.8, 30.9	5-150
22.7	30.8.1	5-155
22.7	30.8.1.3	5-158
25	14.1	5-178
27	21, 22	6-17, 6-29
27.7, 27.8	22	6-42
29	18.1.1, 18.1.2	7-09
29.1	18.1.1	7-08
30	1.9	7-30
30.1	1.9	7-32
32.1	27.2.4	8-36
32.1	27.2	8-37
32.1	27.2.1.3	8-40
32.1	27.2.1.1	8-41
32.1	28.2.1.3	8-42
32.1, 38.2	27.1	8-48
32.2	27.2.3	8-51
32.3	27.3.1	8-56
32.3	27.4	8-56
32.4	27.6.2, 27.6.3.1, 27.6.3.2	8-57
32.4.1	27.6.4	8-58
32.4.2	27.6.1, 5.5	8-59
32.4.4	27.6.5.1	8-61
32.5.1	27.6	8-66
32.5.1	27.6.6.1	8-67
33.1, 33.2, 33.3	27, 28	8-73
33.1	28.3.1	8-79
33.4	27.6.1, 28.4.1	8-82
34.1	28 A.1.2, 28 A.7	8-85
38	1.8	9-07
38	Supplemental Provisions for EDI, Annex 2	9-07
38.1	1.5	9-05
38.2	27.1	9-10

Index